A Specialist Periodical Report

Surface and Defect Properties of Solids
Volume 1

A Review of the Literature Published between January 1970 and April 1971

Senior Reporters
M. W. Roberts, *School of Chemistry, University of Bradford*
J. M. Thomas, *Edward Davies Chemical Laboratories, University College of Wales, Aberystwyth*

Reporters
J. S. Anderson, F.R.S., *University of Oxford*
C. H. Bamford, F.R.S., *University of Liverpool*
L. L. Ban, *Cities Service Co., Cranbury, New Jersey, U.S.A*
C. R. Brundle, *University of Bradford*
G. C. Eastmond, *University of Liverpool*
R. P. H. Gasser, *University of Oxford*
J. Pritchard, *Queen Mary College, University of London*
P. B. Wells, *University of Hull*
J. O. Williams, *University College of Wales, Aberystwyth*

ISBN: 0 85186 250 0

© Copyright 1972

The Chemical Society
Burlington House, London, W1V 0BN

*Printed in Great Britain by Billing & Sons Limited
Guildford* and London

Preface

A few aspects of this Report merit elaboration. The first involves editorial policy, and is concerned with the question of what one should expect from a volume devoted to current trends in a given area of scientific enquiry. Rather than call for one or other of the two extreme types of report, that is, a factual account of recent experiments, or, alternatively, a general interpretative essay setting out to review recent progress, we have deliberately adopted an ambivalent approach. Some of the articles in this first volume are distinctly phenomenological and seek to be comprehensive; others are of a more general nature.

Through the sections dealing with defective solids runs one noticeable thread of coherence, namely the necessity to invoke extended defects in order to rationalize the properties of these solids. It is now absolutely clear that point defects alone fail to account for a wide range of solid-state phenomena and properties, and that both line defects (dislocations) and planar defects are demonstrably real. Surface Chemistry and Catalysis are considered on a broad front, keeping very much in mind that catalysis is concerned with molecular events involving a solid surface whose nature changes during reaction. In this volume, emphasis is laid on the identity and reactions of chemisorbed species on metal surfaces, as studied by exchange and infrared techniques, catalytic selectivity, and the nature of surfaces as deduced from stimulated electron emission.

M. W. R.
J. M. T.

Contents

Chapter 1 Shear Structures and Non-stoicheiometry
By J. S. Anderson

1 Crystallographic Shear	5
2 The Mechanism and Kinetics of Crystallographic Shear	10
3 The Characterization of Compounds with Planar Defects or Superstructures	12
4 Crystallographic Shear in Specific Systems	13
Derivatives of the ReO_3 Structure	13
Tungsten (and Molybdenum) Oxides	13
Niobium Oxyfluoride	18
Derivatives of the MoO_3 Structure	18
Derivatives of the Rutile Structure	21
5 Thermodynamics of Rutile CS Phases	26
6 Ternary Defect Rutile Oxides	32
7 Double Shear and Block Structures	36
Symbolism	37
Ordered Block Structures	39
8 Stoicheiometric Faults, Intergrowths, and Non-stoicheiometric Faults	44
9 Related Types of Non-stoicheiometry	51

Chapter 2 Direct Study of Structural Imperfections by High-resolution Electron Microscopy
By L. L. Ban

1 Introduction	54
2 Electron Optical Considerations	57
3 Instrumentation	65

4 Materials Studied	66
Carbons	66
Carbon Blacks	67
Distribution of graphitic layer spacings	71
Different types of images in heat-treated carbon blacks	76
Carbon Films	80
Pyrolysed Polymers — Glassy Carbon	84
Coal and Char Samples	86
Polymers	89
Amorphous Polymers	89
Macromolecules — DNA	90
5 Summary and Conclusions	94

Chapter 3 The Role of Defects in Solid-phase Polymerization
By C. H. Bamford and G. C. Eastmond

1 Introduction	95
2 Some General Considerations	96
3 Vinyl Polymerization	97
Reaction Kinetics	99
Nature of Reaction Sites	100
Molecular Mobility	102
U.v. Initiation	105
E.S.R. Studies	106
Polymerization at Low Temperatures	112
Topotactic Polymerization	114
The Influence of Additives	117
Effects of Pressure	119
4 Polymerization of Cyclic Monomers	120
5 Co-polymerization	127
6 Concluding Remarks	127

Chapter 4 Structural Imperfections in Organic Molecular Crystals
By J. M. Thomas and J. O. Williams

1 Point Defects	130
2 Dislocations	132

3 Stacking Faults	135
4 Long-range Disorder and Short-range Order	140

Chapter 5 Surface Studies by Photoemission
By M. W. Roberts

1 Introduction	144
Formal Description of the Photoemission Process	146
Photoemission near the Threshold	148
Escape Depth of Photoelectrons	151
2 Experimental Approaches in Photoemission	156
3 Information from Photoemission Studies	157
Alloy Surfaces	158
Studies of Silicon	161
Nickel + Oxygen and Copper + Oxygen Systems	163
Yield	164
Energy Distribution	165
The Aluminium–Oxygen and Aluminium–Water Interfaces	167
Density of States	168

Chapter 6 The Application of Electron Spectroscopy to Surface Studies
By C. R. Brundle

1 Introduction	171
2 Principles of Electron Spectroscopy	172
Summary of the Techniques Involved	172
Chemical Information from Electron Spectroscopy	174
Photoelectron Spectroscopy	174
Energy Loss Electron Spectroscopy	178
Electron Impact Auger Spectroscopy	178
Ion Neutralization (I.N.) Spectroscopy	179
3 Application to Surface Studies	180
Mean Free Path Lengths of Electrons	180
Electron Impact Auger Spectroscopy	183
Ion Neutralization Spectroscopy	187
Photoelectron Spectroscopy	190
Molecular Photoelectron Spectroscopy: Photoemission Studies	190
X-Ray Photoelectron Spectroscopy, ESCA	193
Energy Loss Spectroscopy	200

Chapter 7 Exchange and Equilibration Reactions on Metal Surfaces
By R. P. H. Gasser

1 The Hydrogen–Deuterium Reaction	205
2 Nitrogen Isotope Reactions	211
3 Carbon Monoxide Isotope Reactions	215
4 Oxygen Isotope Reactions	217
5 Exchange of Hydrogen Isotopes in Organic Molecules	218

Chapter 8 Infrared Spectra of Adsorbed Species on Metals
By J. Pritchard

1 Carbon Monoxide	223
2 Reflection Spectroscopy of Chemisorbed Carbon Monoxide	228
3 Other Reflection Spectra	230
4 Hydrogen, Nitrogen, and Nitric Oxide	231
5 Hydrocarbons and other Compounds	233

Chapter 9 Some Aspects of the Selective Action of Metal Catalysts
By P. B. Wells

1 Introduction	236
2 Selectivity and Catalyst Structure	237
3 Selectivity and Reaction Mechanism	244
Consecutive Reactions	245
The Thermodynamic Factor	245
The Kinetic Factor	246
A Selectivity Pattern and its Interpretation	246
The Role of Occluded Hydrogen	249
Unresolved Problems	250
Parallel Reactions	251
Substituent Effects in Selectivity	254
Isomerization of Product before Desorption	254
Reduction of Selectivity by Polymerization	255
4 Selective Poisoning and Shape-selective Catalysts	255
5 Summary	258

Author Index   259

1
Shear Structures and Non-stoicheiometry

BY J. S. ANDERSON

As solid-state theory developed from 1930 onwards, it appeared that the existence and the properties of non-stoicheiometric compounds could be interpreted in terms of point defects—vacancies or interstitial atoms. This outlook is no longer tenable without considerable modification, and this review examines the implications of recent work on one extensive group of substances: those in which the notion of a point defect must be replaced by that of an extended, planar singularity in the structure.

Before doing so, it is desirable to summarize briefly the grounds for modifying the theoretical outlook that has become generally accepted. The power and generality of the statistical thermodynamics of crystals are such that the basic concepts of point defect theory are beyond dispute; moreover, the transport properties of ionic and covalent crystals (self diffusion, chemical diffusion, and ionic conductivity) demand the existence of mobile defects. The native concentration of point defects in a stoicheiometric crystal is given by an expression of the form $A \exp E_d/2kT$, where A depends upon the excess entropy of the defective crystal (its configurational entropy and the changes in vibrational entropy due to the defects) and E_d is the energetic cost of creating a complementary pair of point defects. E_d depends upon a number of factors, including the band gap, and is lower for quasi-covalent crystals (*e.g.* the transition-metal chalcogenides) than for essentially ionic crystals (*e.g.* most of the oxides), but is typically of the order of several electron volts. It follows that the calculated equilibrium concentration of point defects, particularly in the metallic oxides but also in the sulphides *etc.* is extremely small, even at high relative temperatures $\theta = T/T_m$ (T_m = the melting point of the crystal).

By contrast, the apparent defect concentration in non-stoicheiometric compounds, with a chemically significant range of existence, is very high, affecting the occupancy of 0.1—20% of the lattice points of at least one sublattice of the crystal.

In these circumstances, the dilute regular solution theory implied by the basic point defect approach is no longer applicable. Interactions between defects, and between the defects and the 'solvent' crystal lattice, mediated by coulombic forces, electron delocalization, and repulsions, become important. Moreover, the view that non-stoicheiometric compounds involve only the extension of point defect theory—displacement of the intrinsic defect equilibrium in response to changes in the chemical potential of the components [1]—runs into the difficulty that, to account for defect concentrations of the magnitude indicated, the energy of creation of point defects would have to be unacceptably small. Efforts to include interaction effects into a generalized statistical thermodynamic treatment [2] have had only limited success, because the actual structure of non-stoicheiometric compounds and solid solutions shows that some highly specific factors are involved. These depend both on the structure of the parent phase (*e.g.* the types of defect complex found in hyper- [3] and hypo-stoicheiometric fluorite structures,[4] the ordering effects in transition-metal chalcogenides related to the $B8$ and $C6$ structure types [5]) and on the electronic structure of the elements and compounds concerned (*e.g.* the variety of defect structures within the NaCl structure type, as shown by FeO,[6] TiO,[7] and NbC[8]).

Although no simple general description is any longer tenable, the results produced by increasingly strong interaction effects can be summarized schematically, as in Table 1.

There are differences between essentially ionic and essentially metallic structures as regards their non-stoicheiometric behaviour. In the latter, point defects are perturbations of the inner potential; changes in composition change the electron population at the top of the conduction band, *i.e.* usually in the anti-bonding or non-bonding part of the band. Because the local perturbations can be screened by the conduction electrons, long-range interactions are small and it may still be appropriate to speak of isolated point defects. Nevertheless, evidence is accumulating that interactions are strong enough to lead to marked site-preference energy effects, so that there is a considerable degree of correlation between the positions of vacancies (*e.g.* the preferential location of vacant carbon sites in NbC_{1-x} on third neighbour positions), even though this is not strong enough to lead to long-range order. In ionic compounds, coulombic forces and crystal-field interactions exert more profound effects, so that the parent crystal structure around

---

[1] F. A. Kröger, 'Chemistry of Imperfect Crystals', North Holland, Amsterdam, 1964.
[2] J. S. Anderson, *Proc. Roy. Soc.*, 1946, **A185**, 69; 'Problems of Non-stoicheiometry', ed. A. Rabenau, North Holland, Amsterdam, 1970, Chap. 1; *Bull. Soc. chim. France*, 1969, 2203.
[3] B. T. M. Willis, *J. de Physique*, 1964, **25**, 431; 'Thermodynamics and Transport Properties of UO_2 and Related Phases', I.A.E.A. Tech. Rep. Series No. 39, Vienna, 1965; B. E. F. Fender, A. K. Cheetham, and P. E. Childs, to be published.
[4] (*a*) B. G. Hyde, D. J. M. Bevan, and L. Eyring, *Phil. Trans.*, 1966, **A259**, 583; (*b*) M. R. Thornber and D. J. M. Bevan, *J. Solid State Chem.*, 1970, **1**, 536, 545.
[5] Y. P. Jeannin, in 'Problems of Non-stoicheiometry', ed. A. Rabenau, North Holland, Amsterdam, 1970, Chap. 2; M. Chevreton and S. Brunie, *Bull. Soc. France Mineral Crist.*, 1964, **87**, 277.
[6] F. Koch and J. B. Cohen, *Acta Cryst.*, 1969, **B25**, 275.
[7] D. Watanabe, J. R. Castles, A. Jostsons, and A. S. Malin, *Acta Cryst.*, 1967, **23**, 307; D. Watanabe, O. Terasaki, A. Jostsons, and J. R. Castles, 'The Chemistry of Extended Defects in Non-Metallic Solids', ed. L. Eyring and M. O'Keefe, North Holland, Amsterdam, 1970, p. 238.
[8] A. W. Henfrey and B. E. F. Fender, to be published; A. W. Henfrey, D. Phil. Thesis, Oxford, 1970.

Shear Structures and Non-stoicheiometry

Table 1

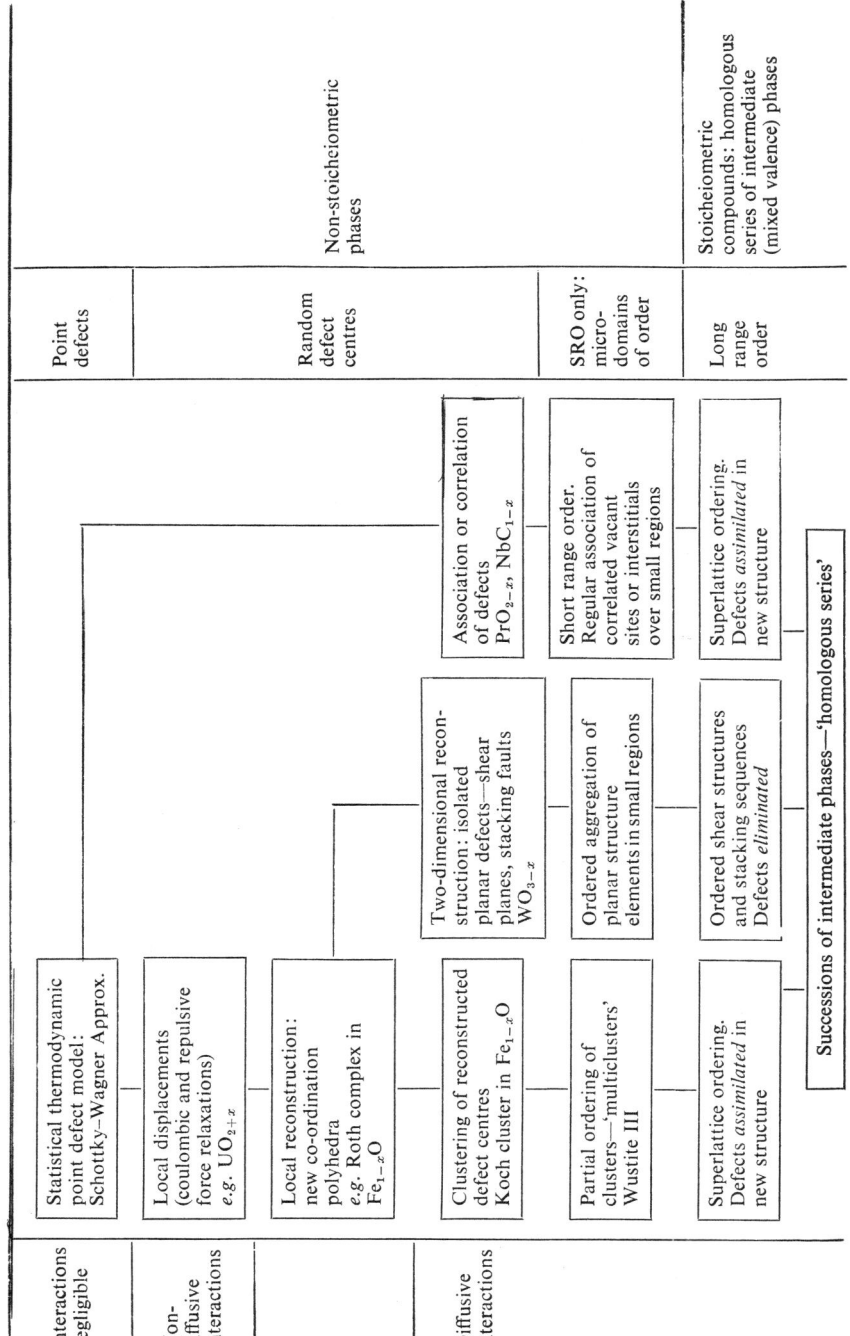

nominal point defects is modified. At significant defect concentrations most 'point' defects are transformed into defect complexes, which may aggregate into defect clusters. Whether this invariably happens cannot be affirmed; the statement is correct for each one of the few defect structures that have been examined in detail.

A second feature shown in Table 1 is that interaction between defect centres, whatever their nature, tends to the formation of ordered structures. It can be maintained that, for attainment of true inner equilibrium at low temperatures, the configurational entropy of any random defect system must be reduced to zero by ordering. In principle, any assemblage of atoms, in any proportions, could order in a sufficiently large unit cell; in practice, non-stoicheiometric crystals either unmix into phases of simple composition and structure or form a succession of intermediate compounds—a homologous series of intermediate mixed valence phases, derived according to some common structural pattern from the structure of the parent non-stoicheiometric compound. In an increasing number of instances, indeed, it has been found that such ordered intermediate phases are invariably formed, and that the apparent existence of non-stoicheiometric phases with wide ranges of composition is illusory. In a number of systems, however, (*e.g.* the transition-metal chalcogenides, CeO_{2-x} etc.) there is a transition from univariant thermodynamic behaviour, with successions of sharply defined phases, at low temperatures, to the bivariant behaviour of a non-stoicheiometric phase at high temperatures. In such cases, the structures of the intermediate phases show how ordering occurs, and it is a reasonable inference that analogous ordering processes operate over a short range within the non-stoicheiometric phase. On that basis, the structure of the non-stoicheiometric crystal could be regarded as not completely random and homogeneous in composition, but involving small regions or microdomains of differing superstructure order (and hence differing composition). The close relationship between the structures concerned implies that they can inter-grow coherently within a single crystal or particle; at the high temperatures for which non-stoicheiometric behaviour represents a stable state, self-diffusion processes are important, so that each microdomain is a dynamically fluctuating region, smaller in extent than the coherent scattering length for X-ray diffraction effects. On this view, put forward particularly by Ariya [9] and Wadsley,[10] a non-stoicheiometric crystal is inherently a hybrid crystal in Ubbelohde's sense; the concept must at present be regarded as non-proven, but has found increasing acceptance,[11] although attempts to formulate more precisely its thermodynamic implications raise a number of problems. Some residual ordering of defect centres in non-stoicheiometric phases can be inferred from the results of careful thermodynamic studies of

---

[9] S. M. Ariya and M. P. Morozova, *J. Gen. Chem.* (*U.S.S.R.*), 1958, **28**, 2647; S. M. Ariya and Y. G. Popov, *ibid.*, 1962, **32**, 2077.
[10] A. D. Wadsley, in 'Non-stoicheiometric Compounds', ed. L. Mandelcorn, Academic Press, New York, 1964, Chap. 3.
[11] B. G. Hyde, D. J. M. Bevan, and L. Eyring, ref. (4*a*).

the ferrous oxide phase $Fe_{1-z}O$ [12] and the praseodymium oxide phase PrO_{2-x} [13] and from thorough X-ray diffraction studies of the La_2O_3—CeO_2 solid solution system.[14]

If, in place of simple vacancies and interstitials, non-stoicheiometric systems must be viewed structurally as based upon defect complexes and a considerable measure of short-range order, an inescapable conclusion is that attempts to interpret their properties in terms of classical point defect theory, without reference to their real structure, lead to quite fallacious results. This has not yet been fully appreciated.

1 Crystallographic Shear

The immediate concern is with those structures in which defects are virtually eliminated by planar singularities and, in particular, with the so-called 'shear structures'.

The term 'crystallographic shear' (which is better not abbreviated to 'shear') was first applied by Wadsley to the structural principle underlying the homologous series of intermediate phases in the oxides of molybdenum, tungsten, vanadium, and titanium, discovered by Magneli and his co-workers. It has proved to be of wider significance, describing the relationship between crystal structures that are based on the same packing of anions, and extending to the defect structure of simple oxides.

If ionic crystals are considered as built up by the linkage of co-ordination polyhedra, e.g. the $[MO_6]$ octahedra in the oxides of the six-co-ordinate cations, crystallographic shear provides a means of altering the anion : cation ratio in some simple parent structure, such as the ReO_3 or rutile type

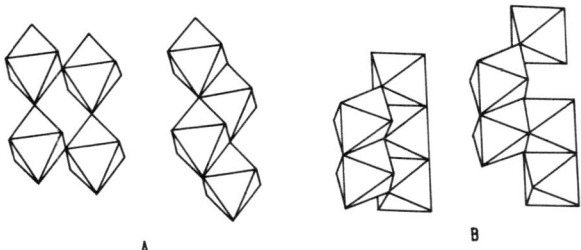

Figure 1 *Conversion of* (A) *apex-sharing* $[MX_6]$ *octahedra in* ReO_3 *structure into edge-sharing, and* (B) *edge-sharing octahedra in rutile structure into face-sharing by a shear displacement*

---

[12] P. Vallet and P. Raccah, *Compt. rend.*, 1964, **258**, 3679; *Mem. Sci. Rev. Met.*, 1965, **62**, 1; B. E. F. Fender and F. D. Riley, *J. Phys. Chem. Solids*, 1969, **30**, 793.

[13] M. S. Jenkins, R. P. Turcotte, and L. Eyring, 'The Chemistry of Extended Defects in Non-metallic Solids', ed. L. Eyring and M. O'Keefe, North Holland, Amsterdam, 1970, p. 36.

[14] D. J. M. Bevan, W. W. Barker, R. L. Martin, and T. C. Parks, 'Rare Earth Research', ed. L. Eyring, Gordon and Breach, New York, 1965, 3, 441.

structures, without the introduction of point defects and without change in cation co-ordination, but with a change in the co-ordination number of certain anion positions. Along some crystallographic plane, the linkage between co-ordination polyhedra is changed, a closer linkage (*e.g.* replacement of apex-sharing by edge-sharing, as in ReO$_3$-type structures, Figure 1A, or edge-sharing by face-sharing, as in rutile-based structures, Figure 1B) eliminating a set of anion sites and conversely.

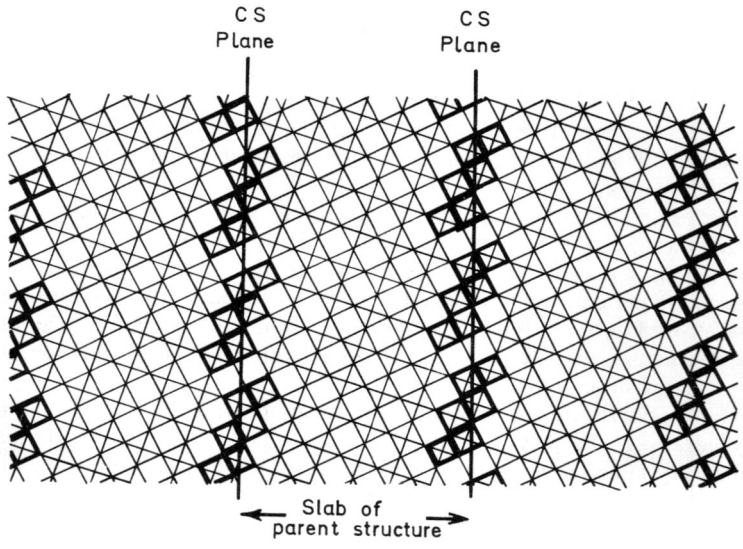

Figure 2 *The* W$_{20}$O$_{58}$ *structure. Recurrent CS planes separate slabs of essentially unaltered* WO$_3$ *(ReO$_3$ type) structure. Diagram shows projection of octahedral* [WO$_6$] *groups down the* b *axis*

The consequences of such planar structural features can be exemplified by the tungsten oxide WO$_{2.90}$ or W$_{20}$O$_{58}$ (Figure 2). WO$_3$ itself has a distorted variant of the ReO$_3$ structure, with [WO$_6$] octahedra linked through all the apical oxygen atoms. In W$_{20}$O$_{58}$, the octahedra are nearly regular and linked as in ReO$_3$ except along regularly recurring planes, the (130) planes of the ReO$_3$ structure, along which there are groups of octahedra (infinite columns in the direction normal to the diagram) which share edges. The slices of structure so formed have a lower oxygen : tungsten ratio (in this instance 8 : 3) than the slabs of unperturbed parent structure enclosed between successive slices of edge-sharing octahedra. Inspection shows that in this, and all, shear structures, the crystallographic planes parallel to the slice of altered composition are of two kinds: *A*, containing metal atoms (or metal and oxygen atoms) and *B*, containing only oxygen atoms. In the parent structure, there is regular alternation $ABABAB..$; at the discontinuity, one oxygen-only sheet is omitted to give the stacking sequence $ABABAABAB.....$ In $W_{20}O_{58}$, type A planes have the composition WO, type B planes the composition O_2. The resulting structure is that which would result if a type B sheet were abstracted from the parent structure, followed by a mutual displacement of the two slabs of structure so as to restore six-fold co-ordination to the cations and continuity to the oxygen sublattice.

In this particular structure, the two type A sheets collapse to form a single sheet (denoted as A_2) with twice the normal density of tungsten atoms). Across the plane concerned, the *crystallographic shear plane* (CS plane), 'normal' and 'interstitial' cation sites switch meanings as between slabs of parent structure separated by the CS plane. This displacement is equivalent to a shear, the direction and magnitude of which is defined by a *shear vector* which approximates to a rational lattice vector of the parent structure, but is not necessarily an exact lattice vector because the new polyhedron linkage in the CS plane may involve significant distortion.[15]

The hypothetical shear process can be represented by the operation of some *displacement vector* $\left\langle \dfrac{a\,b\,c}{u\,v\,w} \right\rangle$ on the {hkl} planes of the parent structure, the CS operation being symbolized as $\left\langle \dfrac{a\,b\,c}{u\,v\,w} \right\rangle$ {hkl}. In order that a set of lattice points may be eliminated, the shear vector must have a component normal to the shear plane; a true mechanical shear, with the displacement lying in the shear plane can produce only an antiphase boundary, without change in stoicheiometry (Figure 3). The crystallographic shear plane and the shear vector together determine the structure, and therefore the composition, of the collapsed slice; the total composition is determined by the slabs of more or less undistorted parent structure. An isolated CS plane constitutes an extended (two dimensional) defect; regular recurrence of the CS operation on some particular set of (hkl) planes generates a new ordered structure. The new repeating unit of the crystal is therefore defined by an enlarged unit cell, of lowered symmetry, with at least one long interplanar spacing, the distance between successive CS planes. Since the structure at the CS planes depends only on their orientation and shear vector, they could in principle be separated by any arbitrary width of parent structure.

If the spacing of CS planes is completely regular, the result is to generate from a parent structure MX_a, a homologous series of stoicheiometrically defined intermediate, mixed valence compounds with the general formula MX_{an-m}; m measures the number of anion sites eliminated by crystallographic shear, n the breadth of the slabs of parent structure, as measured by the number of co-ordination octahedra in unbroken rows between CS planes. Attainment of perfect regularity of CS plane spacing clearly depends upon some long-range interactions, and this problem is considered below. If, on

[15] B. G. Hyde and L. A. Bursill, *Proc. Roy. Soc.*, 1970, **A320**, 147.

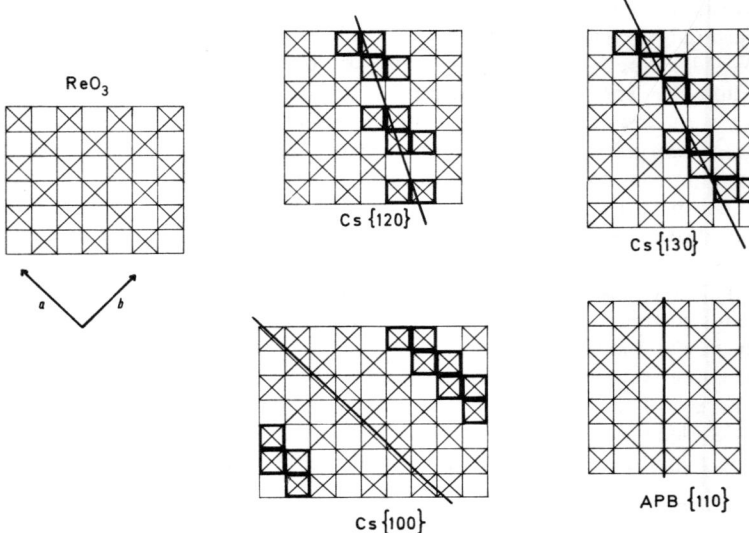

Figure 3 *Crystallographic shear on* {120}, {130}, *and* {100} *in* ReO_3 *structure; shear displacement, with the same displacement vector, on* {110} *produces an antiphase boundary. Diagrams show projection of* [MO_6] *octahedral groups on* (001) *plane*

the other hand, the interaction forces are not sufficient to impose regularity of spacing, then a new type of non-stoicheiometry must result: the composition of a crystal could be a continuous function of the chemical potential of its components, changes in composition being taken up as alterations in the distribution of CS plane spacings. Such a crystal could be regarded from an alternative standpoint: as made up of lamellae of different compositions, in coherent intergrowth—microdomains, effectively infinite in two dimensions but only a few, or even one, unit cells in thickness. With the advent of electron microscopic techniques it has become possible to investigate the microstructure of CS phases at this level, and some results are discussed later, in reference to particular systems; they raise questions as to the stage at which a defined *structure* is recognized as a distinct *compound*, and the distinction between such a structurally defined entity and a *phase* in the thermodynamic sense.

Considerable interest clearly attaches to the thermodynamic definition of the CS phases: whether they display stoicheiometric variability and, in particular, whether at high temperatures thermal fluctuations of composition produce variations in CS plane spacing and a consequential broadened composition range. As will be seen, the evidence from known series of compounds is that CS planes interact strongly enough to impose a high degree of order, so that where the slabs of parent structure are up to 6—8

Shear Structures and Non-stoicheiometry

octahedra wide the compounds are strictly stoicheiometric. In some systems, indeed, they exist only as very high temperature phases. On the other hand, the kinetics of CS plane ordering and readjustment may be very sluggish, so that disorder, at the microdomain level, cannot easily be eliminated.

In a topological sense, crystallographic shear describes the relationship between crystal structures that are based on a common mode of packing of anions and a constant co-ordination of cations, notably octahedral co-ordination. In continuous three-dimensional structures, the progenitor structures are the compounds MX_3, involving co-ordination octahedra linked through all their apices. Of this type, the ReO_3 structure is based upon

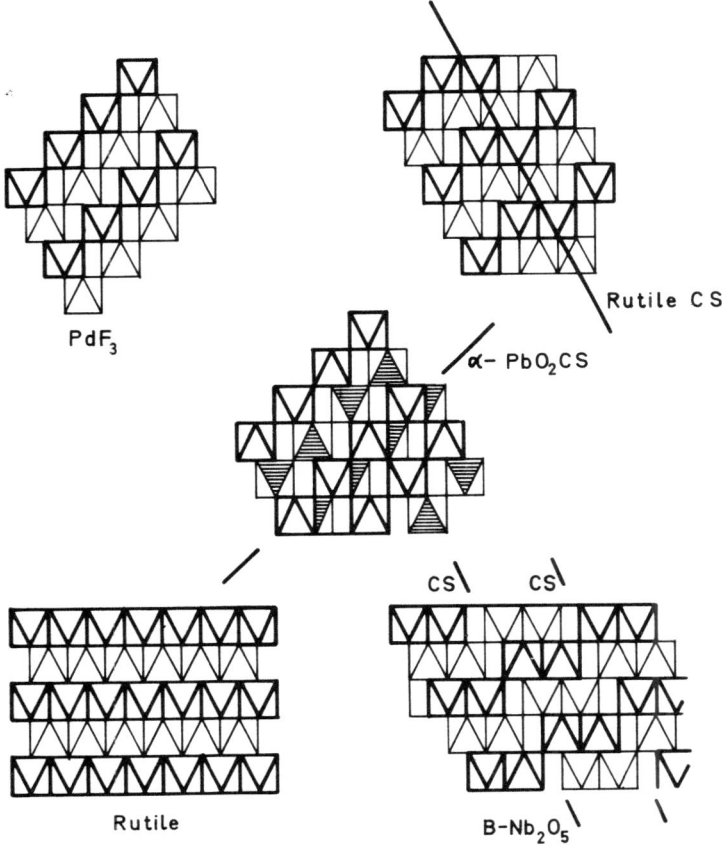

Figure 4 *Possible CS operations in the PdF_3 structure: Derivation o, rutile and α-PbO_2 structures by two different recurrent shears. Relation of B-Nb_2O_5 structure to that of rutile by an expansive shear. Potential extended defects in structures derived from the PdF_3 type*

cubic close-packing and the PdF$_3$ type upon hexagonal close-packing of anions. The operation of recurrent crystallographic shear on the former, involving one or two CS processes, is illustrated in Figure 3; the relations between the structures derived or derivable from the PdF$_3$ type have been discussed by Andersson [16] (Figure 4). These considerations have a predictive value, in foreseeing the possible existence of polymorphic forms (*e.g.* high-pressure structural transformations), of new series of intermediate mixed valence and ternary phases, and of defect structures in the compounds concerned.

2 The Mechanism and Kinetics of Crystallographic Shear

Crystallographic shear has been considered so far in purely geometric terms, but since CS compounds are formed (albeit in an often disordered state) by solid-state reactions and in the reduction of the parent oxides, the transformations involved pose questions as to the mechanism by which reaction and ordering take place. Two stages are involved; how one whole sheet of anion sites can be virtually eliminated to define the locus of collapse of the crystal structure, to form a CS plane, and how CS planes attain a regular spacing. Both steps seem to require extensive co-operative shifts of atomic positions.

Three mechanisms have been proposed for the creation of CS planes. The first [17] presupposes the prior creation of a relatively high concentration of

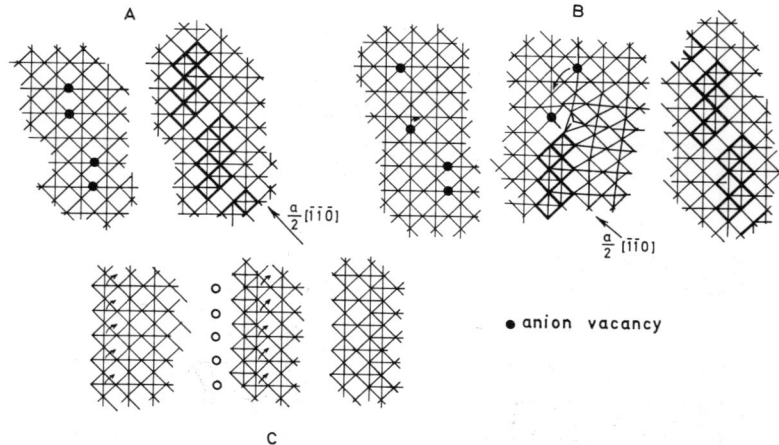

Figure 5 *Proposed mechanisms for formation of CS planes in reduction of oxides:* (A) *Gado,* (B) *Anderson and Hyde,* (C) *Wadsley and Andersson. In* (C), *oxygen atoms are removed and anion sites annihilated over the whole surface*

---

[16] S. Andersson and J. Galy, *J. Solid State Chem.*, 1970, **1**, 576.
[17] P. Gado, *Acta Phys. Hung.*, 1965, **18**, 111.

anion vacancies, which undergo a superlattice ordering into zigzag 'walls' (Figure 5A) along the trace of the CS plane; the crystal then shears across the walls, annihilating the vacant sites. There is little evidence that the requisite high vacancy concentrations can be formed, nor for superstructure ordering as a general precursor stage, presupposing a highly organized migration of vacancies from the surface of the crystal into the bulk. Anderson and Hyde [18] also suggested that random anion vacancies are formed as the initial step but that, at low concentration and at the surface, where vacancies are created, a few of these could aggregate into a vacancy disc which collapses, with simultaneous shear, to form an element of the CS plane (Figure 5B). As long as this is finite, it must be bounded by a dislocation loop; this is a sink for the annihilation of vacancies, which can drain into it *via* stress-facilitated short diffusion paths as the dislocation loop climbs through the crystal, extending the area of CS plane. The model implies that CS planes nucleate at a surface and extend into and across the bulk by growing along their length. It provides a possible way of eliminating very low concentrations of vacancies as isolated CS planes. However, anions are presumably removed from or added to the structure over the whole reactive interface during reactions (*e.g.* reduction or oxidation of an oxide); vacancies are not created at interior sites. It is not clear how the conditions for creation of the CS plane would be established unless the core of the bounding dislocation provided such a preferential diffusion path that, once nucleated, the reaction was effectively localized, and no mechanism is provided for the migration of CS planes, such as is required for the formation and interconversion of ordered structures.

Andersson and Wadsley [19] assumed interstitial cations to be the transient, mobile species. Removal of a layer of oxygen atoms, in a reduction process, displaces a sheet of cations co-operatively into adjacent interstitial sites to form a CS plane at the surface; by a series of repetitive steps this may diffuse into the crystal, along a direction normal to the CS plane (Figure 5C). It would follow that reaction leading to CS planes would occur only at selected orientations on any crystalline particle, and the production of a single CS plane across the middle of a crystal would require the movement of a vast number of atoms (*e.g.* 50% of all the cations to produce a CS plane across a rutile crystal). This is inherently improbable.

None of the interpretations of the reaction process is completely satisfactory. The electron microscopy of CS phases has shown that CS planes can grow along their length and commonly traverse the full width of crystals with perfect planarity. They may be formed (see below) at an exceedingly low anion-deficiency of the crystal, requiring the whole locus of reaction in a macroscopic crystal to be concentrated at a relatively few sites. The Anderson–Hyde mechanism is compatible with this, but leaves the chemistry of reduction and reaction processes unexplained. On the other hand, CS

[18] J. S. Anderson and B. G. Hyde *J. Phys. Chem. Solids*, 1967, **28**, 1393.
[19] S. Andersson and A. D. Wadsley, *Nature*, 1966, **211**, 581.

planes can certainly move laterally, to adjust their spacing, and this is explained only by the Andersson–Wadsley process. Such lateral movement to build up regularly recurrent CS planes requires long-range interactions between them, which are not at all well understood.

3 The Characterization of Compounds with Planar Defects or Superstructures

For several reasons, the identification and differentiation of CS phases and related structures present some special difficulties. As will be seen (later), thermodynamic equilibrium studies can be vitiated by both the slowness of solid-state processes and the extremely small energetic differences between members of a homologous series. Diffraction methods encounter the problem that all the structures of any family are so closely related to that of some parent compound; indeed, the more or less unperturbed parent structure makes up most of the true unit cell of the new compound, and its contributions dominate the diffraction properties, especially for the higher members of any family, for which the true repeat distance between CS planes becomes large.

In single crystal X-ray diffraction work, the reciprocal lattice of the large, true unit cell is frequently very empty, information being sometimes restricted

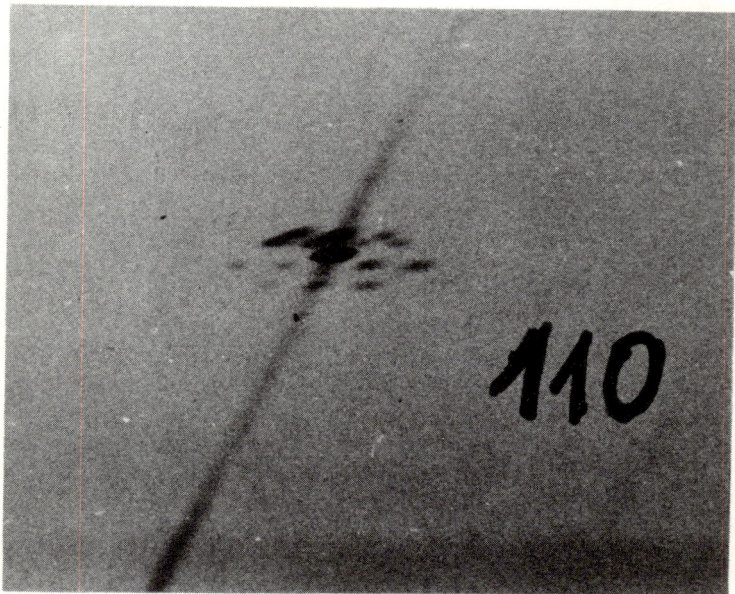

Figure 6 *Fine structure of X-ray diffraction spots from Cr-doped rutile, due to presence of higher CS phases. (After O. W. Flörke and C. W. Lee, ref. 82)*

to the fine structure of diffraction spots assignable to the parent subcell (Figure 6). Moreover, the possibility of coherent intergrowth between members of a homologous series can be masked, in that the diffraction properties are averaged over a macroscopic crystal. Evidence of this and other kinds of disorder is found in the streaked X-ray reflections recorded by several workers. Powder diffraction methods involve yet a further loss of information: only reflections corresponding to the long spacings of the true cell, or symmetry splitting of the subcell reflections, are relevant. The usual Debye–Scherrer technique is therefore quite useless, and even with the high-resolution Guinier method it may be extremely difficult to differentiate between compounds or to establish homogeneity ranges in any system.

Electron diffraction and electron microscopy have become powerful structural tools in this context. Selected area methods enable single crystal diffraction patterns to be obtained from regions about 10^{-4} cm in diameter; coherent diffraction is obtained from small volumes of uniform structure, so that coherent intergrowth structures can give rise to separable superstructure diffraction spots from each of the microdomain components. By high-resolution electron microscopy, the periodicity of the crystal can be directly imaged, and since this is done by forming an image with selected diffracted beams, from a diffraction pattern which contains latent information about the aperiodic features of the structure, the resulting lattice image reveals these also. Direct lattice imaging has therefore developed [20] into a direct means of detecting aperiodicities or changes of periodicity in stacking, *e.g.* variations in the width of the slab of parent structure between CS planes. The importance of evidence derived from electron diffraction and microscopy will be apparent in the survey of some specific systems of compounds.

4 Crystallographic Shear in Specific Systems

Derivatives of the ReO$_3$ structure.—*Tungsten (and Molybdenum) Oxides.* Tungsten trioxide has a distorted form of the ReO$_3$ structure, with the tungsten atom displaced from the centre of each co-ordination octahedron to give unequal W—O bond lengths, in the monoclinic modification stable at room temperature ($a = 7.285$ $b = 7.517$ $c = 3.835$ Å $\beta = 90.90°$ [21]). At high temperatures, in reduced tungsten oxides and in the tungsten bronzes (*i.e.* as electrons are introduced into the non-bonding band of the ReO$_3$ structure, which is empty in WO$_3$ itself), the co-ordination octahedra become more regular, and the structures can be idealized.

It is well known that WO$_3$ loses oxygen readily, in vacuum at high temperatures or by reduction, but the older literature contains discrepant reports of its stoicheiometric variability, which Glemser and Sauer [22] considered to

---

[20] J. G. Allpress, *Materials Res. Bull.*, 1969, **4**, 707, and later references.
[21] S. Andersson, *Acta Chem. Scand.*, 1953, **7**, 154.
[22] O. Glemser and H. Sauer, *Z. anorg. Chem.*, 1943, **252**, 144.

extend as far as the composition $WO_{2.95}$. Magneli [23] showed that there were two compounds, completely characterized by their structures, close in composition to WO_3: the β-oxide $W_{20}O_{58}$ ($WO_{2.90}$) and the γ-oxide $W_{18}O_{49}$ ($WO_{2.722}$). The latter is a tunnel structure, in which anion sites can be formally regarded as eliminated by regularly recurring one-dimensional extended structure elements. $W_{20}O_{58}$ is the prototype of ordered shear phases, of the M_nO_{3n-2} series produced by crystallographic shear on $\{130\}$ ReO_3 and, as such, might be expected to generate a homologous series with differing widths of WO_3 structure between CS planes. This is indeed the case; several distinct phases have now been identified by X-ray diffraction in the composition interval between $WO_{2.90}$ and $WO_{3.00}$: $W_{20}O_{58}$ ($WO_{2.900}$), $W_{24}O_{70}$ ($WO_{2.9167}$), possibly $W_{25}O_{73}$ ($WO_{2.9200}$), $W_{35}Ta_4O_{115}$ [24] [(W,Ta) $O_{2.9487}$] and $W_{40}O_{118}$ ($WO_{2.9500}$).[25] All these phases are formed at high temperatures, showing the stability of CS ordering, and it must be emphasized that the composition corresponding to a given structure is uniquely determined by the spacing between CS planes, which is directly measured ($= c \sin \beta$) from the unit cell dimensions. As is usual, X-ray diffraction yields only a majority-dominated, averaged identification of such structures. Allpress and Gado [26] found that, in well crystallized $WO_{2.90}$, electron microscopy revealed a considerable degree of randomness. The spacing between successive planes, measured on direct lattice images within any one macroscopically homogeneous crystal, varied between the values corresponding to $W_{12}O_{34}$ ($WO_{2.8333}$) and $W_{28}O_{82}$ ($WO_{2.9285}$). In some regions of some crystals (Figure 7) there might be microdomains of uniform composition broad enough to give discrete superstructure diffraction spots; alternatively, the CS plane spacings could be almost randomized, with only one or two consecutive slabs or lamellae of the same width. Thus, single crystals of a preparation that is homogeneous, as determined by X-ray diffraction, may contain one-dimensional fluctuations of composition. It is interesting that the averaged composition of the Allpress–Gado material would be closely represented by the limit set by Glemser and Sauer for the β-phase.

The phase cited as $W_{25}O_{73}$ may not be correctly assigned. Gebert and Ackermann [27] considered it to be $W_{25}O_{74}$ ($WO_{2.960}$), with a structure that, although based on $\{130\}_{ReO_3}$ CS planes, distorted the channels (marked Z, Figure 7) into hexagonal tunnels containing oxygen sites. These authors also give good reason for suspecting the existence of another phase, with composition around $WO_{2.98}$, with an as yet unknown structure.

Although MoO_3 does not have the ReO_3 structure, and gives rise to its own CS derivatives, it generates intermediate compounds of the family

---

[23] A. Magneli, *Arkiv Kemi*, 1950, **1**, 513; *Nova Acta Reg. Soc. Uppsaliensis, Ser. IV*, 1950, **14**, 3; L. Kihlborg, *Advances in Chemistry No. 39*, 1963, 37, and ref. 10.
[24] P. Gado, B. Holmberg, and A. Magneli, *Acta Chem. Scand.*, 1965, **19**, 2010.
[25] P. Gado and L. Imre, *Acta Chem. Hung.*, 1965, **46**, 165; P. Gado and A. Magneli, *Acta Chem. Scand.*, 1965, **19**, 1514.
[26] J. G. Allpress and P. Gado, *Crystal Lattice Defects*, 1970, **1**, 331.
[27] E. Gebert and R. J. Ackermann, *Inorg. Chem.*, 1966, **5**, 136.

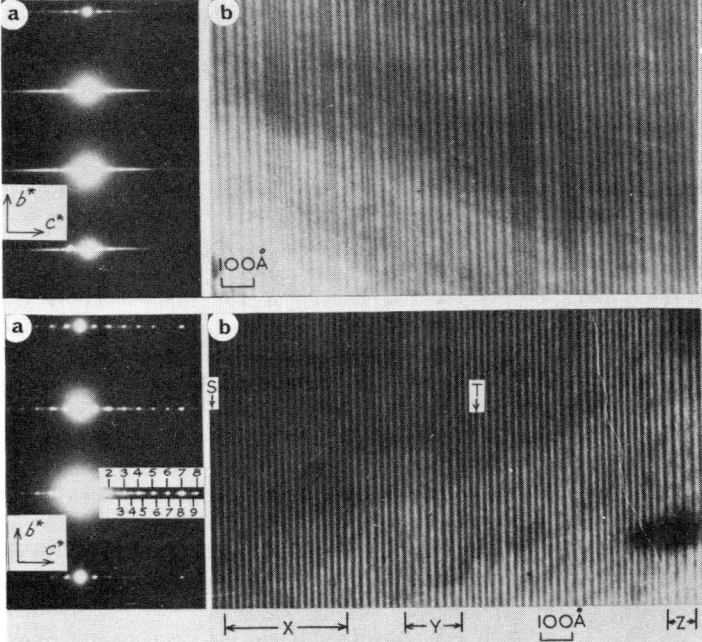

Figure 7 (a) *Electron diffraction patterns* (0kl *reciprocal lattice sections*) *from* $WO_{2.90}$ *showing streaked or doubled sets of reflections parallel to* c* (l *indicates marked*) (b) *Lattice image from same fragment showing CS planes with a range of spacings, Regions* X, Y, Z *are groups of fringes with spacings of* 21 Å ($W_{18}O_{52}$), 24.5 Å ($W_{21}O_{61}$), *and* 27 Å ($W_{23}O_{67}$) *respectively*

M_nO_{3n-1}, based on $\{120\}_{ReO_3}$ CS planes. In the molybdenum oxides proper, only the members Mo_8O_{23} ($MO_{2.875}$) and Mo_9O_{26} ($MO_{2.889}$) are known; no similar phases are known in the binary tungsten oxides, but progressive replacement of molybdenum by tungsten in the mixed oxides extends the series at least to $n = 11$, with $(Mo_{0.85}W_{0.15})_{10}O_{29}$ and $(Mo_{0.5}W_{0.5})_{11}O_{32}$ ($MO_{2.909}$).[28] An oxygen deficiency can be produced also by replacing W^{6+} or Mo^{6+} by Nb^{5+} or Ta^{5+}, and two other members of the M_nO_{3n-1} series, $Nb_2W_{11}O_{38}$ ($MO_{2.923}$) and $Nb_2W_{15}O_{50}$ ($MO_{2.941}$) have been reported[29] (compare the alternative CS structure produced by incorporation of Ta^{5+}). These may not be the only CS phases formed in the system; they not only exemplify the surprising high-temperature stability of CS phases, but emphasize that this type of long-range ordering may be characteristic of high-temperature equilibria: they melt peritectically at about 1629 K and decompose, into a tunnel structure phase and WO_3 (probably an oxygen-deficient

---

[28] B. Blomberg, L. Kihlborg, and A. Magneli, *Arkiv Kemi*, 1953, **6**, 133, A. Magneli, *J. Inorg. Nuclear Chem.*, 1956, **2**, 330.

[29] R. S. Roth and J. L. Waring, *J. Res. Nat. Bur. Stand.*, 1966, **70A**, 281.

solid solution with disordered, widely-spaced CS planes), at 1543 K. There is evidence, from streaking of diffraction spots in X-ray diffraction patterns, that perfect order is difficult to achieve in these materials; there may well be random coherent intergrowth of structures with different CS plane spacings. The discrepancies in the chemistry of the reduced tungsten and molybdenum oxides suggest that, as obtained without the most careful procedure, the ReO_3-like basic skeleton common to all of them results in intergrowths between the various CS phases, the tunnel structures, and domains of WO_3 itself.

Apparent non-stoicheiometry of the intermediate oxides could be a non-equilibrium state. Of considerable interest is the situation in the earliest stages of reduction of WO_3: the defect state that is the precursor of all the ordering processes. A small oxygen deficiency, sufficient to modify the absorption spectrum and electric conductivity, is produced by heating WO_3 crystals in vacuum, or at controlled oxygen fugacities, at 1123—1573 K. Tilley[30] has shown that planar faults, identified by their diffraction contrast properties as $\{120\}_{ReO_3}$ CS planes, are already present in oxide of the composition $WO_{2.994}$; they increase in number as the oxygen deficiency increases, and in $WO_{2.993}$ they tend to cluster in groups at about 40 Å spacing. Thus, even at a concentration of around 200 p.p.m., point defects are largely eliminated from the crystal as extended planar CS type collapsed faults ('Wadsley defects'); these invariably grow in from the edges of the crystal, and finite collapsed vacancy discs or CS discs, terminated by a dislocation ring, have not been observed. The structure with $\{120\}_{ReO_3}$ CS planes is not that of the M_nO_{3n-2} type found in the highest well-defined intermediate oxides; the CS planes apparently interact to build up imperfectly ordered microdomains approximating in composition to $W_{50}O_{149}$. A similar order of magnitude for the effective range of the CS plane ordering interactions is obtained from work on crystallographic shear in rutile. Work by Berak and Sienko,[31] on the optical and electronic transport properties of WO_3 also suggests the formation of isolated $\{120\}$ CS planes at even lower oxygen deficiencies.

It is possible that yet an earlier stage of defect elimination is indicated by some observations of Spyridelis, Delavignette, and Amelinckx.[32] They found that platelets of WO_3, sublimed under slightly reducing conditions at 1673 K, showed linear faults, at very regular spacings of about 160 Å, parallel to the (010) surface of the crystals. They interpreted these as dislocation lines, arising from the removal of a row of oxygen atoms along [100], followed by a collapse process $\frac{1}{2} < 10\bar{1} > \{100\}_{ReO_3}$. This is equivalent to a crystallographic shear of the kind that relates the V_2O_5 structure to the ReO_3 structure, but confined to an exceedingly thin layer at the surface of the crystal. For the spacings observed, the composition of the layer concerned

---

[30] R. J. D. Tilley, *Materials Res. Bull.*, 1970, **5**, 813; J. G. Allpress, R. J. D. Tilley, and M. J. Sienko, *J. Solid State Chem.*, 1971, **3**, 440.
[31] J. Berak and M. J. Sienko, *J. Solid State Chem.*, 1970, **2**, 109.
[32] J. Spyridelis, P. Delavignette, and J. Amelinckx, *Materials Res. Bull.*, 1967, **2**, 615.

would approximate to $W_{92}O_{275}$, but the oxygen deficiency of the crystal as a whole would be very small.

These last observations need to be revised in the light of the later work by Tilley, but it is clear that if the processes of reduction have to follow the sequence

Lines of shear $\| <100>$	→	isolated $\{120\}$ CS planes, clustering at $WO_{2.993}$	→	Ordered $\{130\}$ CS planes in W_nO_{3n-2} compounds

there must be a reorientation, as well as an ordering, of CS planes in forming the first well-defined intermediate phase. Since there are four equivalent $\{130\}$ planes, the CS planes can be established on four orientations. There is little energetic driving force behind the attainment of perfect order over the whole volume of a crystal, and Allpress and Gado find that twinning, irregular domains of different (but permitted by the ReO_3-type subcell) orientation and tangled, kinked CS planes can be left as relics of the nucleation process (Figure 8). The relevance of this observation is that the junctions

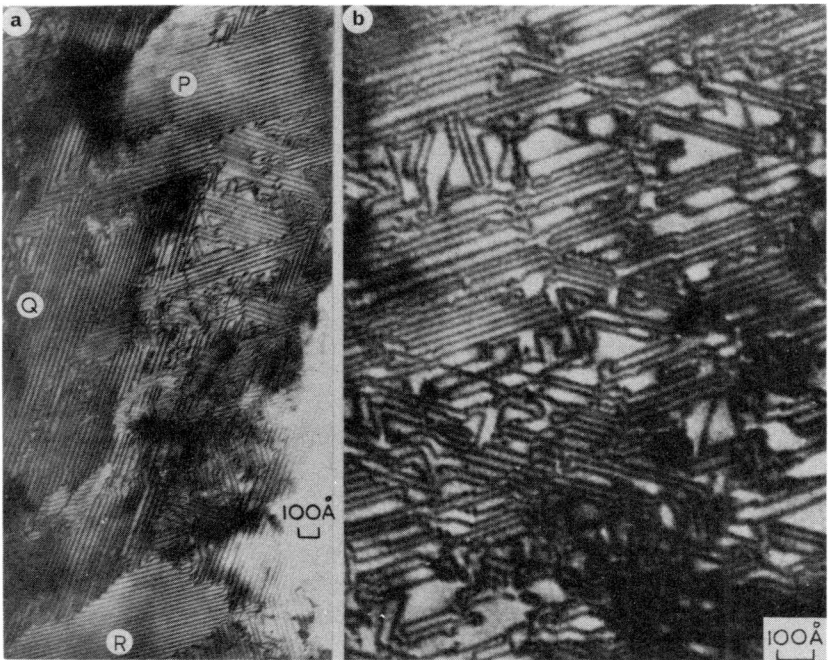

Figure 8 *Lattice image of* $WO_{2.90}$ *showing disordered CS defects. Domains with different orientations* (P, Q, R) *surround an area with kinked and disordered CS planes*

between different {130} CS planes can give rise to improper blocks of fused octahedra—they change the proportion of anion sites eliminated. If the polyhedral regions bounded by CS planes are small, an additional factor is thereby introduced into the stoicheiometric variability.

Niobium Oxyfluoride. NbO_2F has the simple ReO_3 structure with oxygen and fluorine apparently randomized over the anion sites. When it is heated it loses $NbOF_3$: *e.g.* $4NbO_2F \rightarrow NbOF_3 + Nb_3O_7F$. The first chemically detectable stage of the reaction, formation of Nb_3O_7F, a slab structure, involves the introduction of CS planes (shear vector $\frac{a}{2} <110> \{010\}_{ReO_3}$) at every third octahedron linkage (Figure 3). No other CS plane spacing is detectable, and further loss of NbF_5 introduces a second orthogonal set of CS planes ($\frac{a}{2} <110> \{100\}$) which divide the infinite two-dimensional slabs into one-dimensionally infinite columns or blocks, to generate a family of oxyfluorides which are listed (Table 3) under block structures. The same compounds may be more readily prepared by the reaction of NbO_2F with Nb_2O_5 under hydrothermal conditions.

The earliest stages of reaction of NbO_2F again show how the transfer of components of the crystal to the gas (in this case, molecules of $NbOF_3$, creating an imbalance between the populations of anion sites and cation sites) is absorbed within the ReO_3 structure by crystallographic shear rather than by creation of point defects. Heating in the electron microscope, by the electron beam, introduces a mass of planar faults on all the {100} planes.[33] Their diffraction properties are consistent with identification as CS planes of the same type as ultimately line up at close spacings in Nb_3O_7F, which is ultimately formed. They are bounded by dislocations, and grow along their length, as is consistent with the Anderson–Hyde mechanism for the formation of CS planes. As in the reduction of WO_3, completion of the reaction requires the alignment of all CS planes in any one region; this is kinetically difficult, and the Nb_3O_7F shows twinning and evidence of residual disorder. As first observed, the CS planes are isolated and widely separated, up to 1000 unit cell spacings apart; they are already formed when the composition of the crystal has changed to $NbO_{2+\delta}F_{1-2\delta}$, with $\delta \sim 0.001$.

Derivatives of the MoO_3 Structure.—Reference has been made to the ReO_3-type compounds Mo_8O_{23} and Mo_9O_{26}; both of these are high-temperature phases; for the phase relations in the higher molybdenum oxides, see ref. 34. Mo_4O_{11}, in both its modifications, can also be regarded as a CS phase, based on the ReO_3 structure and sheared on the $\{11\bar{2}\}$ planes into slabs three octahedra wide; the shear is not simple, but leaves the slabs connected by a layer of tetrahedral cations. Whether there is any variability in the width

---

[33] L. A. Bursill and B. G. Hyde, *Phil. Mag.*, 1969, **20**, 657.
[34] L. Kihlborg, *Arkiv Kemi*, 1963, **21**, 471.

of the slabs (*i.e.* any metastable range of composition for Mo_4O_{11}) has not been investigated by electron microscope techniques. Below about 973 K, however, the stable structures of the oxygen-richer oxides are derived directly from that of MoO_3 by crystallographic shear.

The structure of MoO_3 has double ribbons of octahedra that share both corners and edges. In a topological sense, this is related to the ReO_3 structure by two successive CS operations on the same orientation, one eliminating anion sites, one creating anion sites.

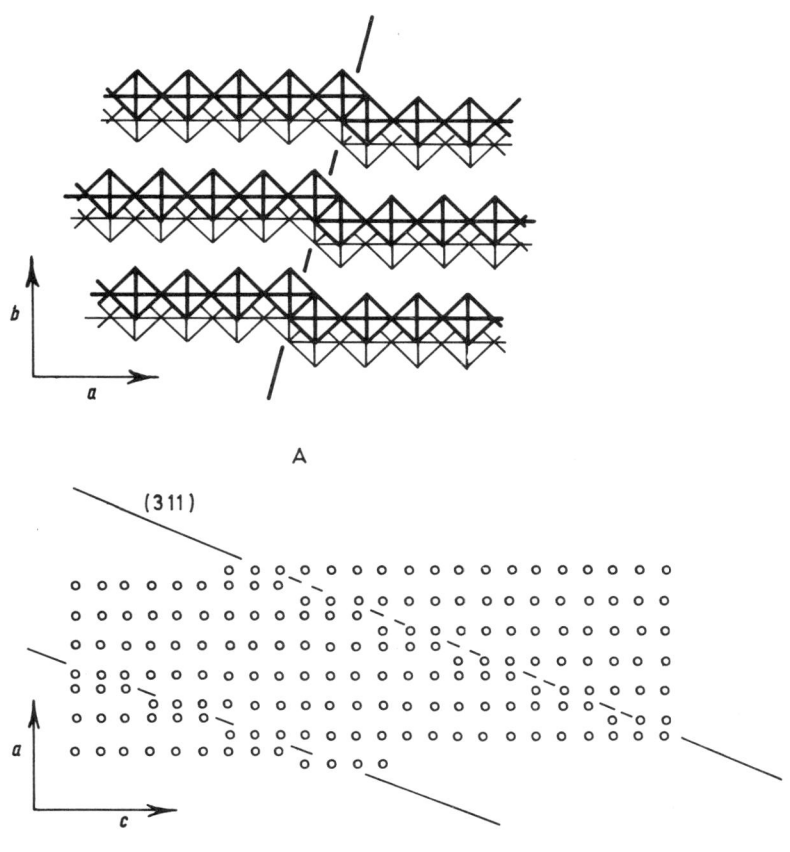

Figure 9 *Crystallographic shear in the MoO_3 structure. (A) a-b section, normal to the double ribbons of $[MoO_6]$ octahedra, showing displacement common to all the CS phases. (B) a-c section of the Mo_nO_{3n-2} structure and the $\{311\}$ CS plane. Only the positions of the metal atoms in the upper levels of the double ribbons of octahedra are shown. In $Mo_{13}O_{38}$ and $Mo_{26}O_{75}$ the edge-sharing in the CS plane involves groups of 2 and 4 octahedra respectively, at each level*

Reduced oxides with closely spaced compositions are formed from this by the introduction of regular, oblique CS planes, with a displacement vector $<\frac{a}{2}+\frac{b}{7}>$, whereby groups of octahedra in adjacent ribbons are made to share edges (Figure 9). The main product of reduction to a composition around $MoO_{2.9}$ is $Mo_{18}O_{52}$, which appears to be the stable form of $MoO_{2.8989}$ below about 923 K, but interconverts with Mo_9O_{26} only sluggishly. In this, a member of a possible Mo_nO_{3n-2} series, the CS planes are $\{311\}_{MoO_3}$, but other orientations are possible, and represented by the occasional single crystal isolated by Kihlborg[35] from preparations of $Mo_{18}O_{52}$: $Mo_{13}O_{38}$ (Mo_nO_{3n-1}), with $\{211\}$ CS planes and $Mo_{26}O_{75}$ (Mo_nO_{3n-3}) with $\{411\}$ CS planes. It is of interest that, in each of these, the slabs of MoO_3 structure are six octahedra wide across the a_{MoO_3} direction, but the repeat distances between CS planes along the unbroken double ribbons of octahedra are long. It may be expected that, as for $W_{20}O_{58}$, electron microscopy will reveal some disorder, and the presence of lamellae corresponding to other members of these homologous series. Some degree of metastable stoicheiometric variability, from this cause, is to be expected.

There is, however, yet an earlier stage in the reduction of MoO_3 that demonstrates the capacity of this structure for ordering by crystallographic shear and its low tolerance for point defects. MoO_3 undergoes curious changes as a result of slight reduction by the electron beam during electron microscopy, readily forming a pattern of domains like a parquet floor.[36] Bursill[37] showed that this arose from an unmixing process, into a matrix of MoO_{3-x} saturated with oxygen vacancies ($x < 0.0025$), enclosing lamellar or cellular domains of a compound in which oxygen vacancies were fairly regularly ordered on a 7×7 superstructure for which the calculated composition is about $Mo_{400}O_{1199}$ ($MoO_{2.9975}$). The domain and matrix structures are fully coherent, but in a structural sense the oxide is already a two-phase system at the composition $MoO_{2.999}$. At slightly elevated temperatures, the oxygen vacancies diffuse and aggregate into discs on $\{120\}$ planes, which collapse (with the same displacement vector as for the lower shear phases discussed above) and grow through the crystal, by climb of their bounding dislocation loop, annihilating the superlattice of oxygen vacancies. With further reduction, the density of these $\{120\}_{MoO_3}$ CS planes increases; an ordered superlattice of vacancies need not be formed as a precursor for the creation of CS planes. It would seem that, within the temperature range investigated, they do not achieve a fully ordered configuration, so that they lead to the formation of an operationally non-stoicheiometric oxide, $MoO_{3-\delta}$ ($0.0025 < \delta < \sim 0.05$), the lower limit being inferred from the observed spacing of about 50 Å between CS planes in the more reduced material; this lower

---

[35] L. Kihlborg, *Arkiv, Kemi*, 1963, **21**, 443.
[36] H. König, *Z. Physik*, 1951, **130**, 483; E. Pernoux and R. Borrelly, *J. Microscopie*, 1963, **2**, 407.
[37] L. A. Bursill, *Proc. Roy. Soc.*, 1969, **A311**, 267.

limit agrees roughly with unpublished observations made by Anderson and Rossell, and reached on different evidence.

Derivatives of the Rutile Structure.—Rutile and the isostructural oxides of other quadrivalent elements have been the object of a great number of investigations, which have left some important unresolved problems about defect structure and non-stoicheiometric behaviour. Three strands of evidence have to be harmonized: phase equilibrium studies, extrinsic electronic conduction due to stoicheiometric variability, and direct structural evidence.

TiO_2 (rutile form) has a very low intrinsic conductivity, corresponding to a wide band gap; it loses sufficient oxygen, when heated in vacuum at moderate temperatures ($< 1273 K$), to acquire n-type conductivity, and one numerous group of investigations has been concerned with the interpretation of the carrier concentration—oxygen pressure relation in terms of point defect theory. At higher temperatures and in reducing conductions the oxygen deficit becomes chemically measurable. Brauer and Littke [38] found that the melting point of rutile was depressed, and an oxygen-deficient product obtained when the oxide was fused in varying ambient oxygen pressures. More recent work [39] indicates the values shown in Table 2. Ehrlich [40] placed the lower composition limit at about $TiO_{1.90}$, but later work has progressively revised this figure upwards; the limit appears to depend critically on the purity of the material and, with very pure rutile, the behaviour in a solid-state galvanic cell at $1273 K$ clearly indicated that $TiO_{1.990}$ was already in a two-phase state.[41] It will be seen later that the phase limit is now hard to define.

Table 2

Oxygen pressure (atm.)	1	0.25	Low (argon atmosphere)
Melting point (K)	2185	2133	2053
Composition	$TiO_{2.000}$	$TiO_{1.985}$	$TiO_{1.894}$

There has, however, been controversy over the point defect description of sub-stoicheiometric rutile within what may be described as the operationally monophasic composition range. Loss of oxygen and development of extrinsic conductivity has been represented alternatively by formation of oxygen vacancies:

$$O_o^x \rightarrow V_o^x + \tfrac{1}{2} O_2$$
$$V_o^x \rightarrow e^- + V_o^{\cdot} \rightleftharpoons 2e^- + V_o^{\cdot\cdot}$$

[38] G. Brauer and W. Littke, *J. Inorg. Nuclear Chem.*, 1960, **16**, 67.
[39] J. P. Coutures and M. Foëx, *Compt. rend.*, 1968, **276** C, 1577.
[40] P. Ehrlich, *Z. Elektrochem.*, 1939, **45**, 362.
[41] C. B. Alcock, S. Zador, and B. C. H. Steele, *Proc. Brit. Ceram. Soc.*, 1967, No. 8, p. 231.

[where O_o^x represents a normally occupied anion site (O^{2-}), V_o^x a vacant anion site with two trapped electrons, and $V_o^{\cdot}$, $V_o^{\cdot\cdot}$ the successive ionization states of the 'relative neutral' vacancy to form an anion vacancy and two electrons in the conduction band], or by formation of titanium interstitials:

$$Ti_{Ti}^x + 2O_o^x \rightarrow O_2 + Ti_I^x$$

$$Ti_I^x \rightleftharpoons Ti_I^m{}^+ + me^-$$

Measurements of the density and cell dimensions of oxides $TiO_{1.983}$ to $TiO_{2.000}$ (quenched from 1673 K) indicated an oxygen vacancy structure,[42] and the thermogravimetric work of Førland [43] fitted the relation x (in TiO_{2-x}) $\propto P_{O_2}^{-\frac{1}{6}}$ as required for fully ionized oxygen vacancies, over the range to $TiO_{1.9232}$ (1467 K, $10^{-13.4}$ atm. O_2). Kofstad [44] also postulated a vacancy model, although he found a change in the pressure dependence (possibly explicable in other ways) at compositions above $TiO_{1.998}$. Hurlen,[45] however, argued in favour of titanium interstitials; the ribbons of edge-linked octahedra in the rutile structure enclose square-columnar tunnels in which there are sites, at two levels, in which a cation would be in slightly expanded octahedral co-ordination. The changes observed in cell dimensions, $a = 4.584$, $c = 2.959$ Å for $TiO_{2.000}$, $a = 4.603$, $c = 2.960$ Å at the lower phase limit,[46] are compatible with the occupation of these sites, which would bring a few Ti atoms into the octahedron face-sharing relationship found at the CS planes in the lower oxides. For Ti^{m+} ions in interstitial sites, the composition should vary as $P_{O_2}^{-(m+1)^{-1}}$, and Tannhauser [47] has, indeed, found the $P^{-\frac{1}{5}}$ dependence, at very high temperatures, expected for fully ionized interstitials. It is, however, difficult to discriminate between the $P^{-\frac{1}{5}}$ and $P^{-\frac{1}{6}}$ (or some apparent law P^{-n} with $\frac{1}{6} < n < \frac{1}{4}$ if oxygen vacancies are not fully ionized). There is no doubt that interstitial sites play a dominating role in the diffusion of cations in rutile. This is highly anisotropic: ions of about the same radius as Ti^{4+} diffuse rapidly parallel to the c axis, i.e. down the tunnels, and for lithium diffusion the anisotropy factor is at least 10^8. The diffusion coefficient of transition-metal ions along c is about 10^{-6} cm^2 s^{-1} at 1273 K, whereas cations much larger than the Na^+ ion (e.g. the lanthanide cations) hardly diffuse with measurable speed up to 1473 K. The much slower diffusion normal to c probably takes place by a different (interstitialcy) mechanism.[48] Oxygen certainly diffuses by a vacancy mechanism, with a higher activation energy than for cation diffusion, and diffuses slower along

---

[42] M. E. Straumanis, T. Ejima, and W. J. James, *Acta Cryst.*, 1961, **14**, 453.
[43] K. S. Førland, *Acta Chem. Scand.*, 1964, **18**, 1267.
[44] P. Kofstad, *J. Phys. Chem. Solids*, 1962, **23**, 1579.
[45] T. Hurlen, *Acta Chem. Scand.*, 1959, **13**, 365.
[46] S. Andersson, B. Collén, U. Kuylenstierna, and A. Magneli, *Acta Chem. Scand.*, 1957, **11**, 1641.
[47] D. S. Tannhauser, *Solid State Comm.*, 1963, **1**, 223.
[48] P. I. Kingsbury, W. D. Ohlsen, and O. W. Johnson, *Phys. Rev.*, 1968, **175**, 1099; J. P. Wittke, *J. Electrochem. Soc.*, 1966, **113**, 193; H. B. Huntingdon and G. A. Sullivan, *Phys. Rev. Letters*, 1965, **14**, 177.

Shear Structures and Non-stoicheiometry

c than normal to c ($D_{1079 K} \parallel c$ 3.2×10^{-16}, $\perp c$ 1.7×10^{-15} cm^2 s^{-1}).[49] It might be expected that oxygen diffusion rates would vary with oxygen pressure in such a manner as to discriminate between the alternative forms of the defect equilibria. In fact, the rate is independent of P_{O_2}, for a reason that indicates the difficulty of categorizing the structure in the very composition range where one might hope to explore the defect rutile phase itself: the level of impurities ($\sim n \times 10^{-4}$) in any but exceptionally pure single crystal specimens is comparable with or greater than the stoicheiometric defect produced by changes in P_{O_2}, thereby shifting the inner defect equilibrium.

These considerations are relevant to the interpretation of the system as a whole, and later work has shown that the interaction between defects, and their elimination by crystallographic shear, introduces a further complication. Magneli and his co-workers [50] showed that a series of mixed valence intermediate phases, with the general formula Ti_nO_{2n-1}, existed between $TiO_{1.75}$ and $TiO_{1.90}$ corresponding to $n = 4, 5, 6 \ldots 10$), with regularly spaced CS planes, on the ($1\bar{2}1$) orientation, in which [TiO$_6$] octahedra shared faces. A definitive analysis of crystal structure was carried out on Ti_5O_9, and could be extended consistently to interpret the other members of the family. $Ti_{10}O_{19}$ was not unambiguously identified, but its existence as a member of the $\{1\bar{2}1\}$ family has since been established by electron microscopy. Between $Ti_{10}O_{19}$ and $TiO_{1.96}$, at least, there was evidence for further superstructure phases, but the system could not be elucidated further by X-ray diffraction methods. In discussing the mechanism of crystallographic shear, Anderson and Hyde [18] pointed out that CS planes other than the $\{1\bar{2}1\}$ plane were to be expected, notably the $\{1\bar{3}2\}$ planes which had been identified by Eikum and Smallman [51] as planar boundaries.

Application of electron microscopy and electron diffraction has shed new light on the composition range $TiO_{1.90}$—$TiO_{2.00}$. Slightly reduced rutile was found to contain planar boundaries, or planar defects with $\{1\bar{3}2\}$ orientation.[52] In material of composition $TiO_{1.986}$ these were randomly dispersed, forming a maze of uncorrelated faults on several (or all) of the $\{1\bar{3}2\}$ orientations. In $TiO_{1.94}$, the diffraction patterns displayed superstructure reflections lying along $<132>$, which can be regarded as arising from the regular recurrence of these planar defects, which can be unequivocally identified as CS planes. A new family of oxides is thereby generated, also represented by the general formula Ti_nO_{2n-1}. The compositions of such oxides can be deduced from the periodicity of the superstructure reflections along $<132>$ rutile, *i.e.* from the cell dimensions. From the standard techniques of fault extinction and fringe contrast observations on isolated $\{1\bar{3}2\}$ boundaries

---

[49] R. Haul and G. Dümbgen, *J. Phys. Chem. Solids*, 1965, **26**, 1; T. B. Gruenwald and G. Gordon, *J. Inorg. Nuclear Chem.*, 1971, **33**, 1151; V. I. Barbanel and V. N. Bogomolov, *Soviet Phys. Solid State*, 1970, **11**, 2160.
[50] S. Andersson and A. Magneli, *Naturwiss.*, 1956, **43**, 495, and ref. 46.
[51] A. Eikum and R. E. Smallman, *Phil. Mag.*, 1965, **11**, 627.
[52] L. A. Bursill, B. G. Hyde, S. Terasaki, and D. Watanabe, *Phil. Mag.*, 1969, **20**, 347.

it was confirmed [53] that the displacement vector at the defects was essentially $\frac{1}{2}<011>$, *i.e.* that they were CS planes with the same shear vectors as in the $\{1\bar{2}1\}$ CS compounds; on an idealized model, removal of one oxygen-only plane and operation of this displacement vector would collapse the structure by one-half of the interplanar spacing. It is known, however, from the structure of Ti_5O_9 (and also for $Mo_{18}O_{52}$ and Nb_3O_7F) that there is distortion at the CS plane; both metal atoms and oxygen atoms are displaced from their ideal sites to an extent that attenuates progressively from the CS planes to the mid-point between them. The effect is to expand the structure slightly at the CS planes, and the collapse factor for the $\{1\bar{2}1\}$ family is 0.34 instead of 0.50. The CS plane spacing is then directly related to the width of the rutile slabs between successive CS planes by the general relation $D = d_{1\bar{2}1} \times (n-0.34)$; $d_{1\bar{2}1} = 1.687$ Å. For the same reason, the displacement vector is not quite a perfect lattice vector: for $\{1\bar{3}2\}$ CS planes it is $\frac{1}{2}<0, 0.90, 0.90>$ rather than $\frac{1}{2}<011>$, giving a collapse factor of 0.45 and a CS plane spacing of $D = d_{1\bar{3}2} \times (n-0.45)$; $d_{1\bar{3}2} = 1.036$ Å. Dimensions so calculated accord with recorded diffraction patterns and with the spacings measured between fringes in direct lattice images.[54] and enable compositions to be deduced.

It has thus been established that the $\{1\bar{3}2\}$ homologous series covers a range of discrete structures from $Ti_{16}O_{31}$ ($TiO_{1.9375}$) to about $Ti_{40}O_{79}$ ($TiO_{1.975}$, with a spacing *ca*. 40 Å between CS planes).

Oxides in this composition range are, however, invariably inhomogeneous. Regions of crystal corresponding to a single n value, *i.e.* of perfectly uniform composition, are rare; a single reduced crystal invariably contains microdomains of different members of the series in random coherent intergrowth. These may be large enough to give coherent diffraction (sharp superstructure spots, possibly with two or more different periodicities overlaid on each other), especially in material with a composition corresponding to the lower members of the family, but may correspond to the intergrowth of isolated lamellae of a particular member. Any crystal particle of physically significant size can thus show a variability of composition but only one dimensional disorder. Such materials may not be in true inner equilibrium, even after annealing; it is certainly to be inferred that the difference in free energy per atom of Ti, for different members of the series, is so small that the driving force behind an ordering process is small. It is of interest that only the members of the $\{1\bar{3}2\}$ CS family with n even have been unambiguously identified, and that in the best ordered crystals, where extensive regions containing only two compositions were found, these invariably differed by $\Delta n = 2$ (*e.g.* $Ti_{18}O_{35} + Ti_{20}O_{39}, Ti_{20}O_{39} + Ti_{22}O_{43}$). The constitution of such substances is a problem that is not amenable to X-ray diffraction methods.

The question then arises as to the nature of rutile in the composition range $TiO_{1.98}$ to $TiO_{2.00}$. That isolated CS planes are already present after a very

---

[53] L. A. Bursill and B. G. Hyde, *Proc. Roy. Soc.*, 1970, **A320**, 147.
[54] (*a*) L. A. Bursill and B. G. Hyde, *Acta Cryst.*, 1971, **B27**, 210; (*b*) J. S. Anderson and R. J. D. Tilley, *J. Solid State Chem.*, 1970, **2**, 472.

slight degree of reduction has been mentioned. Hyde and Bursill [55] have shown that interaction forces between CS planes already operate to bring about incipient compositional segregation in material brought, by coulometric titration at 1273 K, to a known composition of $TiO_{1.9966}$.* Whereas most of the examined areas of such crystals contained only random and zig-zag $\{1\bar{3}2\}$ defects, there were occasional lamellae consisting of groups of 2—25 parallel, regularly-spaced fringes, implying that CS planes were lining up to define slabs of reduced oxide with the generic structure of the $\{1\bar{3}2\}$ family, and at spacings that reflected the energetics of CS plane interactions. This separation depended on the number of CS planes in any group, i.e. on cumulative, long-range interactions which optimize the configurational energy of a whole group. The forces are not wholly repulsive, since the CS planes do not spread out to occupy the whole volume available, and the interaction law must be of a form that leads to a broad shallow minimum in the potential energy curve. For a pair of CS plane fringes, the observed spacing was about 87 Å (corresponding to a local composition of $TiO_{1.988}$ for the lamella); in larger groups of ten or more fringes, the spacing changed progressively from about 63 Å ($TiO_{1.982}$) for the outer slabs to a nearly uniform 38 Å at the centre. If this is taken as the limiting CS plane spacing for an infinite lamella in coexistence with 'rutile', the limiting stable composition of the $\{1\bar{3}2\}$ series is about $Ti_{37}O_{73}$. ($TiO_{1.973}$).

The nature of the interaction forces remains a problem. It is evident that ordering in CS phases is not merely the optimization of repulsions,[56] and detailed consideration casts doubt on the possibility that elastic energy of distortion at the CS planes could generate a potential energy curve of the form required by the evidence (Figure 10). Crystallographic shear has been found only in compounds of transition metals, with d-orbitals available for metal–metal or metal–oxygen bonding, but the evidence of both binary and ternary CS phases shows that either d^0 or d^n configurations can be involved. No simple interpretation on the basis of electron delocalization appears valid.

The picture that emerges divides the composition range between $TiO_{1.75}$ and $TiO_{2.0000}$ into five or six structural ranges.

(I) $TiO_{1.75}$ to $Ti_{1.90}$:

$\{1\bar{2}1\}$ CS phases Ti_nO_{2n-1}, $4 < n < 10$

---

* It must be noted, however, that this small stoicheiometric deficit is comparable with the impurity concentration in high quality single crystals, so that the effects of impurity cations on the inner defect equilibria may not be negligible.

[55] L. A. Bursill and B. G. Hyde, *Phil. Mag.*, 1971, **23**, 3.
[56] S. Amelinckx and J. van Landuyt, 'The Chemistry of Extended Defects in Non-metallic Solids', ed. L. Eyring and M. O'Keefe, North Holland, Amsterdam, 1970, p. 295; J. M. Cowley, *ibid.*, p. 662; P. C. Clapp, *ibid.*, p. 663 and personal communications.

B

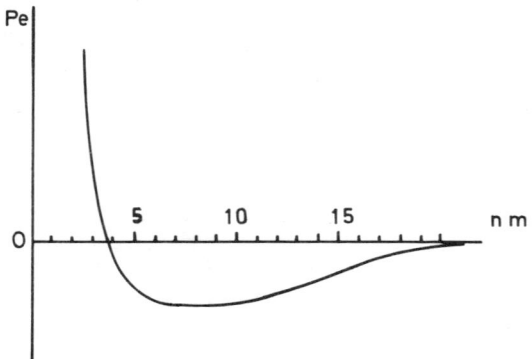

Figure 10 *Potential energy curve for interaction between CS planes, as deduced by B. G. Hyde and L. A. Bursill*

(II) $TiO_{1.90}$ to *ca.* $TiO_{1.9375}$.
a discontinuity between the $\{1\bar{2}1\}$ and $\{1\bar{3}2\}$ CS families. The coexistence of domains of $Ti_{10}O_{19}$ and $Ti_{16}O_{31}$ has been observed, but there is evidence that this composition range involves some new structural principles and cannot be treated as a classical univariant, biphasic coexistence of phases. This point is considered further, below.

(III) $TiO_{1.9375}$ to *ca.* $TiO_{1.9730}$:
$\{1\bar{3}2\}$ CS phases Ti_nO_{2n-1}, $16 < n < \sim 37$

(IV) $\sim TiO_{1.9370}$ to *ca.* $TiO_{1.999}$:
$\{1\bar{3}2\}$ CS planes group in lamellae defining high members of the Ti_nO_{2n-1} family, in coexistence with 'rutile' saturated with random Wadsley defects.

(V) $TiO_{1.999}$ to TiO_{2-x} (where x is certainly $\ll 0.001$):
Wadsley defects on all $\{1\bar{3}2\}$ orientations, at average spacings too great to result in realinement and aggregation.

(VI) TiO_{2-x} to $TiO_{2.00000}$:
Rutile with point defects. There is no evidence to indicate whether this range has any real existence.

5 Thermodynamics of Rutile CS Phases

The closely-spaced compositions of discrete structures in the range $TiO_{1.75}$ to $TiO_{2.00}$ raise interesting questions about the thermodynamics of the system: the increment in ΔG_f° per atom of titanium in traversing the series, the possibility of thermal fluctuations in composition which would relax the stoicheiometric invariability of the compounds and the conditions under which homogeneous, fully characterized members of the series might be formed.

There is no doubt about the high-temperature stability of the CS phases.[57] It was shown by Roy and Kachi that Ti_4O_7—Ti_7O_{13} were each stable in a

[57] G. Andersson. *Acta Chem. Scand.*, 1954, **8**, 1599.

restricted range of oxygen pressures up to 1673 K at least, and that members of the analogous $\{1\bar{2}1\}$ CS phase family V_nO_{2n-1} ($3 < n < 9$) derived from VO_2 were stable up to 1573 K. If, as is probable, their existence range extends to the melting point, the form of the fusion curve at different oxygen pressures [38] would indicate that they undergo a series of peritectic transformations. Thermogravimetric studies [58] of equilibrium with H_2–H_2O mixtures at 1303 K were first interpreted as indicating wide composition ranges for the Ti_4O_7 to Ti_9O_{17} series, but it was pointed out that hysteresis in the system could make the results compatible with the inference that the CS phases were of definite composition. Anderson and Khan [59] found that it was difficult to obtain reversible, hysteresis-free equilibria in either the Ti—O or the V—O systems below 1323 K but that at higher temperatures (1348—1423 K) reversible equilibria with gas buffers could be measured for Ti_4O_7—Ti_6O_{11} and V_3O_5—V_7O_{13}. Within the precision of the composition measurements, and for the increments of oxygen activity used, this work gave no evidence for any stoicheiometric variability. Galvanic cell measurements, with cells of the type Fe, $Fe_{0.95}O|ThO_2,15\%La_2O_3|Ti_nO_{2n-1}+Ti_{n+1}O_{2n+1}$ have been made [60] using adjacent pairs of oxides, but again at temperatures (1000—1300 K) where the inner equilibria of the solids are inevitably sluggish of attainment. Reference has been made [41,61] to the precise work by Zador on the chemical potential of oxygen across the phase field of TiO_{2-n}—ranges (V) and (VI) of the foregoing summary.

All this work was carried out before the structural complexity of the composition range $TiO_{1.90}$—$TiO_{2.00}$, and the existence of the $\{1\bar{3}2\}$ CS phases, were known. The more recent, extremely careful, work by Merritt and Hyde [62] is important because it covers the Ti—O system in the light of the existing structural knowledge. The discovery of the large series of $\{1\bar{3}2\}$ CS phases makes it evident that very fine gradations in the chemical potential of oxygen must be expected, and that a very high density of data, across the composition range, must be obtained if important features are not to be overlooked. This thermogravimetric study of equilibration with H_2–H_2O mixtures was carried out at 1300 K, and later work must show whether some of the conclusions will be modified at higher temperatures, where there is greater mobility in the solid-state equilibration processes. The results show that complete adjustment of solid-state processes is extremely slow. A true stationary composition at a given activity is not operationally obtainable; 'equilibrium' compositions drift slowly and indefinitely in both oxidation and reduction processes, and in no previous work (around that temperature,

---

[58] N. I. Bogdanova, G. P. Pirogovskaya, and S. M. Ariya, *Russ. J. Inorg. Chem.*, 1963, **8**, 401; B. G. Hyde and L. Eyring, *Zhur. neorg. Khim.*, 1965, **12**, 2837.
[59] J. S. Anderson and A. S. Khan, *J. Less Common Metals,* 1970, **22**, 219 (Ti—O system), 209 (V–O system).
[60] I. A. Vasil'eva and E. Y. Shavlova, *Russ. J. Phys. Chem.*, 1969, **43**, 1713.
[61] S. Zador, in 'Electromotive force measurements on high temperature systems', ed. C. B. Alcock, Institute of Metals, London, 1967; Ph.D. Thesis, University of London, 1969.
[62] R. R. Merritt and B. G. Hyde, to be published: personal communication.

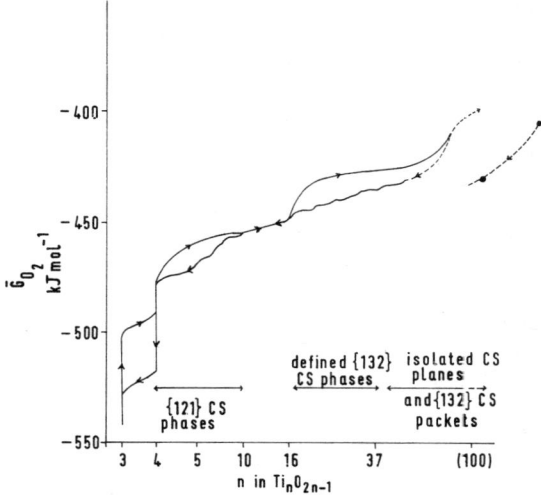

Figure 11 *Chemical potential of oxygen in titanium oxides as a function of composition: schematized from work of B. G. Hyde and R. R. Merritt. Composition axis plotted as log x (in TiO$_{2-x}$) to expand the phase field involving {132} CS defects*

at least) could it be said that stationary state conditions were reached. Corresponding to this, adjustment of composition and structure is attended with considerable hysteresis effects, and the precise course followed in any series of measurements depends upon the extent (and the direction) to which the initial state of the system has been altered. Step changes in $\bar{G}_{O_2}$ were not sharp, and did not correspond either exactly or reversibly to the ideal composition of the CS phases; they were more often observable during reduction than during oxidation processes. The variation of $\bar{G}_{O_2}$ with composition can be schematically represented as in Figure 11; it falls into several distinct regions (A—E below) which can be correlated with the manner in which structural changes are involved in attainment of an equilibrium state.

(A) TiO$_{1.6667}$—TiO$_{1.7500}$. Ti$_3$O$_5$ coexists with Ti$_4$O$_7$. Ti$_3$O$_5$ is not a CS phase, so that interconversion of the oxides is a reconstructive process. There is some hysterises, but the compositions of the phases are absolutely defined and the breadth of the hysteresis loop is not large compared with the changes in $\bar{G}_{O_2}$. The hysteresis can be ascribed to a high activation energy for the nucleation of one structure from the other.

(B) TiO$_{1.7500}$—TiO$_{1.90}$. In this range, covered by the $\{1\bar{2}1\}$ CS phases, a change in composition involves the introduction or elimination of CS planes, and their lateral migration to form the new ordered arrangement with the CS plane spacing of the product phase. There is a large hysteresis, greatly exceeding the height of a unit step in $\bar{G}_{O_2}$ from one phase to the next. Scanning paths within the hysteresis loop can simulate the behaviour of smeared out steps in the $\bar{G}_{O_2}$-composition curves.

(C) $TiO_{1.90}$—$TiO_{1.93}$. In this range it might appear that $\{1\bar{2}1\}$ CS phases and $\{1\bar{3}2\}$ CS phases should coexist; in any case a change of composition across this range implies that the CS planes must pivot round from one orientation to the other. It might be expected that this reorganization would involve considerable kinetic difficulties and crystalline disorder, resulting in irreversibility. In fact, this is the only part of the equilibrium diagram that displays true reversibility. On closer examination, an interesting structural feature emerges. Because of the lower density of sites on $\{132\}$ oxygen-only planes, and as the formulae for the CS plane spacings in $Ti-O_{2n-1}$ show, the CS planes in $Ti_{10}O_{19}$ (the uppermost known member of the $\{1\bar{2}1\}$ family) and in $Ti_{16}O_{31}$ (the lowest known member of the $\{1\bar{3}2\}$ family) are almost identical. To a first approximation, a change in composition between these limits involves no change in the *number* of CS planes per unit volume, but only in their *orientation*. This reorientation appears to be kinetically facile as compared with the introduction and parallel re-shuffle of additional CS planes. How it takes place, and the structure of the oxides in this narrow composition range, remain to be elucidated.*

(D) $TiO_{1.93}$—$TiO_{1.98}$. This is the range of the $\{1\bar{3}2\}$ CS phases, and changes in composition involve a change in the number and spatial density of CS planes. There is again a wide hysteresis loop, and since it is known that any crystal, as prepared, usually contains not a coexisting pair of phases, but lamellae of a multiplicity of compositions, a thermodynamic uncertainty is to be expected. Even if the only members of the series formed are those with n even, the width of hypothetical biphasic regions is very narrow (a change of only 0.0016 for x in TiO_x from $Ti_{34}O_{67}$ to $Ti_{36}O_{71}$). Stepped curves did, indeed, appear in reduction, with changes of only 150—360 cal mol^{-1} (0.65—1.5 kJ mol^{-1}) in $\bar{G}_{O_2}$ across the step, and the compositions at the steps suggested the formation of ordered intergrowths as well as discrete members, of $\{1\bar{3}2\}$ CS structures.

(E) $TiO_{1.981}$—$TiO_{1.986}$. Over this range, in which the structural evidence indicates the progressive aggregation and alinement of $\{1\bar{3}2\}$ CS planes into lamellae of ordered structure, the oxygen potential was a steep function of the composition, and the behaviour appeared almost reversible. The measurements of Merritt and Hyde did not extend above $TiO_{1.986}$, leaving a composition gap and an apparent discordance with the careful measurements of Zador, which related to composition for which the oxygen deficiency was accommodated, altogether or in part, by randomly disordered $\{1\bar{3}2\}$ CS planes. It appears likely that there is, in this region, yet another broad hysteresis loop associated with the progressive introduction of CS planes.

Whatever the interpretation placed upon the hysteresis phenomena, this oxide system is defined to the extent of showing that the differences in ΔG_f° and $\bar{G}_{O_2}$, between different members of the structural series, are exceedingly

---
* See footnote on p. 53.

small. This has wider implications. It means that nearly all the work that has been carried out on the properties of reduced rutile and the intermediate phases relates to uncharacterized, inhomogeneous materials. The extent to which preparations obtained by solid-state reactions can attain uniformity or order, even after prolonged annealing at the temperatures commonly used ($\not> 1273$ K) is very uncertain, and X-ray diffraction methods (especially Debye–Scherrer work on polycrystalline specimens) are incapable of detecting the imperfections and inhomogeneities that have been discussed above.

Considerable interest attaches, however, to the physical properties—especially the electronic conductivity, magnetism, and optical properties—of mixed valence CS phases, so that methods of preparing stoicheiometrically defined and perfectly ordered crystals are potentially important. This may be possible for the $\{1\bar{2}1\}$ CS oxides, which are sufficiently widely separated in their thermodynamic properties. Crystals can be grown from a borate flux in an atmosphere of controlled oxygen activity (H_2—H_2O, CO—CO_2, or H_2–CO_2 gas buffer), to be determined from the thermodynamic data,[63] and the members of the $\{1\bar{2}1\}$ CS family, with $n = 4$—10, have been obtained in this way.[64] Crystals can also be obtained by chemical vapour transport, and since an isolated system should be self-buffering with respect to the oxygen activity, homogeneous crystals should be formed providing that fluctuations in the transport system are small. Anderson and Tilley (unpublished) have grown Ti_4O_7—Ti_8O_{15} in H_2—HCl transport buffers, and Japanese workers have found that $TeCl_4$ is a good transport agent for titanium and vanadium oxide $\{1\bar{2}1\}$ CS phases.[65,66] The HCl-transported material showed coherent intergrowth of domains of compositionally adjacent phases,[54b] each perfect and free from faults, a few hundred Å in extent; crystals prepared by other methods have not been examined for their ultra-structure.

Complex and inconsistent results obtained on microcrystalline preparations of $\{1\bar{2}1\}$ CS phases can therefore be disregarded, but it is now established that Ti_4O_7 crystals, grown either by vapour transport[65] or from a flux at controlled oxygen potential,[67] are metallic at room temperature (resistivity about 10^{-3} ohm cm^{-1}) with a metal $\leftrightarrows$ semiconductor transition at 149 K, attended with a 10^3-fold increase in resistivity. The conductivity appears to be greater parallel to $[010](Ti_4O_7)$, *i.e.* $\|[001](TiO_2)$, that is for directions lying *in* the CS plane, than transverse to the CS planes.

The Hall mobility of electrons in the metallic state is low: 1 cm^2 volt^{-1} s^{-1} at 290 K, 4 cm^2 volt^{-1} s^{-1} at 160 K. This low mobility has been interpreted in terms of small polarons localized around lattice defects.[68] The 149 K transition is accompanied by a sharp decrease in the paramagnetic

---

[63] J. S. Berkes and R. Roy, *J. Appl. Phys.*, 1965, **36**, 3276.
[64] R. F. Bartholomew and W. B. White, *J. Crystal Growth*, 1970, **6**, 249.
[65] K. Nagasawa, Y. Kato, Y. Bando, and T. Takada, *J. Phys. Soc. Japan*, 1970, **29**, 241.
[66] K. Nagasawa, Y. Bando, and T. Takada, *Jap. J. Appl. Phys.*, 1969, **8**, 1262, 1267.
[67] R. F. Bartholomew and D. R. Frankl, *Phys. Rev.*, 1969, **187**, 828.
[68] V. N. Bogomolov and V. P. Zhuza, *Soviet Phys. Solid State*, 1967, **8**, 1904; T. Goto and T. Okada, *J. Phys. Soc. Japan*, 1968, **25**, 289.

susceptibility, but is not an antiferromagnetic transition. There may be a second semiconductor ⇌ semiconductor transition at about 125 K, attended by a small change in structure and an increase in resistivity, but without magnetic effects. Ti_5O_9 and Ti_6O_{11} also show transitions with conductivity changes at about 130 K, but are not truly metallic above the transition temperature since their conductivity increases with temperature. The ~130 K transitions are accompanied by a sharp change in magnetic susceptibility;[69] above them, good Curie–Weiss behaviour is followed but earlier work [70] suggested that the low temperature state is antiferromagnetic. Keys and Mulay consider that the titanium atoms in the CS planes have Ti—Ti distances less than the Mott critical distance R_c^{3d} for effective d-electron overlap, so that there is electron delocalization within the CS plane. Between CS planes, distances are greater than R_c^{3d}, so that thermal excitation is necessary for electron transport. The transitions are then to be regarded as the onset of electron localization, and may be accompanied by some (as yet undetected) structural modification that tends to equalize the Ti—Ti distances. It is not yet clear that the magnetic behaviour is satisfactorily explained, and until enough work has been done on perfect crystals, the problems and inconsistencies are likely to remain. How far the tervalent cations are localized in the CS planes is uncertain; the fine structure of the $K_{\beta_5\beta''}$ emission bands of Ti_4O_7, $Ti_{10}O_{19}$ (supposedly, but this member of the $\{1\bar{2}1\}$ CS family has not been obtained as a homogeneous, well characterized, macroscopic preparation) and $TiO_{1.983}$ does not differentiate between titanium atoms in two discrete valence states,[71] pointing to a single Brillouin zone structure rather than to non-equivalence of atoms in the CS planes and in the rutile slabs respectively. Nothing is known of the properties of the $\{1\bar{3}2\}$ CS phase region of the system and, for reasons that will be apparent in the light of the structural information summarized earlier, it is difficult to assess the great volume of work that has been devoted to 'slightly reduced' rutile. This has long been known as an n-type semiconductor [72] and, as indicated earlier, carrier populations have legitimately been used to determine the stoicheiometric deficit in TiO_{2-x}. Assignments to a rather subtle set of impurity levels close to the conduction band [73] have usually been based on an interstitial point defect model.[74] With the evidence that some, possibly almost all, point defects are subsumed in CS boundaries, the whole subject needs reinterpretation. Only in the most recent work [75] has there been an attempt to take into account the formation and ordering of CS boundaries.

[69] L. N. Mulay and W. J. Danley, *J. Appl. Phys.*, 1970, **41**, 877.
[70] L. K. Keys and L. N. Mulay, *Jap. J. Appl. Phys.*, 1967, **6**, 122; *J. Appl. Phys.*, 1967, **38**, 1466; *ibid.*, 1968, **39**, 598; *Phys. Rev.*, (A), 1967, **154**, 453.
[71] V. I. Chukov and E. E. Vainshtein, *Izvest. Akad. Nauk, S.S.S.R., Inorganic Materials*, 1967, **3**, 910.
[72] W. Kleber, H. Pabst, and W. Schroeder, *Z. Phys. Chem.*, 1960, **215**, 63.
[73] *E.g.* R. R. Hasiguti, N. Kawamiya, and E. Yagi, *J. Phys. Soc. Japan*, 1964, **19**, 573.
[74] *E.g.* R. N. Blumenthal, J. Coburn, J. Baukus, and W. M. Hirthe, *J. Phys. Chem. Solids*, 1966, **27**, 643; R. N. Blumenthal, J. C. Kirk, and W. M. Hirthe, *ibid.*, 1967, **28**, 1077.
[75] *E.g.* C. W. Chu, *Phys. Rev.* (B), 1970, **1**, 4700.

Although the V—O system between V_3O_5 and VO_2 has not yet been studied by electron microscopy, there is little reason to suppose that it will not reveal the same kind of complexity as the Ti—O system. Earlier work on the {121} CS phases was somewhat conflicting, but the carefully prepared polycrystalline materials examined by Kosuge [76] indicate paramagnetic ⇌ paramagnetic transitions for all members, with the transition temperature moving progressively downwards as n increases. More recent work on vapour transport grown single crystals[66] shows that V_3O_5 is an n type semiconductor (activation energy 0.4 eV) down to 130 K, but V_6O_{11} has a metal ⇌ semiconductor transition, with a 10^4-fold change in resistivity, at 177 K, and V_7O_{13} is metallic down to 120 K.

6 Ternary Defect Rutile Oxides

The cation : anion ratio in rutile can be altered by the incorporation of elements differing in valency from the Ti^{4+} cations, as well as by reduction. It has already been pointed out that the tervalent cations of Al, Fe, and Cr are commonly present at the 100 p.p.m. level in high-grade single crystal rutile, and that they may not be without influence on the defect structure and thermodynamics in the interesting region of composition $TiO_{1.998}$ to $TiO_{2.0000}$. At higher concentrations, the size, charge, and d-electron character of substituent cations can, in principle, be used to probe the conditions under which CS planes are formed in oxygen-deficient structures, but if the altervalent ions have to be segregated in the CS planes, it must be borne in mind that the development of an equilibrium structure depends upon solid-state cation diffusion and ordering processes. Whilst the high cation diffusion rates parallel to [001] rutile favour the facile entry of ions of 0.6—0.8 Å radius, subsequent readjustments of the structure depend upon the much slower diffusion parallel to < 100 >.

The first important ternary substituent is hydrogen, which can readily enter the rutile structure [77] and is said to be invariably present in flame-grown Verneuil crystals [78] but can be avoided by using an argon–oxygen plasma torch.[79] Flame-grown crystals have been found to contain hydrogen roughly equivalent to the tervalent impurities, and as it is essentially bound to oxygen, as hydroxide ions, it provides compensation for the anion deficiency that would otherwise involve structural defects. It contributes a time-dependent component to the electrical conductivity, diffusing down [001] by a mechanism analogous to the proton conductivity of ice. It is lost as water—bringing about an equivalent reduction—in vacuum at 873 K, and the formation of n-type conducting rutile by heating in vacuum, in many physical investigations, is probably to be ascribed to this process rather than to the stoicheiometric loss of oxygen itself. Hill has suggested that, in the surface

[76] K. Kosuge, *J. Phys. Chem. Solids*, 1967, **28**, 1613.
[77] A. von Hippel, J. Kolnais, and W. B. Westphal, *J. Phys. Chem. Solids*, 1962, **23**, 779.
[78] G. J. Hill, *Brit. J. Appl. Phys.* [ii], 1968, **1**, 1151.
[79] J. D. Chase and L. J. van Ruyven, *J. Crystal Growth*, 1969, **5**, 294.

step, where two hydroxide ions are converted into a desorbed H_2O molecule and an oxygen vacancy (even if this has only a transient existence) the consequential loss of screening of Ti^{4+} ions may lead to just such a readjustment of cation positions as creates the groups of face-sharing $[TiO_6]$ octahedra that nucleate a CS plane. The extent to which hydrogen can enter reduced rutile, to compensate Ti^{3+} ions, is not clear but, in view of the firm retention of hydrogen, shown by its presence in Verneuil crystals, the question arises whether it may complicate the proper interpretation of the structure and thermodynamics of slightly sub-stoicheiometric oxide equilibrated with H_2—H_2O buffers.

In terms of the point defect model, tervalent cations are regarded as compensated, at least at low levels, by titanium interstitials:

M_2O_3 in TiO_2: $4M^{3+} + 6O^{2-} + Ti^{\times}_{Ti} \rightarrow 4M'_{Ti} + 6O^{\times}_{o} + Ti^{4+}_{I}$;

alternatively, but less favoured, oxygen vacancies may be formed:

$$2M^{3+} + 3O^{2-} \rightarrow 2M'_{Ti} + 3O^{\times}_{o} + V^{\cdot\cdot}_{o}$$

The question is then whether the point defects remain as such, trapped in the vicinity of the randomly distributed M^{3+} cations, whether they cluster in some way, or whether they are transformed by crystallographic shear, as in rutile itself.

Since vanadium forms $\{1\bar{2}1\}$ CS phases like titanium, it has been generally assumed that the Ti–V–O system between M_3O_5 and MO_2 would form solid solutions exactly analogous to rutile itself, but no experimental work is on record. The Ti–Cr–O system does show a close resemblance to the Ti–O system. With 15—33 mol % $CrO_{1.5}$, a family of oxides $Cr_2Ti_{n-2}O_{2n-1}$ is formed,[80] with $\{1\bar{2}1\}$ CS planes. Direct lattice imaging methods [81] have confirmed that these are well ordered, though consecutive members of the series can coexist in coherent domains within a single crystal, and have extended the series to an upper limit with $n = 13$. In a systematic phase study, Flörke and Lee [82] showed that the CS phases are essentially high-temperature phases; they undergo progressive disproportionation in favour of the higher members of the series, i.e. in favour of producing wider slabs of rutile structure, as the temperature is reduced (Figure 12). This tendency to favour more complex patterns of ordering at high temperatures, and the attainment of very perfect order in so doing, is one of the puzzling aspects of crystallographic shear. At lower chromium oxide concentrations, where Flörke and Lee report a rutile solid solution and additional unidentified intermediate phases, electron microscopy shows a close parallelism with the binary system. There is a second homologous series $Cr_2Ti_{n-2}O_{2n-1}$ based on $\{1\bar{3}2\}$ CS planes, of which the $n = 21$ and 22 members were obtained at 1570 K and the $n = 27$—29 members only at lower temperatures; these were well

---

[80] S. Andersson, A. Sundholm, and A. Magneli, *Acta Chem. Scand.*, 1959, **13**, 989.
[81] R. M. Gibb and J. S. Anderson, to be published.
[82] O. W. Flörke and C. W. Lee, *J. Solid State Chem.*, 1969, **1**, 445.

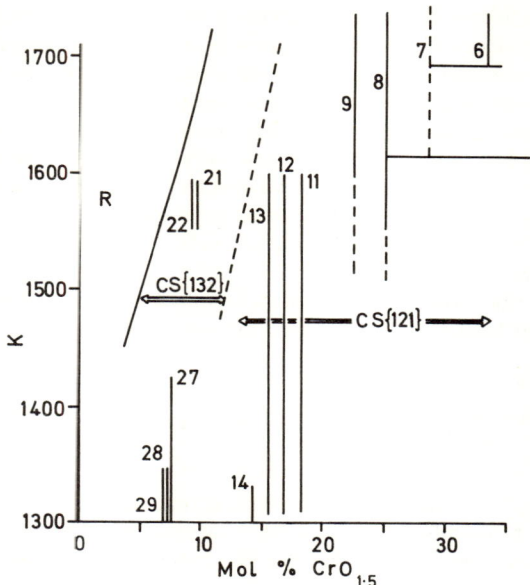

Figure 12 *Equilibrium diagram of the TiO_2—$CrO_{1.5}$ system. High-temperature data from O. W. Flörke and C. W. Lee, ref. 82; bars represent temperature ranges for formation of {121} and {132} CS phases (results of R. M. Gibb and J. S. Anderson). Compositions of these phases indicated by number against each bar, being n in $Cr_2Ti_{n-2}O_{2n-1}$. R = existence range of rutile with random {132} CS defects*

ordered. With 0—5 mol % $CrO_{1.5}$, materials annealed in oxygen showed no planar defects but, as pointed out by Flörke and Lee, low concentrations of chromium may well be incorporated as Cr^{4+} to build up a perfect rutile structure, so that the true anion : cation ratio differs from that of the Cr_2O_3–TiO_2 solid solutions. Slight reduction—e.g. in the electron beam of the microscope—introduced a maze of planar boundaries of {1$\bar{3}$2} orientation which could be confirmed as CS planes. With increasing chromium concentration these tended to aggregate into lamellae of the {1$\bar{3}$2} CS phases, coexisting with a rutile matrix which is traversed by random {1$\bar{3}$2} faults (Figure 13).

There is thus a close correspondence between the Cr^{III}–Ti^{IV} oxides and the Ti^{III}–Ti^{IV} oxides. On the basis of calculations by Sahl,[83] Flörke and Lee argued that Cr^{3+} must enter interstitial octahedral sites rather than the regular octahedral sites, giving a local configuration like that of chromium in corundum structure Cr_2O_3. There is a dimensional misfit, and strain can be relieved by aggregation of these local groupings into a slice of corundum structure which constitutes an isolated CS plane. It might then be expected that other tervalent cations of similar ionic radius would produce similar

[83] K. Sahl, *Acta Cryst.*, 1965, **19**, 1027.

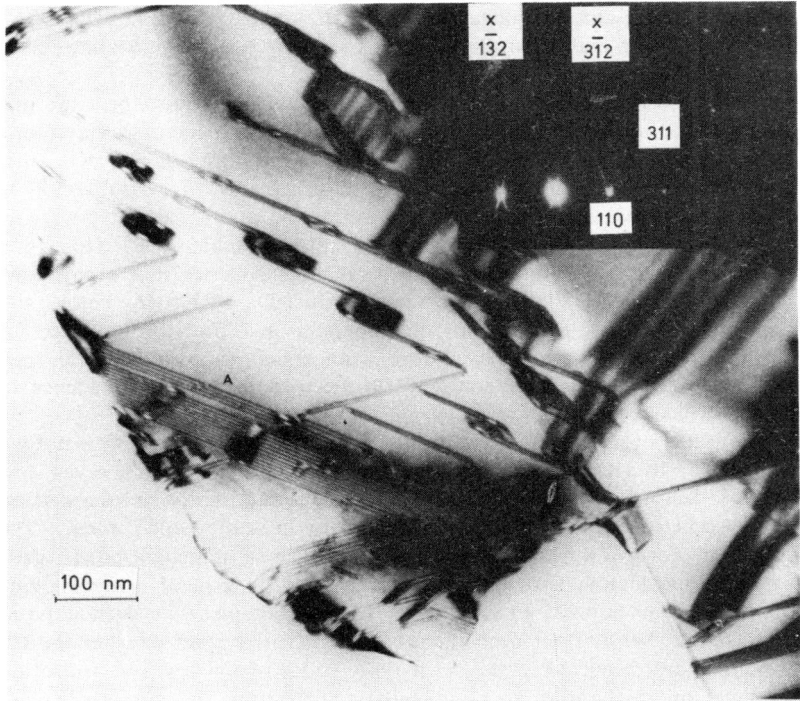

Figure 13 Random {132} CS planes in $Cr_{0.02}Ti_{0.98}O_{1.99}$, on several equivalent orientations. At A, a group of CS planes is clustered at nearly regular spacings

effects, but this is not the case. Gibb and Anderson (unpublished) have found that, within the apparent solid solution range, rutile doped with Al_2O_3 shows no planar boundaries or faults; with more than a few mol % $AlO_{1.5}$, the pseudobrookite phase Al_2TiO_5 is precipitated without the intervention of intermediate phases. With Ga^{3+} and Fe^{3+}, planar defects are formed, but these differ in orientation and displacement vector from the CS boundaries considered above: they are on {210} rutile, with a displacement vector close to $\frac{1}{2}<010>$, but not quite a perfect lattice vector. With iron doping, only random {210} faults are found in dilute systems, while Fe_2TiO_5 is precipitated at higher iron concentrations. In gallium-doped rutile, however, these {210} CS planes* can order themselves, when the oxides are annealed at 1573K, to generate a new homologous series of compounds, $Ga_4Ti_{n-4}O_{2n-2}$. These are rarely perfectly ordered, but lamellae of individual members with $19 < n < 29$ (n odd, only) have been observed. Some observa-

---

* The structure of the (210) boundaries between rutile slabs may differ, strictly, from that described as a CS operation.

tions on naturally occurring rutiles suggest that regular ordering of {210} CS planes can take place in iron-doped samples also, in the long annealing period available for geological specimens.[83a]

As is shown in Figure 4, a crystallographic shear is also possible that would lower the co-ordination number of oxygen and increase the anion : cation site ratio. Since rutile is reported to take a few mol % of $NbO_{2.5}$ into solid solution, and has been shown [84] to accept M : O ratios < 2.06 in Ta_2O_5–TiO_2 solid solutions, it might be expected that these systems would provide evidence of hyperstoicheiometric crystallographic shear. This is not the case. Niizeki [85] found no X-ray diffraction evidence that atoms were located on sites other than normal rutile sites in TiO_2–4%Nb_2O_5; Fender and Collins (private communication) find no neutron diffraction evidence for clustering or any other than a random arrangement of point defects in such materials; Gibb and Anderson find no electron microscopic evidence of planar defects.

The status of defect models of non-stoicheiometry in rutile is thus not yet cleared up. In TiO_{2-x} and $Cr_{2x}Ti_{1-2x}O_{2-x}$ it would appear that the very low intrinsic Schottky–Wagner level of point defects can hardly be exceeded, so that stoicheiometric defects are eliminated by crystallographic shear. The conditions for this appear, however, to involve some unrecognized energetic or electronic factors, so that the doped rutiles, in general, accommodate their non-stoicheiometry in some form (as local complexes or even as point defects) that has not yet been identified. It is not known whether the CS structures are restricted to the rutile-type oxides of the first row transition elements, or whether they are characteristic of the rutile structure *per se* (*e.g.*, possible in the MF_2 fluorides). Some other compounds of this structure, *e.g.* IrO_2, are reported as being non-stoicheiometric, and the thorough study devoted to TiO_2 needs to be extended to other compounds.

7 Double Shear and Block Structures

The compounds and the phenomena considered so far have arisen from either regularly recurrent or irregular crystallographic shear on a single orientation. In the ReO_3 parent structure a complex chemistry of perfected and faulted structures is generated by the operation of two CS operations. In a formal sense, recurrent {100} CS planes create slabs n octahedra wide, as in R-Nb_2O_5 ($n = 2$) or Nb_3O_7F ($n = 3$); a second shear parallel to [001] creates the different slab structure of P-Nb_2O_5 or, normal to [001], the *block structures* to be considered next.

The characteristic of these is that pillars of ReO_3 structure, infinite in one dimension (taken as the b axis), are formed by edge-sharing at their faces with similar (but not necessarily identical) pillars of ReO_3 structure. Metal atoms

---

[83a] P. Buseck and J. S. Anderson, unpublished results.
[84] W. Mertin, R. Gruehn, and H. Schäfer, *J. Solid State Chem.*, 1970, **1**, 425.
[85] N. Niizeki, *Adv. in X-ray Analysis*, 1968, **11**, 482.

Shear Structures and Non-stoicheiometry

retain more or less undistorted octahedral co-ordination but are at two levels, displaced by one half the octahedron diagonal in adjacent blocks. The oxygen sublattice of the ReO_3 structure is thereby left unaltered, but the co-ordination number of oxygen atoms may be 1 at unshared block corners, 2 in the interior of a block, 3 at the edges or 4 at shared corner sites. Changes in the block sizes and in their polyhedron-sharing thereby bring about the most subtle changes in anion : cation ratio. The pillars of ReO_3 structure share faces, but their corners provide three possible types of junction: (a) unshared, so as to create tunnels with tetrahedral cation sites, as in $WNb_{12}O_{33}$ (Figure 14), (b) sharing corner octahedron edges at the same level, as in $M\text{-}Nb_2O_5$ (Figure 14b), or (c) overlapping with blocks at the same level along an edge, as in $Nb_{12}O_{29}$ (Figure 14c). The remarkable character of the chemistry of the block structures comes from the fact that every possible combination of block sizes and block linkages, provided that it satisfies the constraint of preserving the integrity of the oxygen sublattice, represents a possible structure and has a high degree of local short-range order. Every such combination that provides a regular alternation of blocks, no matter how large the repeating unit, must be defined as a distinct compound.

Symbolism.—A block may be defined as having the dimensions $m \times n$ octahedra in plan and infinite in the third (b) direction; the operation of a third CS operation, transverse to [010] and dividing the structure into finite blocks, is conceivable and will no doubt be discovered in due course, but a dispensation of Providence has enabled the sufficiently complex problem of double shear to be explored first. Corner linkages of types (b) and (c) join the blocks at any one level into groups or ribbons of p blocks (p usually 1, 2, or ∞ in simple structures) so that a given set of blocks can be represented by the dimensional symbol $(m \times n)_p$. If p is finite, tetrahedral cation sites are created at the free corners. The composition of such a unit is then given by $M_{mnp+1}O_{3mnp-p(m+n)+4}$. For a structure involving two or more distinct block sizes of types $i, j \ldots$, the total chemical formula is then

$$\sum_i M_{mnp+1}O_{3mnp-p(m+n)+4}$$

It can readily be seen that the same chemical formula may be obtained from different block structures, which represent polymorphic forms with, presumably, very small differences in lattice energy; this is exemplified by probably ten of the fourteen known modifications of Nb_2O_5.

In representing block structures it is convenient, and sufficient, to indicate the rectangle formed by the metal atoms of the bounding octahedra, as is shown in Figure 14 and below. The $[MO_6]$ octahedra are polarized and distorted, but differ little in effective dimensions (diameter ~ 4 Å) for all the known structures. The metric of all block structures is therefore determined by the dimensions of the ReO_3 subcell. This has been of great advantage in applying electron microscope techniques to the structural interpretation of

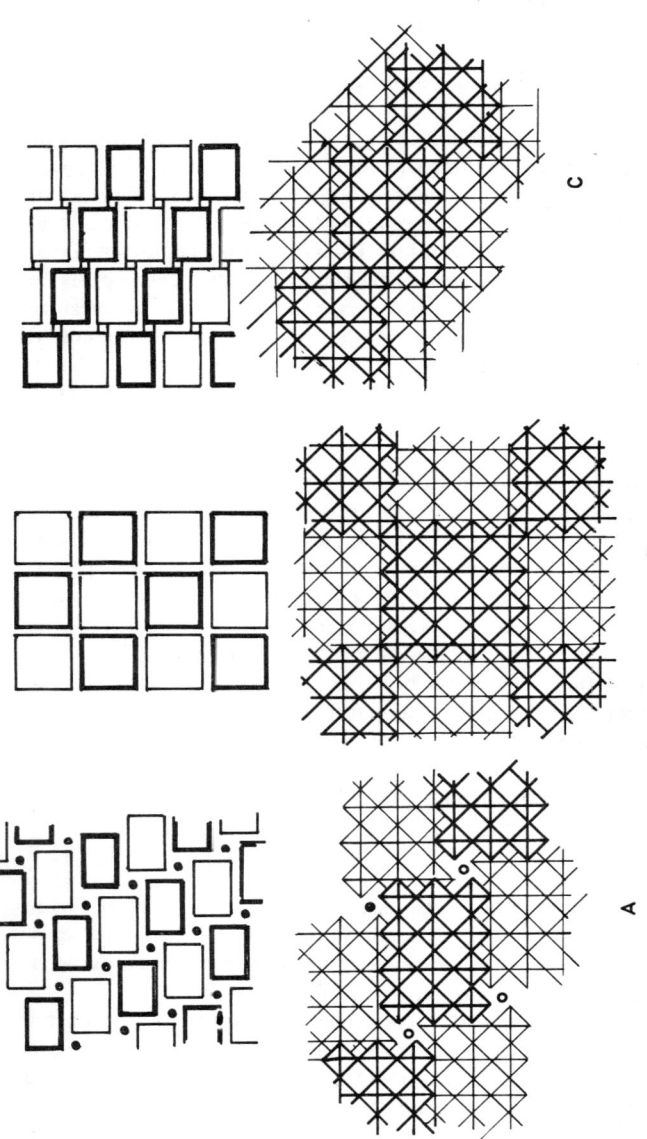

Figure 14 *Block structures, showing modes of linkage between blocks of octahedra at the same level. Lower diagrams: structures drawn out to display [MO$_6$] octahedra; upper diagrams: schematic representation of same structures by rectangles delineating metal atom positions along faces of blocks, with thick and thin outlines to distinguish blocks of octahedra at $z = 0$ and $z = \frac{1}{2}$ respectively*
(a) *Isolated blocks* WNb$_{12}$O$_{33}$, $(3 \times 4)_1$ *with tetrahedral sites*
(b) *Junction through oxygen atoms of corner octahedra.* M-Nb$_2$O$_5$, $(4 \times 4)_\infty$
(c) *Ribbons formed by overlapping edge-sharing between blocks at the same level.* Nb$_{12}$O$_{29}$, $(3 \times 4)_\infty$

block compound chemistry, since measurements of dimensions at the level of the individual unit cell, suffice to furnish an unambiguous identification of the structural units.

Ordered Block Structures.—Formation of block structures shows a high degree of chemical specificity: it is almost restricted (on current knowledge) to oxides and oxide fluorides of niobium. Variations in the anion : cation ratio r (in MX_r) imply either replacement of O^{2-} by F^- (for $M = Nb^V$, $r > 2.500$) or, in the oxides, replacement of Nb^{5+} by cations of lower charge (Nb^{4+}, Ti^{4+} etc., $r < 2.500$) or of higher charge (W^{6+}; $r > 2.500$). Tantalum can partly replace niobium, up to 50 atom % in $Ti_2(Nb,Ta)_{10}O_{29}$, about 40 atom % in $Ti(Nb,Ta)_{24}O_{62}$, and 15 atom % in $H\text{-}(Nb,Ta)_2O_5$,[84]—but tantalum, in the ternary oxides derived from $T\text{-}Ta_2O_5$, has its own structural method of accommodating small changes in stoicheiometry to perfect local order.[86] The crystal structure work of Wadsley and others does not indicate any preferential distribution of quinquavalent and altervalent cations, but from the standpoint of electrostatic charge distribution and local electroneutrality it would seem that cations of lower charge should be preferentially located in the shared corners and possibly the CS plane faces of blocks. This effect is shown in $(W_{0.35}V_{0.65})_2O_5$,[87] which is isostructural with $M\text{-}Nb_2O_5$, with $(4 \times 4)_\infty$ blocks; site occupancy in the corner junction was $(W_{0.12}V_{0.88})$, as compared with $(W_{0.42}V_{0.58})$ in the interior and face octahedra. With this near random cation distribution, there is a close parallel between the Ti–Nb oxides and the binary oxides between $NbO_{2.5000}$ and $NbO_{2.4167}$, but the latter are of particular interest in that they are continuously interconvertible and indicate how the system adjusts itself to stoicheiometric change.

Table 3 lists the block structures that have been obtained as macroscopic crystals and subjected to X-ray crystal structure determination or identified as domains of regular structure, at least as lamellae a few unit cells in thickness, by electron microscopy. There are some outstanding discrepancies, however, between the formulae inferred from crystal structures and the compositions of phases identified in systematic preparations and equilibrium studies. By the usually applied criteria, $H\text{-}Nb_2O_5$ has a range of composition extending to $NbO_{2.489}$ [88] or rather lower,[89] and $NbO_{2.480}$ is reported as biphasic; X-ray diffraction evidence indicated that the $Nb_{25}O_{62}$ structure could be in some way oxygen-deficient, and extended over the range $NbO_{2.472}$—$NbO_{2.478}$ at 1573 K. However, a quite distinct phase of irrational composition, $NbO_{2.483}$, could not only be reproducibly prepared and grown

---

[86] R. S. Roth, J. L. Waring, and H. S. Parker, *J. Solid State Chem.*, 1970, **2**, 445; R. S. Roth and N. C. Stephenson, in 'Chemistry of Extended Defects in Non-metallic Solids', ed. L. Eyring and M. O'Keeffe, North Holland, Amsterdam, 1970, p. 167.
[87] M. Israelsson and L. Kihlborg, *Arkiv. Kemi*, 1968, **30**, 129.
[88] H. Schäfer, D. Bergner, and R. Gruehn, *Z. anorg. Chem.*, 1969, **365**, 31.
[89] A. Burdese, E. V. Tkachenko, and F. Abbatista, *Izvest. Akad. Nauk. S.S.S.R., Inorganic Materials*, 1969, **5**, 1666.

Table 3 Known block structures

Anion: cation ratio	Unit cell formula	Block structure	Compounds
2.4167	$M_{12}X_{29}$	$(3\times4)_\infty$	$Nb_{12}O_{29}{}^a$ $Ti_2Nb_{10}O_{29}{}^b$
2.4490	$M_{49}X_{120}$	$(3\times4)_\infty+(3\times4)_2$	$Ti_5Nb_{44}O_{120}{}^c$
2.4545	$M_{22}X_{54}$	$(3\times4)_\infty+(3\times3)_1$	$Nb_{22}O_{54}{}^d$
2.4595	$M_{37}X_{91}$	$(3\times4)_\infty+(3\times4)_2$	$Ti_3Nb_{34}O_{91}{}^c$
2.4667	$M_{15}X_{37}$	$(3\times5)_\infty$	$MgNb_{14}O_{35}F_2{}^e$
2.4681	$M_{47}X_{116}$	$(3\times4)_\infty+(3\times3)_1$	$Nb_{47}O_{116}{}^f$
2.4800	$M_{25}X_{62}$	$(3\times4)_2$	$Nb_{25}O_{62}{}^d$ $TiNb_{24}O_{62}{}^g$
2.4872	$M_{78}X_{194}$	$(3\times4)_2+(3\times4)_1+(3\times5)_\infty$	$Nb_{39}O_{97}$ $TiNb_{38}O_{97}{}^c$
2.4906	$M_{53}X_{132}$	$(3\times4)_2+(3\times4)_1+(3\times5)_\infty$	$Nb_{53}O_{132}{}^{h,\ c}$
2.5000	$M_{28}X_{70}$	$(3\times4)_1+(3\times5)_\infty$	$H\text{-}Nb_2O_5{}^i$
	$M_{16}X_{40}$	$(3\times4)_\infty$ (2 modes of linkage)	$M\text{-}Nb_2O_5{}^j$ $N\text{-}Nb_2O_5{}^k$
	$M_{10}X_{25}$	$(3\times3)_1$	$PNb_9O_{25}{}^{l,\ m}$
2.5073	$M_{69}X_{173}$	$(3\times4)_1+(3\times5)_\infty$	$WNb_{68}O_{173}{}^n$
2.5085	$M_{59}X_{148}$	$(3\times4)_1+(3\times5)_\infty+(3\times5)_2$	$Nb_{59}O_{147}F^o$
2.5122	$M_{41}X_{103}$	$(3\times4)_1+(3\times5)_\infty$	$WNb_{40}O_{103}{}^n$
2.5161	$M_{31}X_{78}$	$(3\times5)_2$	$Nb_{31}O_{77}F^p$
2.5185	$M_{27}X_{68}$	$(3\times4)_1+(3\times5)_\infty$	$WNb_{26}O_{68}{}^n$
2.5231	$M_{65}X_{164}$	$(3\times5)_1+(3\times5)_2+(3\times6)_\infty$	$Nb_{65}O_{161}F_3{}^o$
2.5294	$M_{34}X_{86}$	$(3\times5)_1+(3\times6)_\infty$	$Nb_{34}O_{84}F_2{}^q$
2.5385	$M_{13}X_{33}$	$(3\times4)_1$	$WNb_{12}O_{33}{}^r$
2.5625	$M_{16}X_{41}$	$(3\times5)_1$	$MoNb_{15}O_{40}F^s$
2.5667	$M_{30}X_{77}$	$(3\times4)_1+(4\times4)_1$	$W_4Nb_{26}O_{77}{}^t$
2.5882	$M_{17}X_{44}$	$(4\times4)_1$	$W_3Nb_{14}O_{44}{}^u$
2.6190	$M_{21}X_{55}$	$(4\times5)_1$	$W_5Nb_{16}O_{55}{}^r$
2.6538	$M_{26}O_{69}$	$(5\times5)_1$	$W_8Nb_{18}O_{69}{}^u$

[a] R. Norin, *Acta Chem. Scand.*, 1963, **17**, 1391; 1966, **20**, 871. [b] A. D. Wadsley, *Acta Cryst.*, **14**, 664, 1961. [c] J. G. Allpress, *J. Solid State Chem.*, 1969, **1**, 66. [d] R. Norin, M. Carlsson, and B. Elgquist, *Acta Chem. Scand.*, 1966, **20**, 2892. [e] M. Lundberg, *J. Solid State Chem.*, 1970, **1**, 463. [f] R. Gruehn and R. Norin, *Z. anorg. Chem.*, 1968, **367**, 209. [g] R. S. Roth and A. D. Wadsley, *Acta Cryst.*, 1965, **18**, 724. [h] R. Gruehn and R. Norin, *Z. anorg. Chem.*, 1967, **355**, 176. [i] B. M. Gatehouse and A. D. Wadsley, *Acta Cryst.*, 1964, **17**, 1545. [j] W. Mertin, S. Andersson, and R. Gruehn, *J. Solid State Chem.*, 1970, **1**, 419. [k] S. Andersson, *Z. anorg. Chem.*, 1967, **351**, 106. [l] R. S. Roth, A. D. Wadsley, and S. Andersson, *Acta Cryst.*, 1965, **18**, 643. [m] H. L. Levin and R. S. Roth, *J. Solid State Chem.*, 1970, **2**, 250. [n] J. G. Allpress and A. D. Wadsley, *J. Solid State Chem.*, 1969, **1**, 28. [o] R. Gruehn, *Naturwiss.*, 1967, **54**, 645. [p] A. Astrom, *Acta Chem. Scand.*, 1966, **20**, 969. [q] S. Andersson, *Acta Chem. Scand.*, 1965, **19**, 401. [r] R. S. Roth and A. D. Wadsley, *Acta Cryst.*, 1965, **19**, 32. [s] J. Galy and S. Andersson, *Acta Cryst.*, 1968, **B24**, 1027. [t] S. Andersson, W. G. Mumme, and A. D. Wadsley, *Acta Cryst.*, 1966, **21**, 802. [u] R. S. Roth and A. D. Wadsley, *Acta Cryst.*, 1965, **19**, 38.

by chemical vapour transport [90] but had its analogues in the mixed valence ternary systems incorporating Ti^{4+},[91] V^{4+},[92a] and Al^{3+}.[92b] It was proposed that this was a defective intergrowth of the $Nb_{28}O_{70}$ and $Nb_{25}O_{62}$ structures. The $Nb_{47}O_{116}$ structure was found [93] to be systematically oxygen-deficient, in

[90] R. Gruehn and H. Schäfer, *J. Less Common Metals*, 1966, **10**, 152; R. Norin, *Z. anorg. Chem.*, 1967, **355**, 176.
[91] R. Gruehn and H. Schäfer, *Naturwiss.*, 1963, **50**, 642.
[92a] J. L. Waring and R. S. Roth, *J. Res. Nat. Bur. Stand.*, 1965, **69A**, 119; [b] R. Norin, *Acta Chem. Scand.*, 1969, **23**, 1210.
[93] R. Gruehn and R. Norin, *Z. anorg. Chem.*, 1969, **367**, 209.

$NbO_{2.464\pm0.001}$, and Burdeze *et al.* concluded that $Nb_{22}O_{54}$ and $Nb_{12}O_{29}$ were also inherently oxygen-deficient, invariably giving biphasic preparations at their ideal compositions. The current status of each of these compounds will not be discussed here; the literature evidence is conflicting on many points (see papers quoted and references therein; also ref. 94). It must be borne in mind that structures are determined on highly selected specimens: 'good' crystals for single crystal X-ray work, which are neither themselves analysed nor necessarily fully representative of the phase assembly from which they are picked, or selected areas from individual crystals in electron diffraction and microscopy. There is a link of inference and extrapolation between the structure and the composition. The evidence is clear that, as prepared, these binary and ternary oxide systems have stoicheiometric faults at one or more of three levels: point defects or defect complexes; extended defects; or domain intergrowth in some form that imposes a modification of the usual Phase Rule conditions for equilibrium and co-existence.

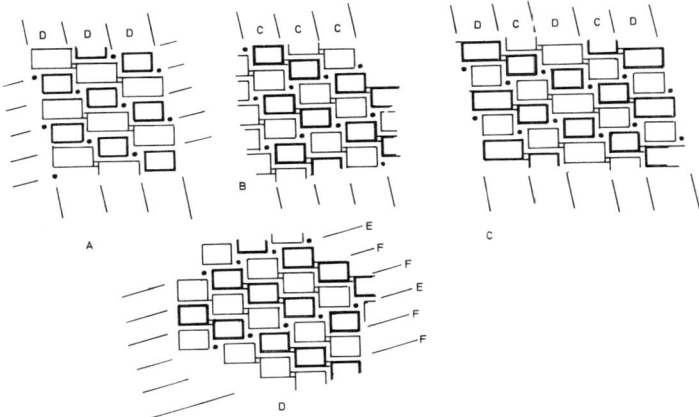

Figure 15 *Schematic structures of* (A) $M-Nb_2O_5$ *and* (B) $Nb_{25}O_{62}$, *showing stacking sequence of columns 16.8 Å wide (D columns) and 15 Å wide (C columns) respectively.* (C) *Regular intergrowth stacking* -D-C-D-C-D- *in* $Nb_{53}O_{132}$. (D) *Regular intergrowth of geometrically compatible rows of* $M_{12}O_{29}$ *and* $M_{25}O_{62}$, *with tetrahedral sites (E rows) and without tetrahedral sites (F rows), as in* $Ti_5Nb_{44}O_{120}$. After J. G. Allpress, Table 3, ref. *c*

This problem has not yet been fully resolved, but a good deal of information has been derived from direct lattice imaging methods in work by Allpress on the Ti–Nb–O and W–Nb–O systems and by Hutchison, Browne, Nimmo, and Anderson (unpublished) for the binary niobium oxides.

All the block structures can be regarded as involving the stacking of units

---

[94] A. D. Wadsley and S. Andersson, 'Perspectives in Structural Chemistry', ed. J. D. Dunitz and J. A. Ibers, Wiley, New York, 1970, Vol. 3, p. 1.

of a few standard sizes in regular *rows* and *columns*. It can be seen from Figure 15 that the long dimension of the (4 × 3) and (5 × 3) blocks found in most of the compounds between $MO_{2.4176}$ and $MO_{2.5385}$ defines one or other of two characteristic column widths, varying only by a small amount for different ways of linking blocks in adjacent columns and denoted as *C* and *D* type columns respectively in Figure 15. A one-dimensional lattice image, using only the reflections on a reciprocal lattice row normal to the columns, consists of a set of fringes spaced at a distance equal to the column width. Thus H-Nb_2O_5, which could be represented by the stacking sequence -*D-D-D-D*- gives fringes at 17 Å spacing; $TiNb_{24}O_{62}$, with stacking sequence of -*C-C-C-C*- fringes at 15 Å spacing (Figure 16 a and b). In Figure 16c, the repeat distance of 32 Å embraces two fringes (spaced at 15 Å and 17 Å), except for a lamella of triple fringes with a 47 Å (= 15+15+17 Å) repeat. The former corresponds to the stacking -*C-D-C-D*-, which leads to the composition $TiNb_{52}O_{132}$; this is the structure found for the $NbO_{2.483}$ phase. The latter implies the stacking -*C-C-D-C-C-D*-, corresponding to $TiNb_{38}O_{97}$.

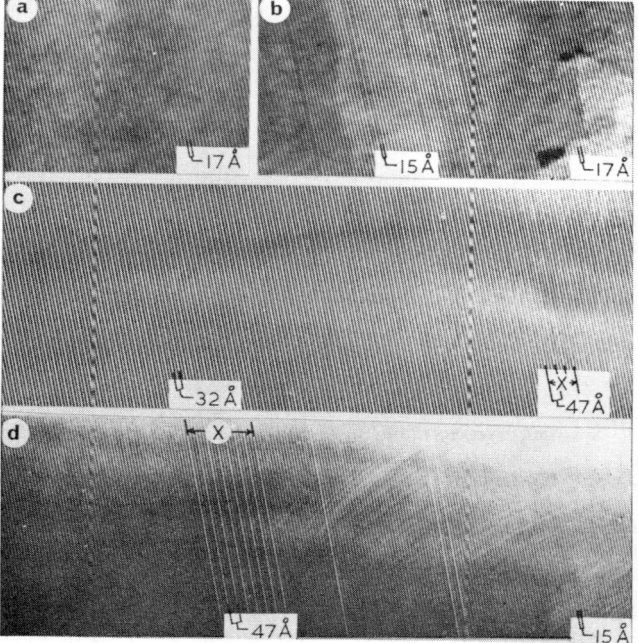

Figure 16 *One-dimensional lattice images showing the type* C *and* D *columns of Figure 15.* (a) H-Nb_2O_5. (b) *From a preparation* TiO_2+40 Nb_2O_5: H-Nb_2O_5 *structure with occasional, isolated intergrowth of lamellae of* $TiNb_{24}O_{62}$. (c) $TiNb_{52}O_{132}$ *with a small domain* X *of* $TiNn_{38}O_{97}$, *three unit cells thick.* (d) $TiNb_{24}O_{62}$ *with isolated* 17 Å *(type* D, H-Nb_2O_5) *lamellae and a domain* X *of* $TiNb_{38}O_{97}$. After J. G. Allpress, Table 3, ref. *c*

Shear Structures and Non-stoicheiometry

In a direction transverse to these C or D type columns, the structures can be dissected into rows of blocks which again conform closely to one or other of two widths, depending on whether they are defined by lines of tetrahedral sites (E) or lines of edge-sharing between blocks only (F). Within a column of given breadth, e.g. the C columns of the $M_{25}O_{62}$ structure, differences are possible in the stacking sequence of rows. Thus stacking of rows in $WNb_{12}O_{33}$, $Nb_{25}O_{62}$ (or $TiNb_{24}O_{62}$), and $Nb_{12}O_{29}$ is: -E-E-E-E-, -E-F-E-F-, and -F-F-F-F respectively. Mixed stacking sequences are again possible, representing intergrowths of the $M_{25}O_{62}$ and $M_{12}O_{29}$ structures, and are revealed in lattice images by virtue of the difference in breadth of the $E(10.5$ Å$)$ and F (10 Å) rows. Thus Figure 17a shows the regular 21 Å double fringes ($=2\times10.5$ Å) of $TiNb_{24}O_{62}$ and Figure 17c domains of triple fringes ($10.0+10.5+10.5$ Å) of the -E-F-F-E-F-F- stacking of $Ti_3Nb_{34}O_{91}$ and quadruple fringes of -E-F-F-F-E-F-F-F- stacking of $Ti_5Nb_{44}O_{120}$.

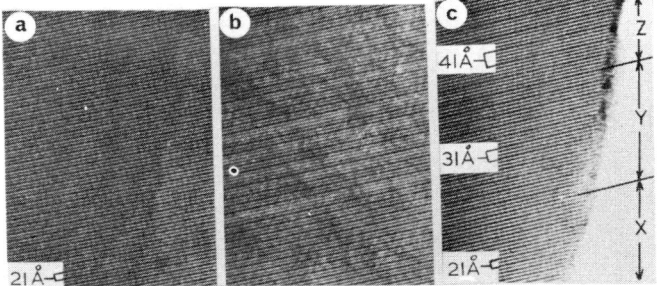

Figure 17 *One-dimensional lattice images, showing the type* E *and* F *rows of Figure 15.* (a) *Fully ordered* -E-F-E-F *stacking in* $TiNb_{24}O_{62}$. (b) *Randomized sequences of rows representing non-stoicheiometric (metastable) composition.* (c) *Ordered domains of* -E-F-E-F *stacking* (X, $TiNb_{24}O_{62}$), -E-F-F-E-F-F- *stacking* (Y, $Ti_3Nb_{34}O_{91}$) *and* -E-F-F-F-E-F-F-F- *stacking* (Z, $Ti_5Nb_{44}O_{120}$). *After J. G. Allpress, Table 3, ref. c*

It will be noted that the columns and rows of blocks are demarcated by planes through tetrahedral sites and corner junctions of blocks, and close to the shear planes that bound them. They are thus directions with a higher than average density of atoms, and it has been an acceptable working hypothesis that such structural features correspond to the contrast maxima in lattice images, especially for very thin crystals which approximate in their diffraction properties to phase gratings.[95] In two-dimensional lattice images, using all reflections needed to give resolution at the 4—6 Å level, similar considerations apply. Theoretical calculations of fringe image contrast distribution is at an early stage of development, but the dimensions and symmetry of the image accord with the known structures and the distribution of contrast within the unit cell can be correlated with the distribution of

---

[95] J. M. Cowley and A. F. Moodie, *Proc. Phys. Soc.*, 1960, **76**, 382.

tetrahedral sites, shared corners, and shear planes. (Figure 18). It has been possible, in this way, to confirm that the -C-D-C-D stacking pattern proposed by Allpress, rather than the -C-D-D-C stacking proposed by Gruehn, is the correct structure for $M_{53}O_{132}$.

8 Stoicheiometric Faults, Intergrowths, and Non-stoicheiometric Faults

The foregoing comments are relevant in that high-resolution electron microscopy provides the only direct method of investigating the inconsistency

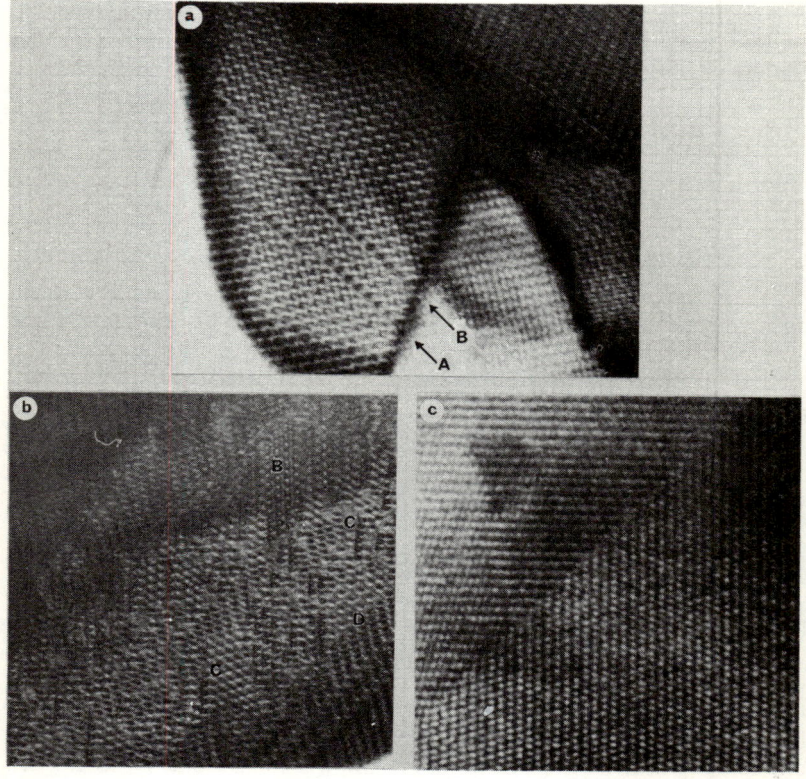

Figure 19 *Faults and extended defects producing no, or insignificant change in stoicheiometry.* (a) *Insertion of a row of* (4×4) *blocks* (= M-Nb$_2$O$_5$) *in a crystal of* H-Nb$_2$O$_5$. *Sidestepping of intergrown row involves wrong sized blocks with possible local change of stoicheiometry.* (b) *Heavily faulted* H-Nb$_2$O$_5$ *with twinning, inserted rows of* M-Nb$_2$O$_5$, *domains and strips of* N-Nb$_2$O$_5$ (B, C) *and a new modification,* H'-Nb$_2$O$_5$ (D). (c) *Monoclinic* Nb$_{12}$O$_{29}$ *with twinning, but no change in stoicheiometry*

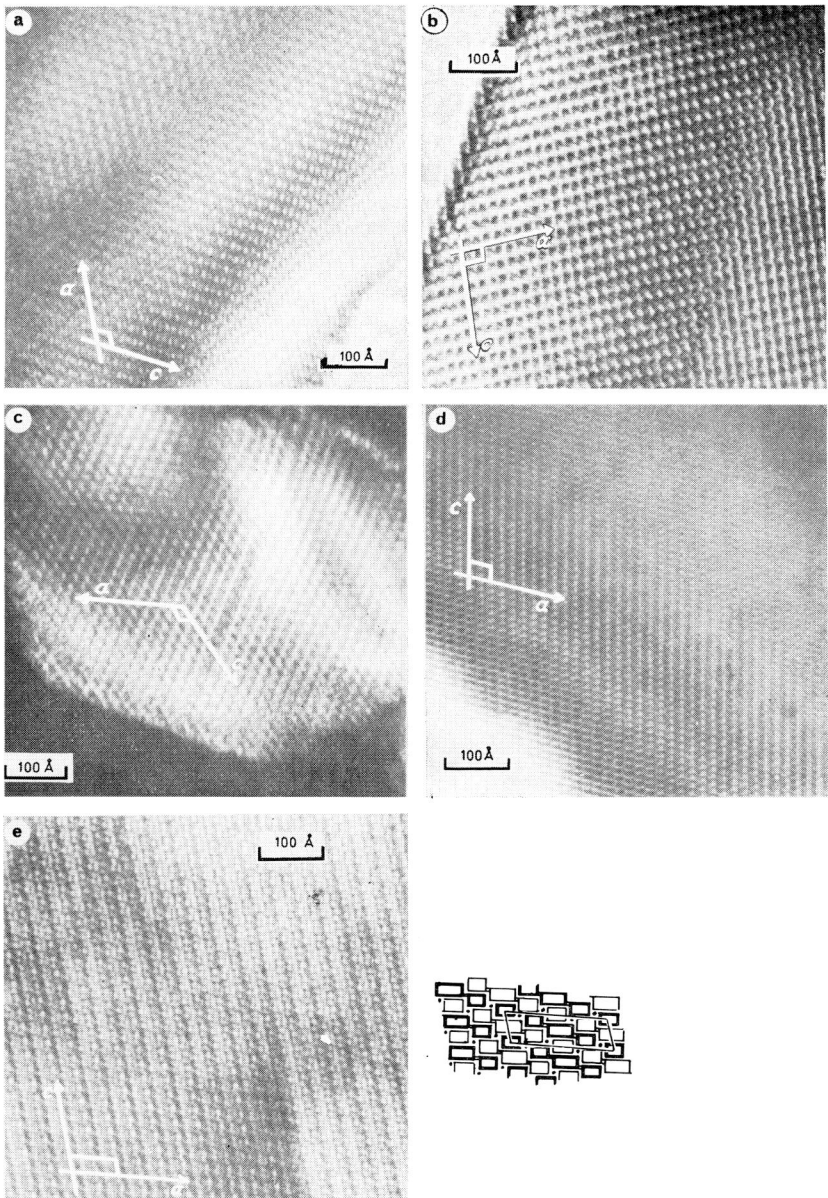

Figure 18 *Two-dimensional lattice images* (a) H-Nb_2O_5, $(3\times4)_1+(3\times5)_\infty$. (b) $Nb_{25}O_{62}$. (c) $Nb_{22}O_{54}$, $(3\times3)_1+(3\times4)_\infty$. (d) $Nb_{12}O_{29}$ $(3\times4)_\infty$. (e) *The intergrowth structure of* $Nb_{53}O_{132}$. *The dimensions, symmetry, and contrast distribution in one unit cell of the lattice image correlate directly with the block structure.* Micrographs from J. L. Hutchison and J. S. Anderson

between the ideal structures and the chemical evidence for these compounds. The first class of faults revealed are those that produce no stoicheiometric defect: twinning, incorporation of other polymorphs (*e.g.* the recurrently twinned orthorhombic form of $Nb_{12}O_{29}$, *etc.*). These are relevant to the present discussion only in so far as they indicate the capacity of the block structures for coherent faulting and intergrowth; some examples are shown in Figure 19.

Non-stoicheiometric faults can be introduced by insertion of rows or columns of blocks of different size, or with a different linkage, subject always to the constraint that the fault is geometrically compatible with the matrix structure. As with the CS phases considered earlier, these extended defects may be isolated, or may constitute coherent lamellae of a regular structure. They indicate fluctuations of composition within the crystal: in the Ti–Nb and W–Nb oxides, locally different concentrations of the minor component; in the niobium oxides, differing distributions of Nb^{4+} and Nb^{5+}. Thus Allpress, Sanders, and Wadsley observed,[96] in one region of a crystal of $W_5Nb_{16}O_{55}$, 125 fringes 18.4 Å wide, proper for the $(5 \times 4)_1$ structure, 4 15 Å fringes corresponding to the $(4 \times 4)_1$ structure of $W_3Nb_{14}O_{44}$, and 2 22 Å fringes that must be attributed to a $(6 \times 4)_1$ structure—an unknown compound $W_7Nb_{18}O_{66}$. The total composition of this 0.25 μm strip of crystal would be $MO_{2.6187}$, as compared with $MO_{2.6190}$ for the perfect crystal, representing an enrichment of tungsten in one part at the expense of another, but also a perfect reorganization of the structure, in two dimensions, to accommodate the local composition without random defects. Such fluctuations are probably frozen-in, out-of-equilibrium states; in the ternary oxides they can arise from incomplete homogenization, if diffusion of the minor component is the

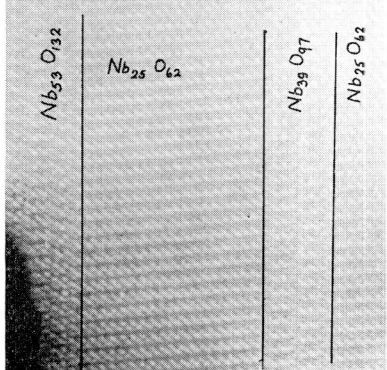

Figure 20 *Faults producing stoicheiometric variability.* $NbO_{2.490}$: *region of crystal with wide domains of* $Nb_{25}O_{62}$ *and fully coherent lamellae of* $Nb_{39}O_{97}$ *and* $Nb_{53}O_{132}$ *ordering*

[96] J. G. Allpress, J. V. Sanders, and A. D. Wadsley, *Acta Cryst.*, 1969, **B25**, 1166.

rate-limiting step. As is discussed below, it is not at all certain that this is the case. Moreover, lamellae and isolated sheets of the intergrowth structures have been found in the non-stoicheiometric composition ranges of the binary niobium oxides, which are electronic conductors and which can adjust the distribution of Nb^{4+} and Nb^{5+} by electron transfer. Figure 20 shows an area of $NbO_{2.490}$, in the reported non-stoicheiometric range of $H\text{-}Nb_2O_5$, after annealing at 1300 °C, with lamellae of $Nb_{25}O_{62}$ and $Nb_{53}O_{132}$. It might then be supposed that the migration of shear planes, to rearrange the block structure, is the determining rate process, but this is also uncertain. Andersson [97] found that MoO_3 vapour, at 10^{-2} to 10^{-3} atm partial pressure, reacted with $H\text{-}Nb_2O_5$ at 1073 K within a few minutes to form $MoNb_{12}O_{33}$ and $Mo_3Nb_{14}O_{44}$, and with $TiNb_{24}O_{62}$ at 1173 K to give complete conversion into '$Ti_6Mo_{19}Nb_{144}O_{429}$,' (the $M_{13}O_{33}$ phase) within 15 min; decomposition of $Mo_3Nb_{14}O_{44}$ to $MoNb_{12}O_{33}$ at 1223—1273 K was equally rapid. These reactions involve a complete rearrangement of block sizes and movement of shear planes, and it must be inferred that the co-operative movements involved can be fast. No ordered intergrowths or random faults have been found in the reduction of $H\text{-}Nb_2O_5$ or in the preparation of intermediate oxides from $NbO_2 + Nb_2O_5$. The only intermediate structures observed are those of $Nb_{12}O_{29}$ and $Nb_{25}O_{62}$ even after annealing periods of 100 h at 1273 K; intergrowth structures are formed only at higher temperatures (> 1573 K).

Deviations from ideal composition can therefore arise from coherent intergrowth of strips or domains of a variant structure, but this interpretation does not suffice to account for all the facts. In slightly reduced $H\text{-}Nb_2O_5$, preparations with compositions down to $NbO_{2.495}$, at least, frequently show no evidence of 'wrong' block sizes; $WNb_{12}O_{33}$ appears to span a composition range without including microdomains with (5 × 3) blocks, needed to form the intermediate intergrowth structures.[97a] As has been mentioned, the lower oxides can be oxidized to Nb_2O_5, forming polymorphs (Schäfer's Ox I to Ox VI forms [98a] which differ from each other and from $H\text{-}Nb_2O_5$ but, on the X-ray diffraction evidence, bear a strong structural relation to the block structures of the original lower oxides. Of these, only the form derived from the $(4 \times 3)_\infty$ structure of $Nb_{12}O_{29}$ has, as yet, been examined by electron microscopy.[98a] In it, the ribbons of (4 × 3) blocks are distorted by faulting, but retained as the basis of the structure, with only very small regions converted into the $H\text{-}Nb_2O_5$ structure. There must be some structural method, other than block reorganization, for accommodating a change in anion : cation ratio. Wadsley and Andersson considered that some variability might arise from a variable occupancy of the tetrahedral sites. This could lead only to metal-deficient composition whereas (e.g. in $NbO_{2.495}$) the

---

[97] S. Andersson, *Z. anorg. Chem.*, 1969, **366**, 96.
[97a] J. G. Allpress, private communication; J. G. Allpress and R. S. Roth, *J. Solid State Chem.*, 1971, **3**, 209.
[98] H. Schäfer, *Angew. Chem. Internat. Edn.*, 1966, **5**, 40.
[98a] J. M. Browne, J. L. Hutchinson, and J. S. Anderson, unpublished results.

problem is that of a metal excess. Moreover, in $Nb_{12}O_{29}$ there are no tetrahedral sites; oxidation can only affect the population of cation sites within the blocks. In the present state of knowledge only two interpretations seem possible. Either some localized defects, possibly point defect cation vacancies or oxygen vacancies, can be tolerated or sets of sites can be created or eliminated by a third CS operation transverse to the infinite extension of blocks, but not necessarily parallel to the $(h0l)$ plane.

There is some evidence suggesting that a third CS direction is possible. In a study of quenched Ti–Nb oxides, Allpress[99] has interpreted lattice images from partially disordered material as indicating not only jogs, or parallel displacement of CS planes from a change in block sizes, but also steps in the CS planes (displacements normal to the b axis; Figure 21). Moreover, electron micrographs of materials quenched in the course of chemical reaction or ordering not infrequently show contrast effects strongly suggestive of families of oblique CS planes producing stepped blocks (see, for example, ref. 99, Figures 12 and 13). The nature of stoicheiometric variability is thus not fully interpreted.

Table 4 *Thermodynamics of niobium oxides* NbO_2—Nb_2O_5

Ratio $H_2O : H_2$	Solid phases	Assignment	$\bar{G}_{O_2}$ (1573 K) kcal mol^{-1})
0.167	$NbO_{2.000}$—$NbO_{2.024}$	NbO_2	
0.167	$NbO_{2.024}+NbO_{2.417}$	$NbO_2+Nb_{12}O_{29}$	−87.6
0.207	$NbO_{2.417}+NbO_{2.453}$	$Nb_{12}O_{29}+Nb_{22}O_{54}$	−86.3
0.412	$NbO_{2.453}+NbO_{2.464}$	$Nb_{22}O_{54}+Nb_{47}O_{116}$?	−82.0
0.412—0.740	$NbO_{2.464}$—$NbO_{2.467}$	$Nb_{47}O_{116}$ variable	
0.740	$NbO_{2.467}+NbO_{2.472}$	$Nb_{47}O_{116}+Nb_{25}O_{62}$	−78.3
0.740—2.51	$NbO_{2.472}$—$NbO_{2.478}$	$Nb_{25}O_{62}$ variable	
2.51	$NbO_{2.478}+NbO_{2.483}$	$Nb_{25}O_{62}+Nb_{53}O_{132}$?	−70.7
< 4.02	$NbO_{2.483}+Nb_{2.489}$	$Nb_{53}O_{132}+Nb_{28}O_{70}$ defective	−67.8
4.02	$NbO_{2.489}$—$NbO_{2.500}$	$Nb_{28}O_{70}$	

Although these systems can discriminate between the structures appropriate to very fine scale changes in anion : cation ratio, the energetic differences must be small. How far do the structurally different compounds behave as thermodynamically distinct species? Schäfer et al.[88] equilibrated niobium oxides with H_2O–H_2 mixtures (0.167 < $H_2O : H_2$ < 4.02) at 1300 °C and found that transformation proceeded in steps, as shown in Table 4. Even at 1573 K, the stage taken as gas–solid equilibrium was attained very slowly. Around 1273 K, the first approach to equilibrium was rapid, but thereafter, composition changes and adjustment of equilibrium continued for extremely long periods.[99a] At that temperature, only the H-Nb_2O_5 structure, the $Nb_{25}O_{62}$ structure and the $Nb_{12}O_{29}$ structure of the final reduction product could be

[99] J. G. Allpress, *J. Solid State Chem.*, 1970, **2**, 78.
[99a] K. M. Nimmo and J. S. Anderson, unpublished results.

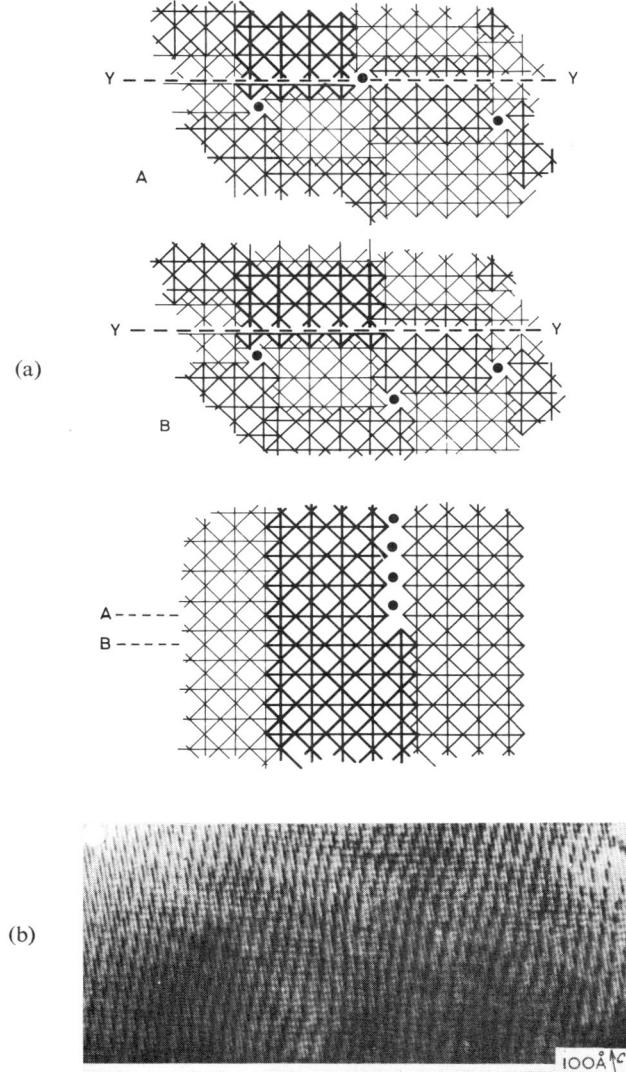

Figure 21 Displacement or jogs due to change in block sizes along the length of the blocks (i.e. along the b axis of the crystal) (a) schematic. (b) One-dimensional lattice image from a quenched sample of $TiO_2+26Nb_2O_5$ showing fringe displacement and contrast attributable to such displacements. Micrograph from J. G. Allpress, *J. Solid State Chem.*, 1970, **2**, 78

detected at any stage by electron diffraction and microscopy, and there was no discontinuity or step corresponding to biphasic equilibria involving $Nb_{25}O_{62}$. The behaviour is very similar to that found in the study of reduced TiO_2, with extreme hysteresis effects involved in the adjustment of CS plane spacings.[62] There is an apparent conflict with the evidence quoted earlier, for the ready reorganization of CS planes, and there are problems in applying the usual phase equilibrium considerations to all these compounds, with their special aptitude for coherent intergrowth.

None of the earlier papers on non-stoicheiometry of Nb_2O_5 has been cited in the foregoing section, since earlier work must be regarded as completely superseded. There is, however, a considerable body of work on electronic conductivity which illustrates the difficulties of deducing defect structures from indirect data. Most of this work relates to 'α-Nb_2O_5', now broadly identifiable as H-Nb_2O_5, which behaves as an n-type metal excess semiconductor; a given specimen is sufficiently reversible in behaviour for its conductivity at 1273 K to be used as a direct-reading oxygen pressure meter down to about 10^{-18} atm.[100] The oxygen pressure dependence of conductivity has been determined and analysed as a continuous function of the stoicheiometric defect x in $NbO_{2.5-x}$ by a number of workers [101–104] on the assumption that a single phase ('α-Nb_2O_5') spanned the composition range from $NbO_{2.5000}$ to about $NbO_{2.42}$, i.e. to the stage of reduction now known to involve reconstruction to $Nb_{12}O_{29}$. Although results scatter, and varying degrees of hysteresis and slow equilibration are reported, the structural transformations that occur in this composition range are not evident as breaks in the conductivity or in the activation energy. At higher oxygen pressures (10^{-6}—1 atm), corresponding to very small deviations from stoicheiometry, the conductivity approximates to a $P_{O_2}^{-\frac{1}{4}}$ dependence; in CO–CO_2 buffer mixtures (e.g. 10^{-10} atm or less) at 1373 K, i.e. for compositions below about $NbO_{2.49}$, results have been fitted to a $P_{O_2}^{-\frac{1}{6}}$ relation. This has been interpreted on a point defect model with oxygen vacancies; at low stoicheiometric defect, reduction takes the form

$$O_o^x \rightleftharpoons \tfrac{1}{2}O_2 + V_o^{\cdot} + e^-$$

and with increasing vacancy concentration there is a progressive lowering of ionization energy for the reaction

$$V_o^{\cdot} \rightleftharpoons V_o^{\cdot\cdot} + e^-$$

so that beyond about $NbO_{2.49}$, double-ionized vacancies predominate. Blumenthal et al. found that for the most reduced material, the pressure

---

[100] J. J. Oehlig, H. Le Brusq and F. Marion, *Compt. rend.*, 1968, **266**, 1774.
[101] R. F. Janninck and D. H. Whitmore, *J. Chem. Phys.*, 1962, **37**, 2750.
[102] P. Kofstad, *J. Phys. Chem. Solids*, 1962, **23**, 1571.
[103] R. N. Blumenthal, J. B. Moser, and D. H. Whitmore, *J. Amer. Ceram. Soc.*, 1965, **48**, 617.
[104] W. K. Chen and R. A. Swalin, *J. Phys. Chem. Solids*, 1966, **27**, 57.

dependence was roughly $P_{O_2}^{-\frac{1}{2}}$, and Kofstad [105] rationalized the result in terms of an increasing transfer of niobium atoms to interstitial positions, in the improbable Nb^{2+} charge state. It is hard to assess the significance of these studies of electronic properties, but two conclusions can be drawn. The first is that since measurements of this type, even on well-characterized single crystal substances, invariably show scatter due to impurity effects, surface states, *etc.*, they may be inadequate to discriminate between valid and false models for defect structure and conduction mechanism. More important, it appears that electronic properties of these and other CS phase systems are not very sensitive to the actual structure of the crystals.

Janninck and Whitmore reported the carrier mobility as 0.2 cm^{-2} volt^{-1} s^{-1}, and independent of the composition; this suggests a hopping electron or polaron mechanism. There may be a significant ionic component of the conduction, by oxygen ion transport. In H-Nb_2O_5, the oxide ion transference number does not exceed 0.05 at 1273 K [106] but it appeared to be substantially higher in TiO_2-doped Nb_2O_5, of a composition and history suggesting that it was largely $TiNb_{24}O_{62}$. Self-diffusion measurements, by isotope exchange with Nb_2O_5 at very small stoicheiometric deviations from the ideal composition give values for D (in cm^{2}s^{-1}) increasing from 1.5×10^{-10} at 1 atm O_2 to 5×10^{-10} at 10^{-2} atm O_2, following a $P^{-\frac{1}{4}}$ law.[107] The behaviour is in accord with an oxygen vacancy diffusion mechanism. This evidence for the transport mechanism is clearly relevant to an understanding of the transformation processes during oxidation and reduction of the block structures.

9 Related Types of Non-stoicheiometry

The essential feature of the crystallographic shear phases is that they are built up of slabs of more or less defect-free structure, joined at planar singularities which are two-dimensional elements of a different composition. In the CS phases, the singularity is a switch of cation sites from 'normal' to 'interstitial' across some plane. Other planar structure elements can be envisaged as delimiting slabs of constant type but variable thickness, and leading to many of the features that have been discussed in relation to CS phases. A new type of slab structure, conforming to this specification, has been identified in the tin tungsten bronzes of low tin content,[108] and may well prove to be the basis of other low bronzes, such as the cadmium and lead bronzes described by Hagenmuller. These compounds may be represented as SnW_nO_{3n+1}, and are based upon slabs of only slightly modified WO_3 structure, from 6 to probably at least 18 [WO_6] octahedra in thickness, joined through sheets of [SnO_6] octahedra (Figure 22a). All the conditions

---

[105] P. Kofstad. *J. Less Common Metals*, 1968, **14**, 153.
[106] R. Elo, R. A. Swalin, and W. K. Chen, *J. Phys. Chem. Solids*, 1967, **28**, 1625.
[107] W. K. Chen and D. A. Jackson, *J. Chem. Phys.*, 1967, **47**, 1144.
[108] I. J. McColm, R. Steadman, and A. T. Howe, *J. Solid State Chem.*, 1970, **2**, 555; R. Steadman, R. J. D. Tilley, and I. J. McColm, *ibid.*, in publication.

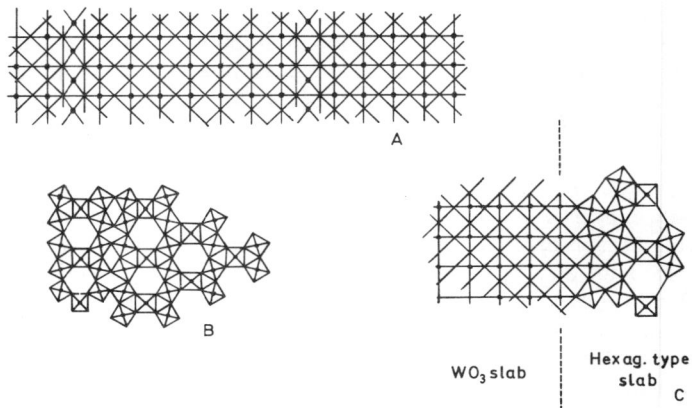

Figure 22 *Structures of tin tungsten bronzes, schematic* (A) SnW_nO_{3n-1}, *with slabs of* WO_3 *structure n octahedra wide* ($6 < n < 19$) *separated by sheets of* $[SnO_6]$ *octahedra.* (B) Sn_xWO_3 *with hexagonal tunnels; the slices containing tunnels can be joined by layers of two thicknesses.* (C) *Possible intergrowth of the two structures, as is found by electron microscopy*

for intergrowth of lamellae of different thickness are fulfilled and, as with the CS phases, regular recurrence of the bounding sheets can be imposed only by long-range interaction effects. As grown by chemical vapour transport, the crystals are generally highly ordered, with the intercalation of occasional lamellae of wrong width, but occasional almost random stacking

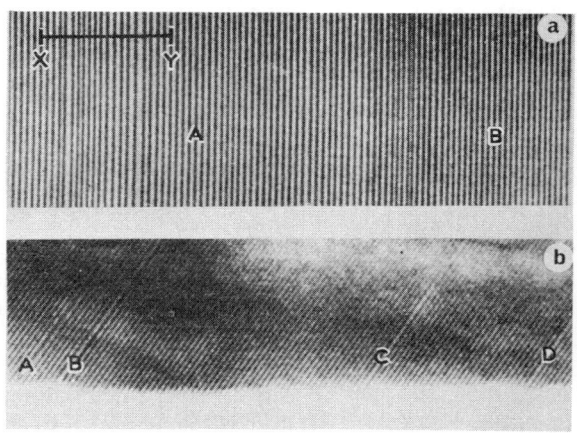

Figure 23 *Lattice images from tin tungsten bronzes.* (a) *Faults in* SnW_nO_{3n+1} *type. At A, 32.5 Å fringes from* WO_3 *slabs of 9 octahedra wide; at B, ordered 8-octahedra slabs; at X—Y disordered lamellae 7, 8, 9, 10, 11 and 12 octahedra wide.* (b) *Faults in hexagonal tunnel type* $Sn_{0.17}WO_3$. *C, D lamellae of 7-octahedron* WO_3 *slab type. A, B planar faults not corresponding to lamellae of identified structure.*

is found (Figure 23). A second structure type in this group of compounds is related to the known hexagonal tungsten bronzes, but with the hexagonal tunnels (which are empty) arranged in sheets (Figure 22b). These are true bronzes, Sn_xWO_3, their relevance for the present discussion is that the evidence of direct lattice imaging shows that lamellae of this type can be spliced coherently, *i.e.* with no intolerable distortions, into the structures of the former type. In these intergrowth structures each lamella may be perfectly ordered; the crystal as a whole can have infinitely variable composition without introducing localized defects.

The further extension of this principle is a matter for speculation; it could underlie other classes of well-authenticated non-stoicheiometric phases. Thus, in the transition-metal chalcogenides, it is familiar that variable composition phases (*e.g.* the TiS_{2-x} and $Fe_{1-x}S$ phases) provide a transition between the NiAs ($B8$) and CdI_2 ($C6$) structure types. In all of these, the general principle is that alternate sheets of cations are fully populated and incompletely populated (with occupancy fractions 0 to 0.5 or 1.0). In some of the systems [5] there are discrete, intermediate phases with a well-defined and ordered arrangement of filled sites and empty sites in the partially populated cation layers. It has been generally assumed, on the basis of the occupancy numbers derived from X-ray crystallographic work, that in the (high-temperature) non-stoicheiometric phases this order is lost and that the cation population of the incomplete layers is variable. This is equivalent to introducing point defect vacancies into the NiAs structure, or interstitials into the CdI_2 structure, in alternate layers. It is possible that the true structure is an irregularly recurrent stacking anomaly of the type considered in this review: each layer itself conforming to a rational, ordered pattern, but with 'wrong' layers aperiodically distributed. Only the application of techniques that can detect either local changes in slab dimensions (as with CS phases and the tin bronzes), or local changes in atomic packing density (as just proposed), at the unit cell level, will resolve such questions.

The concept of crystallographic shear has provided a new insight into an astonishingly versatile chapter of transition-metal chemistry, but many questions must be left unanswered. As at present known, it operates within the most ionic compounds of a restricted and compact group of elements within the Periodic Table. Why this should be, and whether it is of wider applicability is not known. It does not provide a complete and final answer to the understanding of non-stoicheiometric compounds.

* It has now been established (L. A. Bursill, B. G. Hyde, and D. K. Philip, *Phil. Mag.*, 1971, **23**, 1501; R. M. Gibb and J. S. Anderson, *J. Solid State Chem.*, in press) that the reorientation of CS planes, *e.g.* from {1$\bar{3}$2} to {1$\bar{2}$1} rutile is indeed a perfectly continuous transition. With change of composition in the rutile structure, co-operative displacements of cations progressively pivot the CS planes round, through all intermediate {h$\bar{k}$l} orientations in the [1$\bar{1}$1] zone. In principle, every orientation could generate a complete homologous family of compounds. The transition actually takes place spontaneously in such a way that the normal spacing between CS planes remains nearly constant, *i.e.* change of composition does not involve creating or eliminating CS planes, but only reorients them. The same type of continuous transformation takes place in the WO_{3-x} CS phases (J. G. Allpress, personal communication, to be published).

2
Direct Study of Structural Imperfections by High-resolution Electron Microscopy

BY L. L. BAN

1 Introduction

The value of the electron microscope in studying the solid state has been well established. Since its first reported development (in 1932),[1] the electron microscope has undergone many improvements. Some of these were introduced by individual researchers who designed and modified instruments for their own specific applications. Other improvements, particularly in recent years, have been implemented by the microscope suppliers in satisfying the diversified needs of many different scientific disciplines. However, despite rather rapid advances in instrumental design, it is only in the last ten years that a really significant improvement in resolution has been achieved. The Rayleigh resolution limit of approximately 2 Å, which was estimated thirty years ago,[2,3] is now achieved and even exceeded with the imaging of crystal lattices. The resolution in the range of 2 Å is brought about, in part, by the reduction of spherical aberration and by the reduction of fluctuations in the objective lens current and high voltage (accelerating potential). The latter two problems no longer restrict the achievement of the theoretical resolution limit in most modern-day electron microscopes. However, the technical difficulties in overcoming spherical aberration continue to be a problem.[4]

From a practical standpoint, the most common limitations on reaching the best possible resolution are: (*i*) environmental–mechanical and electromagnetic disturbances in laboratories and insufficient cooling of the magnetic lenses; (*ii*) the sample is not sufficiently thin, so that chromatic distortions, which are the result of increased inelastic scattering of electrons, will decrease the resolution; (*iii*) insufficient protection of the sample from contamination caused by low molecular weight hydrocarbon molecules that can exist as free radicals,[3] the sources of which are the oil-diffusion-pumped vacuum system, rubber gaskets, and vacuum greases.

With only limited resolving power, the kinematical and dynamical diffrac-

---

[1] M. Knoll and E. Ruska, *Ann. Phys.*, 1932, **12**, 607; E. Bruche and H. Johannson, *Naturwiss*, 1932, **20**, 353.
[2] M. Knoll and E. Ruska, *Z. Physik.*, 1932, **78**, 318.
[3] E. Ruska, in 'Advances in Optical and Electron Microscopy', ed. R. Barer and V. E. Cosslett, Academic Press, London and New York, 1966, Vol. 1, p. 115.
[4] A. Septier, ref. 3, p. 204.

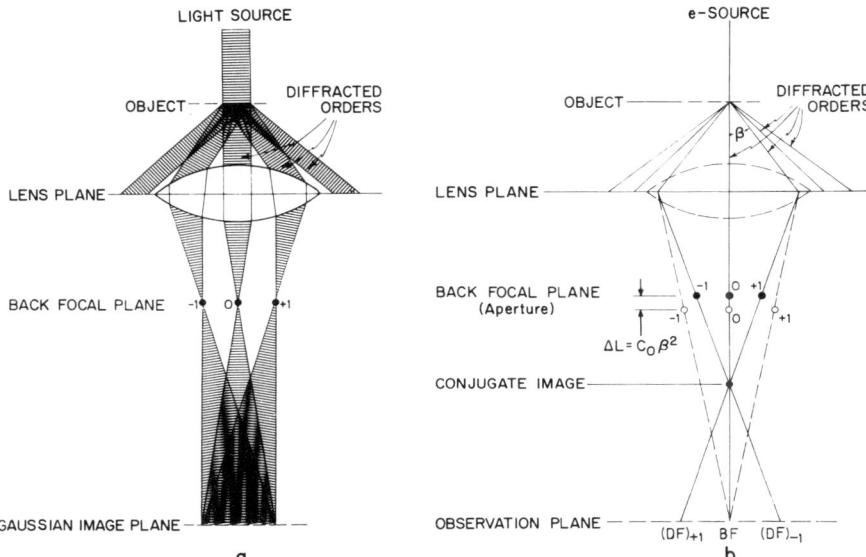

Figure 1 *Diagrams describing image formation:* (a) *light microscope;* (b) *electron microscope*

tion theories were used with good success by many of the early electron microscopists to explore the diffraction contrast details in the images of defective thin crystals.[5-7] The present-day trend is to consider the many beam-diffraction effects in the images, this approach being especially necessary in the case of high-voltage electron diffraction and microscopy.[8]

The direct resolution of crystal lattices was first achieved by Menter.[9] He used axial electron illumination and resolved the (20$\bar{1}$) planes in copper and platinum phthalocyanine with a spacing of 12.0 ± 0.2 Å. These images were interpreted in terms of the theory of Abbe [10] for the light microscope. This theory indicates that image formation of a grating (amplitude or phase grating) depends on the angular aperture (or numerical aperture) being large enough to allow at least the first-order spectrum of the Fraunhofer diffraction pattern to pass through the objective lens. The interference of the zero- and

[5] P. B. Hirsch, A. Howie, and M. J. Whelan, *Phil. Trans.*, 1960, **A252**, 499; A. Howie and M. J. Whelan, *Proc. Roy. Soc.*, 1961, **A263**, 217; 1962, **A267**, 206.
[6] P. B. Hirsch, A. Howie, R. B. Nicholson, D. W. Pashley, and M. J. Whelan, 'Electron Microscopy of Thin Crystals', Butterworth, London, 1965.
[7] R. D. Heidenreich, 'Fundamentals of Transmission Electron Microscopy', Interscience, New York, 1964.
[8] R. Uyeda, *Acta Cryst.*, 1968, **24A**, 175; G. Dupouy, in 'Advances in Optical and Electron Microscopy', ed. R. Barer and V. E. Cosslett, Academic Press, London and New York, 1968, Vol. 2, p. 167; C. J. Humphreys, J. S. Lally, L. E. Thomas, and R. M. Fisher, 'Electron Microscopy', ed. P. Favard, Société Francaise de Microscopie Electronique, Grenoble, 1970, Vol. 1, p. 91.
[9] J. W. Menter, *Proc. Roy. Soc.*, 1956, **A236**, 119.
[10] M. Born and E. Wolf, 'Principles of Optics', Pergamon, Oxford, 1965.

higher-order diffracted waves in the image plane reproduces the object grating (Figure 1a). However, in the case of the electron microscope, the image must be defocused so that there will be a correct superposition of the waves a short distance away from the Gaussian image plane [7] (Figure 1b). This required amount of defocus corrects the effect of spherical aberration on the diffracted waves at higher angles. Crystal lattice imaging was further developed by Dowell,[11] Komoda,[12, 13] and Watanabe et al.,[14] who, by using the tilted illumination method, resolved lattice spacings as small as 1.27 Å (e.g. {220} planes of copper). However, tilted illumination is only useful when large single crystals are available. For small randomly-oriented crystals, the tilting method selects for correct imaging only those crystals whose electron diffraction patterns at the back focal plane are moved closer to the optical axis, where spherical and chromatic aberrations are less. Imaging of other crystals, whose diffraction patterns move further away from the optical axis will be affected appreciably more by these optical defects.

Yada and Hibi,[15] using pointed cathodes and axial illumination, obtained crystal lattice images below 2 Å. In doing so, they proved that by improving the coherence of the electron beam one can approach the resolution of the tilted illumination method with a gain in symmetrical resolution in the image plane. Yada and Hibi also produced three beam interference images showing spacings of 1.02 Å due to {200} simultaneous reflections in gold. However, this achievement primarily indicates the very high mechanical stability of their instrument. It is not a true test of the ability of the objective lens to resolve these small spacings because both reflections are at the same distance from the optical axis and are therefore affected by the same amount of spherical aberration. Lattice images of some light elements were also imaged by Yada and Hibi. These include the (10.0) planes of graphite which were also resolved by the present author.[16]

The imaging of single molecules and atoms with the transmission electron microscope is considerably more difficult than the imaging of crystal lattices. There are many reasons for this, the most important being the resolution limit of the instruments. In addition, there are sample preparation problems and radiation damage in the case of organic molecules because of the very high inelastic scattering cross-section of low atomic number elements.[7] Nevertheless, attempts at imaging molecular structures [17, 18] are being made.

[11] W. C. T. Dowell, Proc. Internat. Conf. on Mag. and Cryst., Kyoto, 1961, 175; W. C. T. Dowell, 'Electron Microscopy', ed. S. S. Breese, Academic Press, New York, 1962, Vol. 1, KK12; W. C. T. Dowell, *Optik*, 1963, **20**, 535.
[12] T. Komoda, *Optik*, 1964, **21**, 93.
[13] T. Komoda, 'Electron Microscopy', ed. R. Uyeda, Maruzen, Tokyo, 1966, Vol. 1, p. 29.
[14] M. Watanabe, H. Shinagawa, and K. Shirota, ref. 13, 33.
[15] K. Yada and T. Hibi, ref. 13, p. 25; *Jap. J. Appl. Phys.*, 1967, **6**, 1007; 1968, **7**, 178; *Jap. J. Electron Microscopy*, 1969, **18**, 266.
[16] L. L. Ban and W. M. Hess, 9th Biennial Conf. on Carbon, Columbus, Ohio, Battelle's Laboratories: Defense Ceramic Inf. Centre, 1969, p. 162.
[17] R. D. Heidenreich, *Jap. J. Electron Microscopy*, 1967, **16**, 23.
[18] H. Fernandez-Moran, *Jap. J. Electron Microscopy*, 1967, **16**, 65; H. Fernandez-Moran, ref. 13, p. 13.

Direct Study of Structural Imperfections

The success of imaging single atoms is strongly dependent on a reduction of the spherical aberration of the objective lens.[4,19] However, the imaging of heavy atoms such as gold or uranium appears possible for present-day high-resolution microscopes operated at 100 kV. One of the limitations is the sample preparation, which requires a monolayer of light atoms supporting the heavy atoms, with the latter being separated by about 2 Å or more. The problem with this type of preparation is that light atoms produce a high background of inelastically scattered electrons in the image which would tend to smear the Airy discs of neighbouring atoms, *i.e.*, there would be a general lowering of contrast. However, by energy selection in a high-resolution microscope, this problem could be overcome. Another problem is determination of the right amount of defocus, which can be either positive or negative. As the diffracted waves from the object spread out into different zones of the objective lens, they obtain different phase relations which shift in a complex manner with changes in focus. This change in phase relationship is governed by the spherical aberration of the objective lens.[20-22] Because of these intrinsic characteristics relating to phase shifts of diffracted waves as a function of spherical aberration and focus, the electron microscope is considered to be a phase contrast microscope. Lattice imaging is achieved by this phase contrast mechanism. In the case of crystal lattices, however, there is a sharply defined zone in the objective lens which is simply related to the spacings through Bragg's law. This does not apply in the case of single atoms.

The subject matter of this article covers the application of high-resolution electron microscopy to the study of crystal lattice images in carbons. Some results on the imaging of polymer microstructure are also included.

2 Electron Optical Considerations

Electrons passing through the object are scattered both elastically and inelastically. Included among the elastically scattered electrons are the Bragg diffracted electrons (in the transmission or Laue case) which are theoretically related by the Ewald sphere of reflection (Figure 2). Among the inelastically scattered electrons are those that are interacting with the atoms of the object. Single atomic excitations and collective excitations, *e.g.* plasma losses in metals,[7,23] as well as collective losses in semiconductors, insulators, and organic substances, including biological materials,[24] can also occur. Although their importance cannot be overlooked, these latter type interactions will not be covered here.

---

[19] E. Zeitler, 'Advances in Electronics and Electron Physics', ed. L. Marton, Academic Press, New York and London, 1968, Vol. 25, p. 277.
[20] O. Scherzer, *J. Appl. Phys.*, 1949, **20**, 20.
[21] R. D. Heidenreich and R. W. Hamming, *Bell System Tech. J.*, 1965, **44**, 207.
[22] C. B. Eisenhandler and B. M. Siegel, *J. Appl. Phys.*, 1966, **37**, 1613.
[23] D. Pines, 'Elementary Excitations in Solids', Benjamin, New York, 1963.
[24] C. D. Johnson and T. B. Rymer, *Nature*, 1967, **213**, 1045; A. V. Crewe, M. Isaacson, and D. Johnson, *Nature*, 1971, **231**, 262.

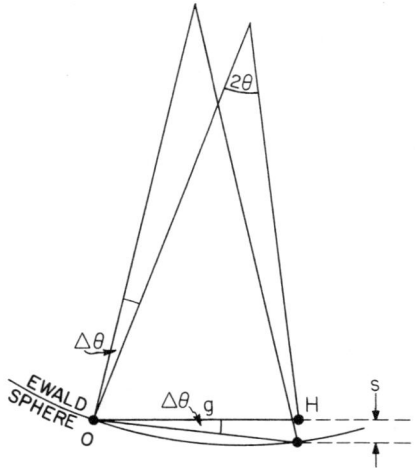

Figure 2 *Ewald sphere of reflection with reciprocal lattice vector $\vec{g}$, reciprocal lattice point* H, *and resonance error* s *indicating deviation of a crystal from the correct Bragg condition (in the Laue case) by* $\Delta\theta$ *(dispersion surfaces are not shown)*

The amplitude distribution of the diffracted waves in the back focal plane of the objective lens is the Fourier transformation of the object potential distribution. The intensity distribution in this plane is the Fraunhofer diffraction pattern of the object. The Fourier transformation of the diffraction pattern produces the image which is modified by the aberrations of the optical system (Figure 1b). The aberrations change the phase of the diffracted waves (Bragg or other). These effects will be covered in detail since the image contrast that is produced is strongly dependent on them. From a practical standpoint, electron microscopy is really a study of contrast (or visibility), without which there is only empty resolution.

Contrast, C, in a lattice image can be defined as:

$$C = \frac{I_{max} - I_{min}}{I_{max} + I_{min}}$$

where I_{max} and I_{min} are the intensities of the electron beams in the fringe pattern. It is necessary to point out that the image is only a representation of the period of the lattice planes, *i.e.* only the relative lattice period is observed.[7] Contrast in the two-beam dynamical theory is given, assuming correct Bragg alignment of the crystal with the illuminating beam, by:

$$C_{dyn} = \frac{2 \sin 2\pi t}{t_0 \sin \chi} \qquad (1)$$

where t is the crystal thickness and t_0 is the extinction distance; $t_0 = \lambda V_a/V_g$, where λ is the electron wavelength, V_a is the accelerating voltage and V_g is the Fourier coefficient of the operating reflection, where $\vec{g}$ indicates a reciprocal lattice vector for a set of $\{hkl\}$ planes (Figure 2). V_g is equal to:

$$V_{hkl} = 48F^*(hkl)/\Omega \text{ volts}$$

where $F(hkl)$ is the structure factor and * indicates the complex conjugate, Ω is the volume of the unit cell in Å^3. χ in equation (1) is the phase shift in the diffracted beam, which depends on the atomic composition of the object, objective lens properties, disturbances such as mechanical vibrations, a.c. hum, and stray fields, and is equal to the following polynomial:[17]

$$\chi = \gamma + \frac{|K|}{2}\Delta L \beta^2 - \frac{|K|}{4}C_0\beta^4 + \chi_{ast} + \chi(t) \tag{2}$$

where $\gamma = (Z/137)(v/c)$ is a phase term,[25] which is a function of atomic composition with atomic number Z, v/c is the relativistic velocity factor. For light atoms, such as carbon, this first phase term can be neglected, but it becomes important with heavy atoms. K is the wave vector with $|K| = 2\pi/\lambda$; ΔL is the defocus term, the only term that is in direct control of the operator of the instrument; β is the diffraction angle ($\beta \approx \lambda/d$, where d is the interlayer spacing), and is equal to 2θ Bragg angle (Figure 2). $C_0 = C_s f$ is the spherical aberration coefficient of the objective lens, where C_s is the spherical aberration constant and f is the focal length. χ_{ast} is the phase shift due to astigmatism:[26]

$$\chi_{ast} = \frac{|K|}{2}Z_{ast}(\sin \alpha)\beta^2 \tag{3}$$

where Z_{ast} is the astigmatism of the lens and α is a reference angle in the plane of the lens. This term indicates that astigmatism has an effect of different focus change in different directions. The last term in equation (2), $\chi(t)$, is due to stray fields, a.c. hum, and high-frequency vibrations; it can be eliminated by careful selection of microscope position in laboratories. The astigmatism can also be made small by carefully correcting the objective lens with the magnetic stigmator which is standard equipment on most high-resolution instruments.

For maximum phase contrast in equation (1) we should have:

$$\sin \chi = \pm 1 \quad \text{or} \quad \chi = \frac{\pi}{2}(2n-1) \tag{4}$$

where n is a positive or negative integer or zero; thickness should be $t_0/4$ or less. If we assume that $\gamma \approx 0$ and $\alpha \approx 0$ in equations (2) and (3) respectively,

[25] E. Zeitler and H. Olsen, *Phys. Rev.*, 1964, **136A**, 1546.
[26] R. D. Heidenreich, W. M. Hess, and L. L. Ban, *J. Appl. Cryst.* 1968, **1**, 1.

C

then by equating (2) and (4) we obtain the defocus necessary for a set of lattice planes with a spacing of d:

$$\Delta L = \frac{d^2}{2\lambda}(2n-1) + \tfrac{1}{2}C_0\frac{\lambda^2}{d^2} \tag{5}$$

In this formula, ΔL is always positive for small crystal spacings, and if we take $n = 1$, the first-order diffraction, then the amount of focus is simply a value which causes the bright-field (BF) and dark-field (DF) images to coincide in the conjugate image plane (Figure 1b), with the right phase difference (the BF image is related to the zero-order beam and the DF image to one of the first-order beams). In the image plane, the BF image rotates with change in focus, but the DF image both shifts and rotates with change in focus. This also means that the DF image is displaced from the BF image in the image plane when the image is in Gaussian focus (paraxial focus) by the amount $C_0\beta^3$. This brings about a rather useful relationship. By marking the movement of the DF image with a change in focus (objective aperture removed) the type of $\vec{g}$ can be determined from the following relationship:

$$\Delta X = C_0\beta^3 \pm \Delta L\beta \tag{6}$$

where ΔX is the change in the position of the DF image with respect to that of the BF image with focus change ΔL. However, one must have only axial illumination for the above expression to be meaningful. $\Delta L = C_0\beta^2$ when there is complete superposition of the DF and BF images. This relationship could be useful for determining the nature of dislocations when high-resolution images are not available.

The phase shift, $\chi_{\Delta r}$, across a wave front which is diffracted by a crystal of size Δr, is given by:

$$\chi_{\Delta r} = \tfrac{1}{4}C_0\beta^4 + C_0\beta^2\frac{\Delta r}{f} \tag{7}$$

where f is the focal length of the objective lens. The size Δr is measured in the direction perpendicular to the lattice planes which are diffracting. The second term in the above relationship is very small and can be eliminated, because there is coherent interference across the narrow crystal areas used in electron microscopy within a recorded image; only the first term is used in the phase polynomial (2). To determine the best focus for spacing d, formula (5) is used, but the same level of contrast (with a reversal of the black and white 'lines') is shown in the image at the subsequent intervals of focus $\Delta L = d^2/\lambda$. However, the size of the crystal is limited since the diffracted image, (DF) (Figure 1b), will move off the BF image, as pointed out before.

The other consideration for contrast and coherency is the electron source. To obtain coherent diffraction from a crystal the source has to be coherent across the crystal which it illuminates.[7,26] Within reasonable limits, the source is spatially coherent to length $\Delta Y = \lambda/2\beta_s$, where β_s is the angle of

Direct Study of Structural Imperfections

illumination. For maximum interference and good contrast $\Delta Y > Nd$, where N is the number of lattice planes with spacing d. It is also important to consider the defocus values that are limited with source coherency, since the image formed with defocus is the sum of all images that are formed by all incident wave fronts with a total angle of β_s. Then, the minimum useful limit on focus steps with a given β_s is $\Delta L < d/2\beta_s$. The above restrictions for the source necessitate the use of pointed single crystal filaments in the Schottky-emission mode.[27]

In using formula (1) it is assumed that the crystal is exactly, or very nearly exactly, aligned with the Bragg angle to the electron beam. However, most often the above assumption is not correct; then the intensities and contrast in the two-beam dynamical case are modulated accordingly. If the crystal is misoriented from the Bragg angle then the transmitted and diffracted intensities, I_t and I_g respectively, are given by the following terms:[12]

$$I_t = \frac{1}{2(1+w^2)}\left[1+2w^2+\cos\left(2\pi\sqrt{1+w^2}\cdot\frac{t}{t_0}\right)\right]$$

$$I_g = \frac{1}{2(1+w^2)}\left[1-\cos\left(2\pi\sqrt{1+w^2}\cdot\frac{t}{t_0}\right)\right]$$

(8)

where $w = t_0 s$ is the deviation parameter in which s is the resonance error in the dynamical theory (Figure 2). If the incident beam is normalized to unity, $I_t + I_g = 1$, if there is no absorption, inelastic scattering, plasma excitation, or core excitation present during the passage of the high-energy electrons through the crystal.[7] Contrast in the lattice image is given, taking into account the deviation above, by:

$$C_{\text{dyn}} = \frac{1}{1+w^2}\left\{\left[1+2w^2+\cos\left(2\pi\sqrt{1+w^2}\cdot\frac{t}{t_0}\right)\right]\times\left[1-\cos\left(2\pi\sqrt{1+w^2}\cdot\frac{t}{t_0}\right)\right]\right\}^{\frac{1}{2}}$$

(9)

From equation (9) it is seen that, if $s = 0$, there is maximum contrast for the lattice image if $t = t_0/4$, as was also pointed out earlier. By varying s and t, the contrast is varied in a complicated manner; one can arrive at a 'rocking curve' by keeping t fixed and varying s.[6]

The simplest thickness variation is that of a wedge crystal, such as a cubic crystal lying with the bisector of the wedge angle normal to the optic axis, a case which was examined by Heidenreich and Sturkey[28] for determining the Fourier coefficients of inner potential for MgO crystals. In this case the intensities become, when the Bragg alignment condition is met, the following:

---

[27] T. E. Everhart, *J. Appl. Phys.*, 1967, **38**, 4944.
[28] R. D. Heidenreich and L. Sturkey, *J. Appl. Phys.*, 1945, **16**, 97.

$$I_t = \cos^2 \frac{\pi t}{t_0} \quad \text{and} \quad I_g = \sin^2 \frac{\pi t}{t_0} \tag{10}$$

The intensity variation with thickness introduces the well-known 'extinction fringes' which are the consequence of the interaction of transmitted and diffracted waves in the crystal. The spacing of the fringes, x_t, is:

$$x_t = \tfrac{1}{2} t_0 \cot \tfrac{1}{2} \gamma_w \tag{11}$$

where γ_w is the wedge angle. If there is deviation from the Bragg angle, t_0 has to be modified and used above: $t_0' = t_0/(1+w^2)^{\frac{1}{2}}$. As was pointed out by Sturkey,[29] the above formulation for obtaining a V_g is rather doubtful, but as a rule can be used for a sparsely populated reciprocal lattice. However, there should also be further modification of t_0 by taking into account thermal diffuse scattering,[7] relativistic effects, and absorption.[6,7,30] If the lattice planes that are responsible for the fringes are resolved, then, according to the two-beam dynamical theory,[31] the fringes shift with $d/2$ as thickness changes by $t_0/2$ (Figure 3). This shifting pattern will become more complicated as the angle of alignment deviates from the correct Bragg angle, a case which was examined in great detail by Hashimoto et al.,[32] not only for the wedge-shaped crystals, but also for bent plate-shaped crystals. From the above studies, one can expect complicated bending patterns around dislocation images. The contrast in and around dislocated areas is affected by the strain field, $\vec{R}$, which affects the phase, χ_R, of the diffracted wave front from that area: $\chi_R = 2\pi \vec{g} \cdot \vec{R}$. The above phase shift enters into the phase polynomial (2), and will thus affect the defocus value for correctly imaging the dislocation area. Once the correct defocus is determined, the intensities and contrast of the lattice images at and near dislocations can best be understood by again considering the modified extinction distance since there is systematic deviation from the correct Bragg angle around the dislocation. The strain field around the dislocation complicates the resolution and location of dislocations, since there will apparently be terminating lattice planes at areas where they are not terminated (e.g. only distorted). This phenomenon will be demonstrated later in images of graphitic carbon black. The above effect was also considered theoretically by Cockayne et al.[33] As the angle of alignment is varied for a set of lattice planes it is also likely that there will be a position when simultaneous diffraction, $\vec{g}$ and $-\vec{g}$, occurs. In this instance (i.e. the three-beam case) the extinction distance has to be modified:[7] $t_{\pm g} = \lambda V_a / V_g \sqrt{2}$, with strictly axial illumination. This effect will also be illustrated in the lattice images of graphitic carbon black.

[29] L. Sturkey, *Proc. Phys. Soc.*, 1962, **80**, 321.
[30] H. Hashimoto, A. Howie, and M. J. Whelan, *Proc. Roy. Soc.*, 1962, **A269**, 80.
[31] H. Niehrs, *Z. Physik*, 1954, **138**, 570.
[32] H. Hashimoto, M. Mannami, and T. Naiki, *Phil. Trans.*, 1961, **A253**, 459, 490.
[33] D. J. H. Cockayne, I. L. F. Ray, and M. J. Whelan in 'Electron Microscopy', ed. D. S. Bocciarelli, Tipografia Poliglotta Vaticana, Rome, 1968, Vol. 1, p. 129; D. J. H. Cockayne and M. J. Whelan, 'Electron Microscopy', ed. P. Favard, Société Francaise de Microscopie Electronique, Grenoble, 1970, Vol. 1, p. 29.

Direct Study of Structural Imperfections

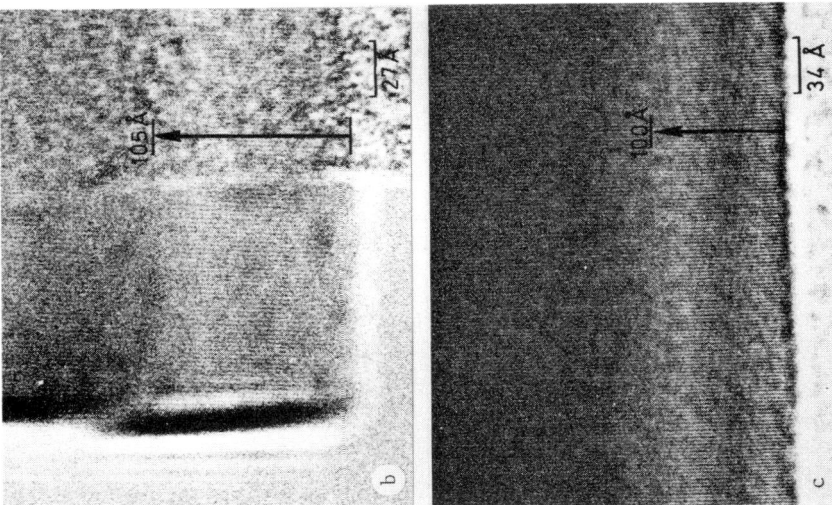

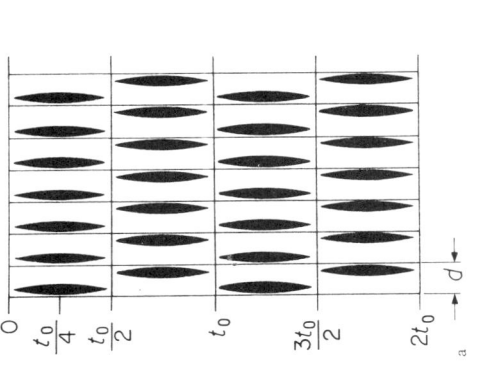

Figure 3 (a) Diagrammatic representation of the lattice image of a wedge crystal when the correct Bragg angle is satisfied; images of lattice planes shift by d/2 as thickness changes by $t_0/2$.; (b) (200) lattice image (d = 2.10 Å) of MgO crystal indicating shifting of the lattice image at approximately 105 Å. The lattice image is not discontinuous as in diagram (a) because it is most likely that the Bragg angle is not satisfied; (c) (00.2) lattice image of a needle-shaped ZnO crystal (d = 2.60 Å) showing the shifting of the lattice image as thickness increases laterally to approximately 100 Å, also with a probable deviation from the correct Bragg angle

Let us now examine the possible contrast produced in the images of molecular detail, such as single macromolecules and the images of superpositioned molecules, or detail from randomly oriented, small short-range structures. The contrast can be treated kinematically;[17] that is, the amplitude of diffracted waves is the sum of the amplitudes from the individual atoms at the back focal plane:

$$C_{kin} = \lambda[1-(v/c)^2]^{-\frac{1}{2}} |f(\beta)| t \left(\frac{\sin \pi t/2\beta^2}{\pi t/2\beta^2}\right) T(N_1) \sin \chi \qquad (12)$$

where $|f(\beta)|$ is the atomic scattering amplitude, the term in brackets is the relativistic correction, $T(N_1)$ is an integral of the phase spread across the shape transform of N_1 atoms (representing only one type of atom) and the other terms have their usual meanings. For obtaining reliable information on the shape and dimensions of a molecular object the thickness should be monomolecular and, for best results, with well-separated molecular segments. If the object is composed of superpositioned molecules then the image formed does not represent the true structure of the object. The latter case was examined by Thon,[34] who by combined electron microscopy and optical diffraction analysis of the images, proved that the defocused and Gaussian focused images of carbon films are complicated by the simultaneous existence of positive and negative phase contrast. By applying Scherzer's [20] phase contrast relationship, Thon developed the following expression, which takes into account positive and negative phase contrast effects:

$$a = \lambda \left[\frac{\Delta L}{C_0} \left(\pm \frac{\Delta L^2}{C_0^2} + \frac{(2n-1)\lambda}{C_0}\right)^{\frac{1}{2}}\right]^{-\frac{1}{2}} \qquad (13)$$

where a is the object period and the other terms have their usual meanings. This expression, of course, is the solution of equation (5) for a in terms of $\beta \approx \lambda/a$. For one defocus value there is a range of a values that depend on the value of n, ($n = 0, \pm 1, 2, 3 \ldots$) giving positive phase contrast if n is even and negative phase contrast if n is odd; for $\Delta L = 0$, at Gaussian focus, n can only have positive values. Contrast can be called positive when object regions having a higher inner potential (or which are thicker than their surroundings) are reproduced in the images as darker areas. Such object regions delay the phase of the electron wave more than their surroundings. Object regions with a leading phase, on the other hand, appear lighter in the images. With negative phase contrast the opposite is true. Then the object details with trailing phase are lighter, while those with a leading phase are darker than their surroundings. One can see the complexity of an image of thick, or even thin, short-range order objects when positive and negative phase contrast exists simultaneously. Overcoming this problem will be demonstrated in a subsequent section of this chapter in the case of imaging the structure of evaporated carbon films.

[34] F. Thon, *Z. Naturforsch.*, 1965, **20a**, 154; F. Thon, *Siemens Rev.*, 1967, **34**, 13.

3 Instrumentation

Two different transmission magnetic electron microscopes (Philips EM-200 and EM-300) were employed in the high-resolution studies described here. Both microscopes have been installed under special environmental conditions to facilitate high-resolution microscopy on a daily, practical basis. Each instrument is centrally positioned in a relatively large room in an area that is essentially free from stray magnetic fields (below 0.001 Oe). Line voltage is regulated within 1 % and each instrument is mounted on a vibration isolator. With this type of mounting, the centre of gravity of the microscope is centred on a concrete isolator block weighing about 11.5 tons. The block and microscope are supported on four large steel springs which have a very low natural frequency (3 Hz). In addition to the elimination of vibration, great attention has also been given to specimen contamination. Contamination has been effectively reduced, to below 2 Å min^{-1}, with the following measures: (*i*) the temperature of the cooling water of the magnetic lenses was kept low (≈ 283 K), (*ii*) dry pre-purified nitrogen, rather than air, was bled into the vacuum system when breaking the vacuum in the microscope, (*iii*) a molecular sieve is used with the mechanical pump, and (*iv*) a special three-blade platinum cooling device surrounded the specimen (and the objective aperture, if one was used). The cooling device is an improvement over

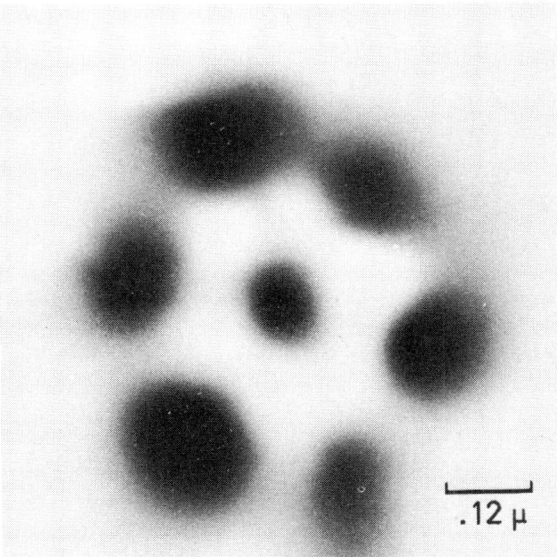

Figure 4 *A typical temperature-field emission image of a tungsten pointed single crystal source before complete thermal saturation, as seen in the electron microscope. Emission pattern indicates* (110) *orientation (middle dark spot). Dark regions are indicating high work function areas, i.e. closely packed crystal planes*

the conventional two-blade type. It is cooled with liquid nitrogen and provides a larger surface area in the vicinity of the specimen for trapping the various contaminants.

It was emphasized before that another important factor in performing high-resolution microscopy on a consistent basis is the electron source. Pointed single crystal tungsten filaments (≈ 1000 Å tip radius) have been employed in this study (Figure 4), used in the Schottky-emission mode. Beam currents of 1–2 μA were sufficient from such a source for recording the images at high magnification.

To facilitate fine 'tuning' for phase contrast imaging, a continuous fine focus control was added to the existing focus controls (which are stepwise), so that the best possible Gaussian focus can be reached. This can be accomplished with ± 50 Å accuracy, and from this point the required defocus can be adjusted with the stepwise focus controls. The images were also corrected for astigmatism by using the continuous focus control, on evaporated carbon films. A convenient magnification was selected which helps to reach the required defocus at low illumination levels. This magnification was 2.3×10^6 on the viewing screen using a $10 \times$ binocular microscope. With this selected magnification, the recorded magnification was 2.95×10^5 on photographic plates (Kodak medium contrast emulsion developed in D19). An average density of 1.6 is used in most of the images, which required on the average 10^{-11}—10^{-12} A cm^{-2} electron density, which is approximately 10^8—10^7 electrons cm^{-2} s^{-1}. The microscopes were operated in most instances with 100 kV electrons and without objective lens apertures. Objective apertures were used only in the case of dark-field imaging or for bright-field imaging not requiring lattice resolution.

4 Materials Studied

Carbons.—Carbons exist in many diverse morphological forms which make them ideal objects for electron microscope study. Because of the advent of high-resolution electron microscopy, the microstructure of these carbons can be viewed (in conjunction with X-ray diffraction) in a completely new light.

Warren,[35] from his atomic radial distribution studies using X-rays, indicated that carbon blacks are not composed of small three-dimensional graphite crystals. Instead, he proposed that they are composed of small graphite-like layers with the same atomic positions within the layers as in graphite. He further inferred that the layers are parallel but rotated around the c-axis. The theory of the diffraction of X-rays by random layer lattices, developed by Warren[36] and applied by Biscoe and Warren[37] to the study of carbon blacks, showed that the X-ray diffraction patterns of these carbons are composed of (00.l) three-dimensional and (hk) two-dimensional reflections. The

---

[35] B. E. Warren, *J. Chem. Phys.*, 1934, **2**, 551.
[36] B. E. Warren, *Phys. Rev.*, 1941, **59**, 693.
[37] J. Biscoe and B. E. Warren, *J. Appl. Phys.*, 1942, **13**, 364.

position of the (00.*l*) reflections indicated that the layers were further apart than in graphite. Crystal size along the *c*-axis, L_c, was determined from the (00.*l*) reflections, notably from the (00.2) and (00.4), using the Scherrer equation. Crystal size in the plane of the layers, L_a, was derived by Warren [36] from the (*hk*) reflections, assuming 60° rhombus shape layers with sides N_a, where N is the number of unit cells with $a = 2.46$ Å, which is the side of the hexagonal unit cell.

The work of Franklin [38] showed that the various types of carbon differ from one another only in the manner in which they resemble or differ from graphite, and not on the basis of different crystallographic structural modifications. She separated carbons into two groups and called them graphitic and non-graphitic. Those carbons whose graphite-like layers lie in parallel groups but oriented rather randomly around the *c*-axis were called non-graphitic. Those carbons whose graphite-like layers have some mutual orientation (similar to graphite's *ABA* or *ABCA* orientation) were called graphitic. Those non-graphitic carbons which formed graphitic carbons when heated to high temperatures (≈ 3270 K), Franklin called graphitizing carbons; those which on similar heating do not reach the graphitic state were called non-graphitizing. Franklin attributed these differences to the earliest stages of formation and subsequent carbonization from the starting materials. She also proposed a cross-linking mechanism of disorganized carbon atoms which restrict the movement of single and groups of layers into the graphitic order. The work of Warren and Franklin showed the direction for further research into the study of carbon structure and a large literature on the published findings now exists.[39, 40]

More recent developments in analysing the atomic radial distribution of carbons indicated that the layers are much larger than one would obtain from the two-dimensional (*hk*) reflections using Warren's formulations.[39]

Electron microscopy of carbons proceeded with the assumption that the findings of the *X*-ray L_a and L_c measurements were correct. The electron microscope was used to locate these crystallites, or simply to see how many of the crystallites were possibly present in an aggregate (primary unit) of these carbons. The locating of dislocations was also a very important development, but the dislocation studies were primarily limited to graphites.

Carbon Blacks. One of the first to attempt to resolve and locate crystallites in carbon black was Hall,[41] who formed dark-field images with the (00.2) reflections by using the displaced aperture method. He reported that the images of very large thermal black particles showed a concentric orientation

---

[38] R. E. Franklin, *Acta Cryst.*, 1950, **3**, 107; 1951, **4**, 253; R. E. Franklin, *Proc. Roy. Soc.*, 1951, **A209**, 196.
[39] S. Ergun, in 'Chemistry and Physics of Carbon', ed. P. L. Walker, Marcel Dekker, New York, 1968, Vol. 3, p. 211.
[40] W. Buland, in 'Chemistry and Physics of Carbon', ed. P. L. Walker, Marcel Dekker, New York, 1968, Vol. 4, p. 1.
[41] C. E. Hall, *J. Appl. Phys.*, 1948, **19**, 271.

of crystallites, with the graphitic layers parallel to the surface. At that time, however, he was not able to determine whether a similar orientation existed for small particle size carbon blacks. Later electron microscope studies [42] on treated carbon blacks (*e.g.* oxidation and high temperature graphitization) supported the findings of Hall. It was not until the application of high-resolution dark-field microscopy,[43] however, that Hall's hypothesis was shown to be valid for all types of carbon black in their initial, untreated form. These studies were based on high-resolution (≈ 3 Å) diffracted-beam images from both the (00.2) and (10) reflections. These images enabled detection of small differences among different types of carbon black, which were not resolvable by Hall's method or the subsequent oxidation and graphitization studies based on bright-field imaging. The diffracted beam method indicated a somewhat more orderly graphitic layer structure for channel carbon blacks (made from a gas feed stock) in comparison to oil furnace blacks of the same particle size. The difference is apparently related to the greater high-temperature heat history of the channel blacks during their manufacture.

Using high-resolution phase contrast microscopy, the graphitic layer lattice was resolved in a heat-treated carbon black (2873 K) by the interference of the (00.2) beam with the 0 beam (Figure 1b).[26] These images were the first to show the very extensive bending of the graphitic layer planes. The imaging of layer planes in every type of carbon black, including commercial non-graphitized types, also became possible with further improvements in instrumentation.[16, 44] It was shown that the graphitic layers are quite extensive and are distorted in a random pattern. Hence, the crystal model based on X-ray diffraction must be considered incorrect. It is now also apparent that the dark-field images, which were interpreted as crystallites, actually represent parallel segments within the larger, distorted layer network. This type of structure may be called paracrystalline.[45] The primary units of carbon black can be considered as carbonized molecular crystals [46] or single graphitic paracrystals. One can assume that distortions in the layer network are produced during conformation and carbonization of the large aromatic layers or molecules around growth centres. The primary paracrystalline units appear to be formed by the coalescing of hydrocarbon droplets, before carbonization begins at higher temperatures.[47]

In the phase contrast imaging of carbon, a different portion of the para-

[42] F. A. Heckman, *Rubber Chem. Technol.*, 1964, **37**, 1245.
[43] W. M. Hess and L. L. Ban, ref. 13, p. 569; W. M. Hess, L. L. Ban, F. J. Eckert, and V. Chirico, *Rubber Chem. Technol.*, 1968, **41**, 356.
[44] L. L. Ban and W. M. Hess, Proc. 26th Annual EMSA, ed. C. J. Arceneaux, Claitor's Pub. Div., Baton Rouge, 1968, p. 256.
[45] R. Hosemann and S. N. Bagchi, 'Direct Analysis of Diffraction by Matter', North Holland, Amsterdam, 1962.
[46] J. D. Brooks and G. H. Taylor, in 'Chemistry and Physics of Carbon', ed. P. L. Walker, Marcel Dekker, New York, 1968, Vol. 4, p. 243.
[47] H. B. Palmer and C. F. Cullis, in 'Chemistry and Physics of Carbon', ed. P. L. Walker, Marcel Dekker, New York, 1965, Vol. 1, p. 265.

Direct Study of Structural Imperfections

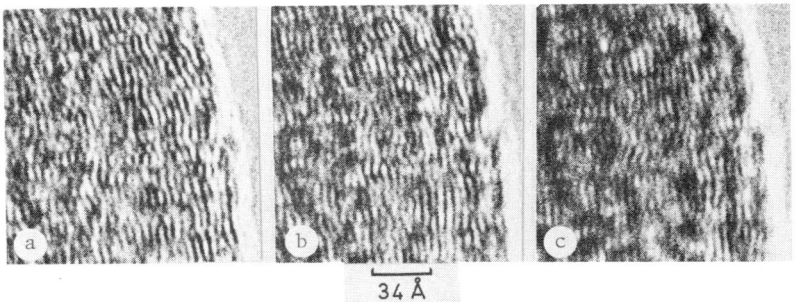

Figure 5 *Defocusing series with* $\Delta L = 250$ Å *differences between images* a, b, *and* c *of a thermal carbon black particle indicating different lattice detail in each image*

crystalline network is shown with changes in focus. This is especially noticeable for the large particle size carbons (Figure 5). However, the focal setting with best contrast and resolution is determined by equation (5). The contrast is dependent on the orientation of the layer planes with respect to the electron beam, in the same manner as regular crystals, and is determined by the combination of orientation and lattice spacing variation. That is, there is a random variation of extinction distance, since the structure factor and the volume of the effective unit cell is fluctuating throughout the material [the extinction distance for graphite (00.2) lattice planes is 380 Å with $d_{002} = 3.35$ Å].

The image corresponding to the (hk) reflections (without layer type detail) is in the central portion of the particles and also changes with defocus. The reason for this change with focus is the same as in the case of carbon film structure, which was discussed earlier in connection with Thon's work.[34] Work is now being carried out to extract useful information from these images.

Images are also observed in which some of the layers appear to emanate from one or two layers (Figure 6). There is a problem of interpretation here in that uncertainty exists regarding the actual position of the layers.[7] It is not possible at this stage of our understanding of phase contrast images to tell which lines represent the true layers, *i.e.* where the carbon atoms are positioned. However, this uncertainty is not too discouraging since the black and white lines seem to be complementary. Hence, on a relative basis they represent the layer structure. It follows then that, if there are layers which emanate from others, it is possible to propose a type of bonding at these sites different from the trigonal and π-bonded graphite. However, this bond would have to occur in long narrow rows, possibly as long as some of the layers in the images (*e.g.* up to 500 Å or more). One can only speculate upon its nature. It could well be the tetrahedral or diamond-like bond which was proposed by others, most notably by Ergun and his co-workers.[39]

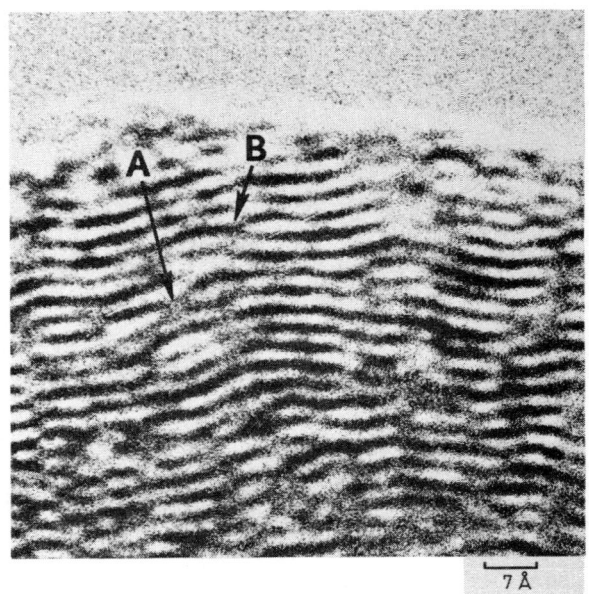

Figure 6 *A segment of a carbon black particle in which layers seem to emanate from one another: four at A and three at B.*

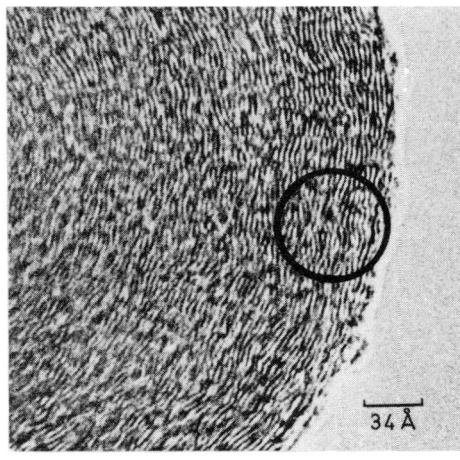

Figure 7 *Segment of a carbon black particle on which the encircled zone indicates the size of the area from which optical diffraction* (OD) *is recorded*

However, the type of imaging illustrated in Figure 6 could also be indicative of layers which are similar to the non-basal dislocations found in graphites.[48]*
(i) *Distribution of graphitic layer spacings.* There is a range of interlayer spacings associated with the distorted layers in the images of carbon blacks. The distribution of spacings has been measured by optical diffraction [49, 50] (OD) of the high-resolution electron micrographs (photographic plates). The optical diffractometer that was used consists of a 3-metre optical bench with a laser source (He–Ne, $\lambda = 6328$ Å, 1 mW). There were two possible approaches available for sampling distribution of interlayer spacings: (*i*) to study a small number of primary paracrystalline units in detail by varying the focus in small increments around the value given by equation (5) and record the different images or (*ii*) examine a number of different paracrystals of varied morphology (*e.g.* spherical, clustered, fibrous) and record each image only once using equation (5), or at $\Delta L \approx 750$ Å for 3.44 Å. The second approach was employed, because of the obvious advantage of being able to sample a greater variety of layer groupings around different size growth centres. In sampling the areas to be analysed, an opaque mask with a circular opening of ~ 2 mm diameter is positioned over the negative images on the photographic plates (Figure 7). By using short exposures, the higher order diffraction rings of the aperture and the weak transfer of mask opening around the diffraction spots were filtered out of the patterns, along with the general phase contrast effects reported by Thon.[34] The statistically correct number of samples was determined experimentally as the number of areas where additional sampling did not significantly change the arithmetic averages (Ar.Av.) of the measured spacings in the following ranges: $a = 3.09$—3.43, $b = 3.48$—3.74, $c = 3.80$—8.24 Å. Range *b* approximates the possible spacings obtained by *X*-ray diffraction for non-graphitized carbons. The two extremes in the spacings (3.09 Å and 8.24 Å) were determined experimentally from the observed maximum and minimum spacings measured during the experiments. The measurements were carried out on photographic prints which were calibrated with the OD patterns of the (111) lattice images of gold with spacings of 2.36 ± 0.02 Å and also a graphitized ISAF type carbon black with $d_{002} = 3.44$ Å (Figure 8a, b). All OD spots which correspond to layer spacings were measured in the wedge-shaped pattern outside the mask's high-order diffraction rings (Figure 9). The spots represent spacings between single layers as well as groups of layers which can also diffract to single

---

* Figure 6 indicates two possible forms of non-basal dislocations: *a*, edge dislocation with Burgers vector perpendicular to the layer planes and the dislocation in the planes; *b*, screw dislocation with Burgers vector and dislocation line perpendicular to the layer planes, in this case, the Burgers vector appears to be 3.6 Å [*c*/2 <00.1 >].

[48] J. M. Thomas and C. Roscoe, in 'Chemistry and Physics of Carbon', ed. P. L. Walker, Marcel Dekker, New York, 1968, Vol. 3, p. 1.
[49] C. A. Taylor and H. Lipson, 'Optical Transforms', Cornell University Press, Ithaca and New York, 1965.
[50] L. L. Ban and W. M. Hess, 10th Biennial Conf. on Carbon, Columbus, Ohio, Battelle's Laboratories, Defense Ceramic Inf. Centre, 1971, p. 159.

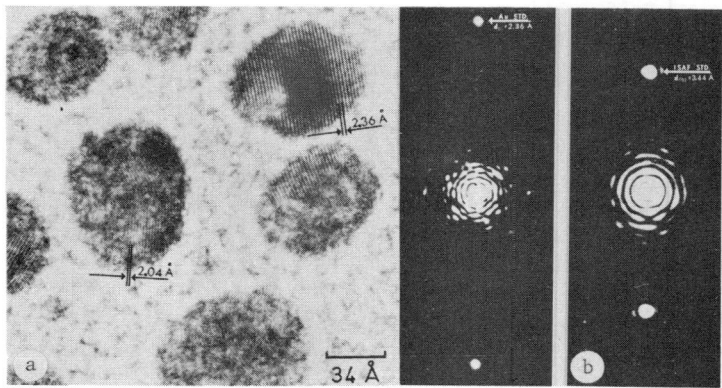

Figure 8 (a) *Lattice image of gold aggregates formed by vacuum evaporation on carbon film and used for magnification calibration of the images;* (b) *the OD pattern of the* (111) *image* (d = 2.36 Å) *is shown along with the OD pattern from a straight segment of graphitized ISAF paracrystal (see Figure 15 for the lattice image)*

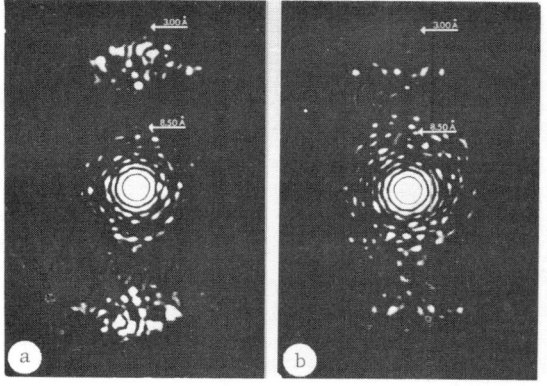

Figure 9 *Typical OD patterns of carbon blacks:* (a) *MT control;* (b) *ISAF control*

spots. The angle of the wedge is inversely proportional to the radius of the growth centres and directly proportional to the degree of bending of the graphitic layers. The measuring accuracy was 0.04 Å at 3 Å and 0.2 Å at 8 Å. The decrease in accuracy with increasing spacings is the result of the reciprocal relationship between distances on the electron optical images and in the OD pattern. The OD pattern is the Fourier transform of the image, and the intensities are the Fraunhofer patterns.

Significantly different distributions of layer spacings were obtained for MT and ISAF type carbon blacks (Figures 10 and 11) both in the normal and heat-treated states (900—1400 °C in 100 °C increments for 1 h). These

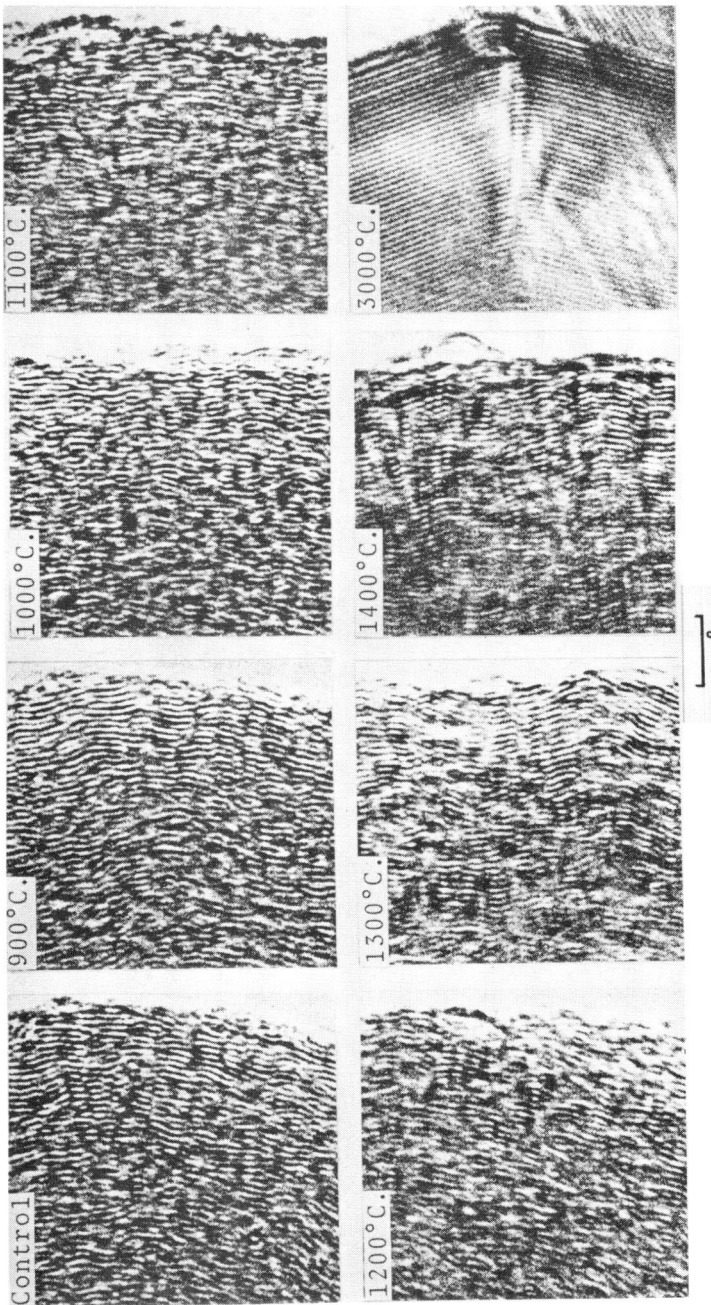

Figure 10 Segments of MT carbon black paracrystals at different heat treatment temperatures

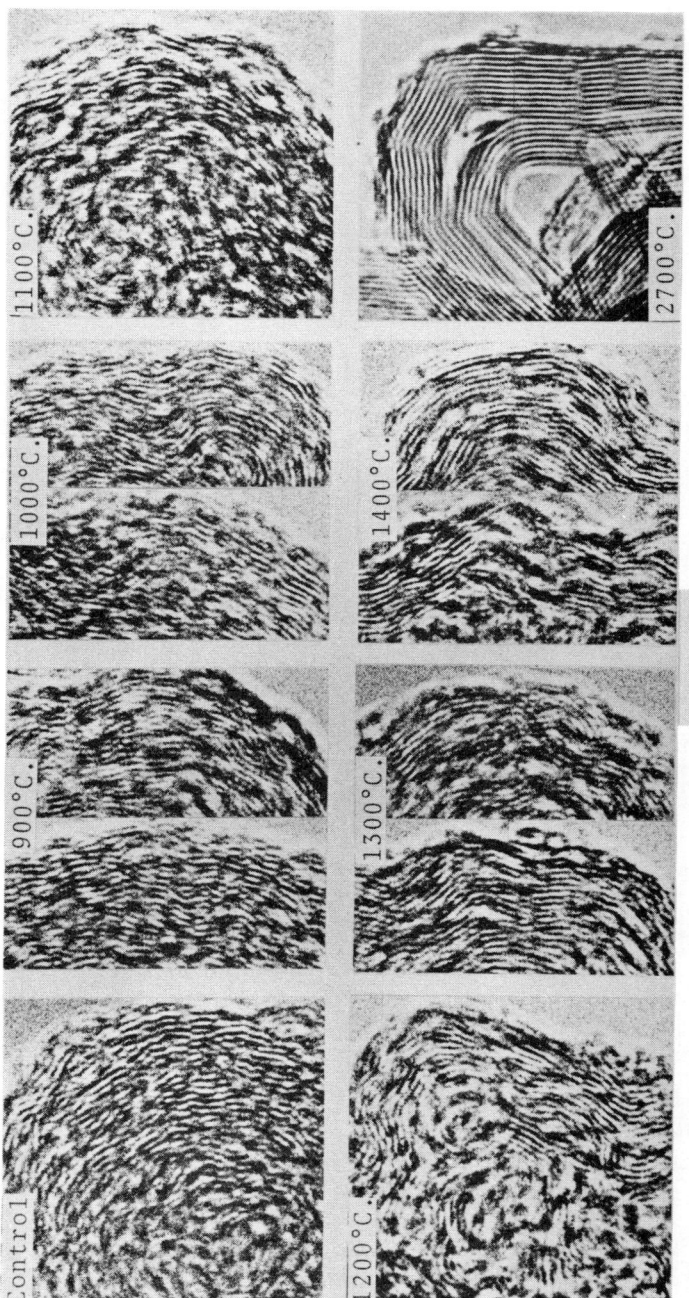

Figure 11 *Segments of ISAF carbon black paracrystals at different heat treatment temperatures*

Direct Study of Structural Imperfections

two carbon blacks were produced under different conditions. The MT thermal black is made from a hydrocarbon gas by means of a cyclic process and its formation time is longer than that of ISAF. The latter black is produced from oil by a continuous operation in the high-temperature furnace and has much shorter residence time. As graphitization of the MT black proceeds, the changes in the distribution of layer spacings indicate a general annealing out of the layer distortions until, at 1200 °C, there is a sudden

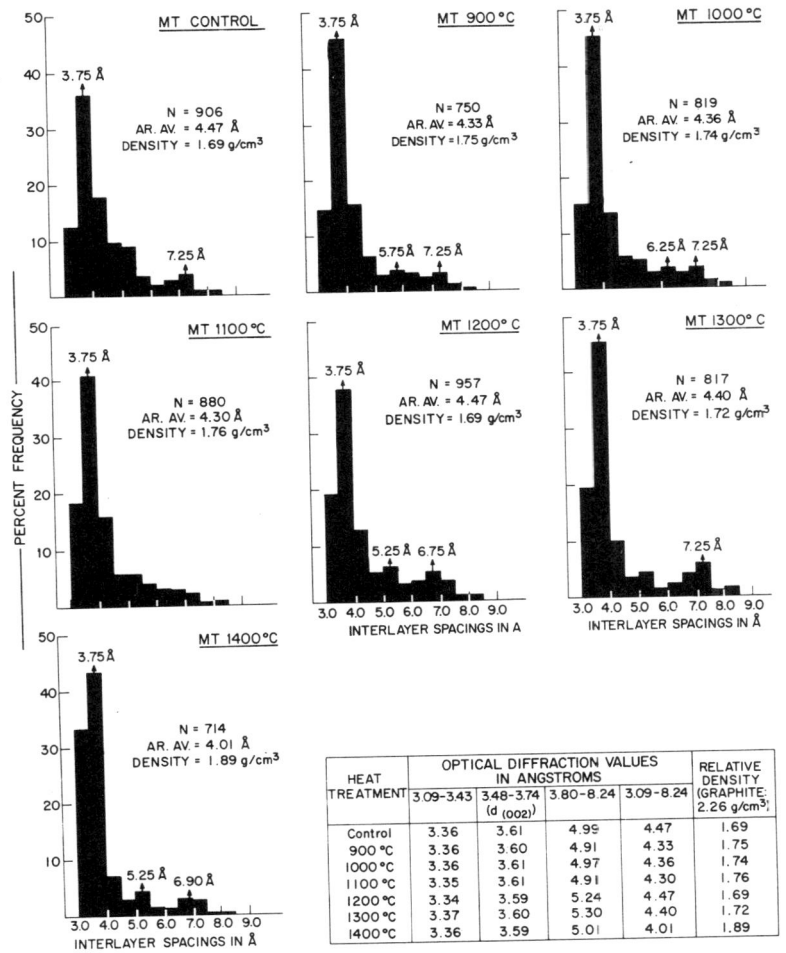

Figure 12 *Histograms of the interlayer spacings in MT carbon black. Number of diffraction spots* (N), *arithmetic averages* (Ar.Av.) *and relative densities are indicated*

increase in the frequency of distortions (Figure 12). It is possible that at this temperature the layers receive a sufficient amount of energy to initiate rearrangement. It is not known whether or not layer growth proceeds along with straightening of the layers. OD information on the ISAF sample indicates both annealing effects and the introduction of different types of distortions (Figure 13). At 900 °C this carbon showed a pronounced change as great as that of the MT at 1200 °C. Densities (relative to the hexagonal graphite value at 2.26 g cm^{-3}) were also calculated for both carbons using the arithmetic averages of the spacings, and are recorded in Figures 12 and 13.

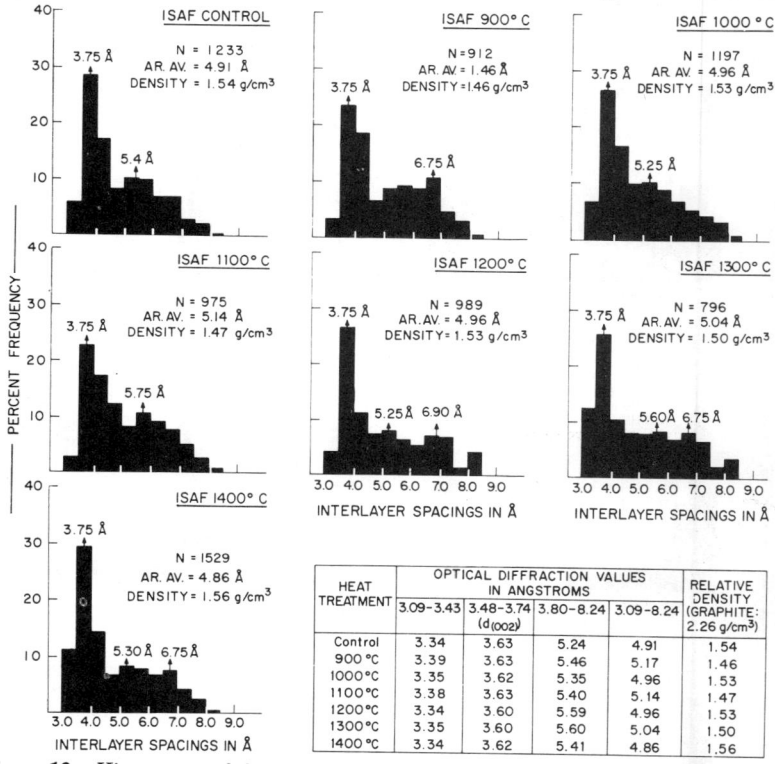

Figure 13 *Histograms of the interlayer spacings in ISAF carbon black. Number of diffraction spots* (N), *arithmetic averages* (Ar.Av.) *and relative densities are indicated*

(ii) *Different types of images in heat-treated carbon blacks.* The 3000 °C MT image (Figure 10) shows a portion of a polygonized particle in which the layer planes continue from one face to another. At the corner, a cavity has formed with a terminated layer on one side. This is similar to, but not the same as, a non-basal edge dislocation found in graphite, since the cavity formed at the corner of the polygon has lowered the energy of the dislocation.

Direct Study of Structural Imperfections

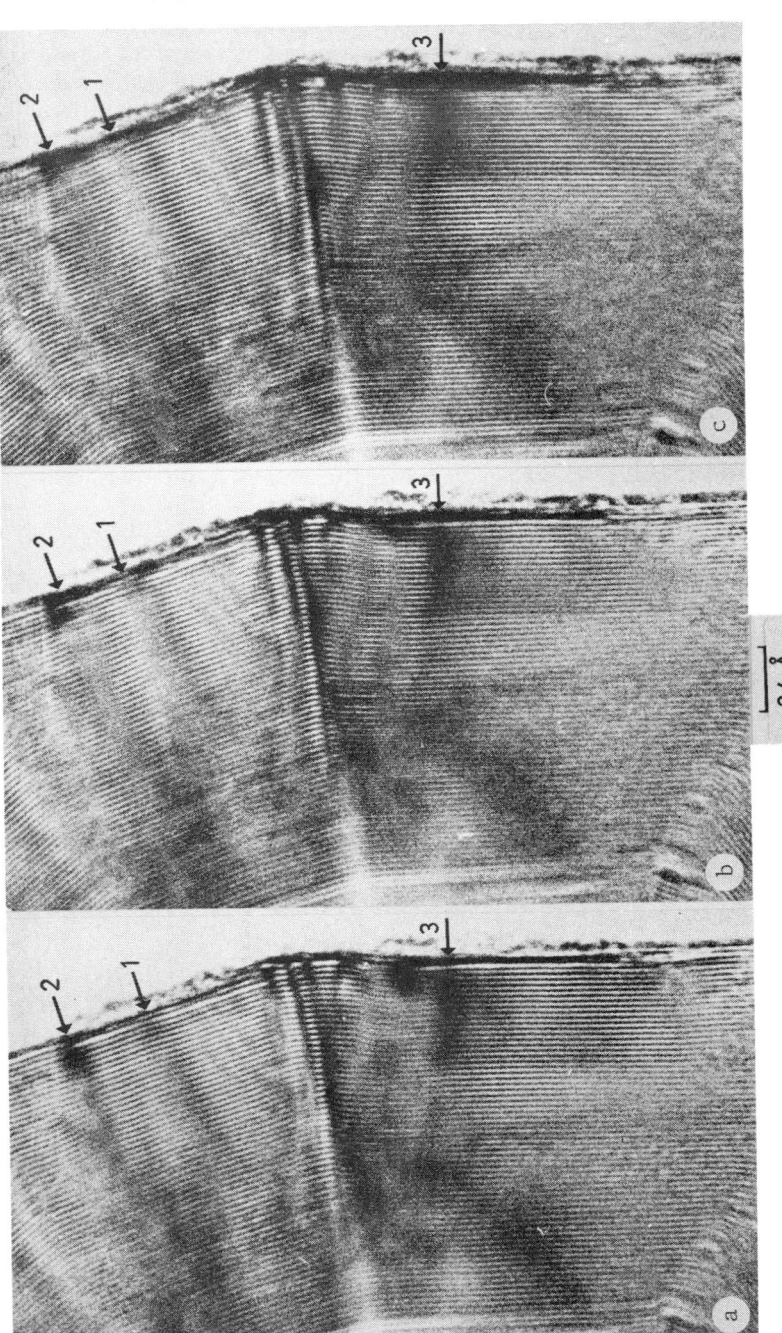

Figure 14 Defocusing series of a segment of a graphitized carbon black with $\Delta L = 250$ Å differences between images. Area marked 1 shows the appearance of non-basal edge dislocations in (b) and the shifting its position in (c). Areas marked 2 and 3 indicate three-beam interference

The 2700 °C ISAF image (Figure 11) shows a portion of a primary unit (single paracrystal). The layers have large dimensions and show the formation of polygonized layer groups in addition to bending or kinking. A large cavity has formed between the inner polygonized wall and the outer surface of the bent layer group. There is also a small interstitial layer appearing in this image. However, its exact location is uncertain at a single focal setting. This is readily apparent in Figure 14, where the same portion of a graphitized segment is imaged at three different defocus values. Dislocation images appear and shift position as the focus is changed. Three-beam interference images with half spacings may also appear and will change their appearance as focus is changed. (To see these spacings best, it is advisable to view the photographic images at a small angle looking along the layer planes). In determining the position of a dislocation by a focal series, there is still an uncertainty in its location by at least one unit dislocation. In Figure 14(a), the dislocation [marked 1 in 14(a), (b) and (c)] is not visible. In the next focus change, (14b), $\Delta L = 250$ Å over (a), the dislocation becomes visible, but a further change (14c), $\Delta L = 250$ Å over (b) indicates a shift in its posi-

34 Å

Figure 15 *Heat-treated (graphitized at 3000 °C) ISAF showing polygonized and bent layer groups around growth centres which most likely become hollow with graphitization*

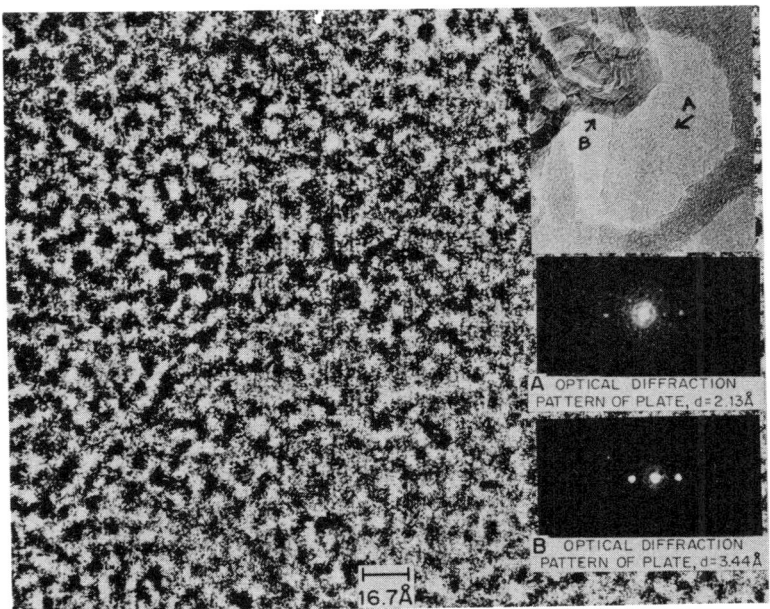

Figure 16 (10.0) *lattice image of a graphite flake (area A in insert) supported over a hole in a carbon film, which also supports the graphitized ISAF (area B in insert) for magnification calibration*

tion. This image also suggests the possibility of an interstitial loop [51] which is being viewed edge on, *i.e.* rotated 90°.

A portion of the graphitized ISAF with many different types of polygonized faces is shown in Figure 15. The most likely structure appears to be a closed shell, with an empty core. The angle between the polyhedron faces varies, but there is no definite twinning with the types of angle that are found in graphites.[52] The three-dimensional (10.0) image, which can be resolved in graphite (Figure 16), is missing in carbon blacks, even for those that are highly graphitized. The interlayer spacings in these graphitized samples show a distribution around a mean of about 3.44 Å. This large average spacing would be expected because of the large number of distortions. However, the arithmetic average value for highly graphitized blacks is nearly always 3.44 Å when a large number of measurements are taken.[53] On straight segments, with at least 10—20 layers in the image, one consistently obtains the 3.44 Å

[51] S. Amelinckx, in 'Solid State Physics', ed. F. Seitz and D. Turnbull, Academic Press, New York, 1964, Supplement 6.
[52] C. Baker, L. M. Gillin, and A. Kelly, Second Conference on Industrial Carbon and Graphite, 1965, Society of Chemical Industry, London 1966, p. 132.
[53] P. A. Marsh, T. J. Mullens, and D. Price, Proc. 28th Annual EMSA, ed. C. J. Arceneaux, Claitor's Publ. Div., Baton Rouge, 1970, p. 50.

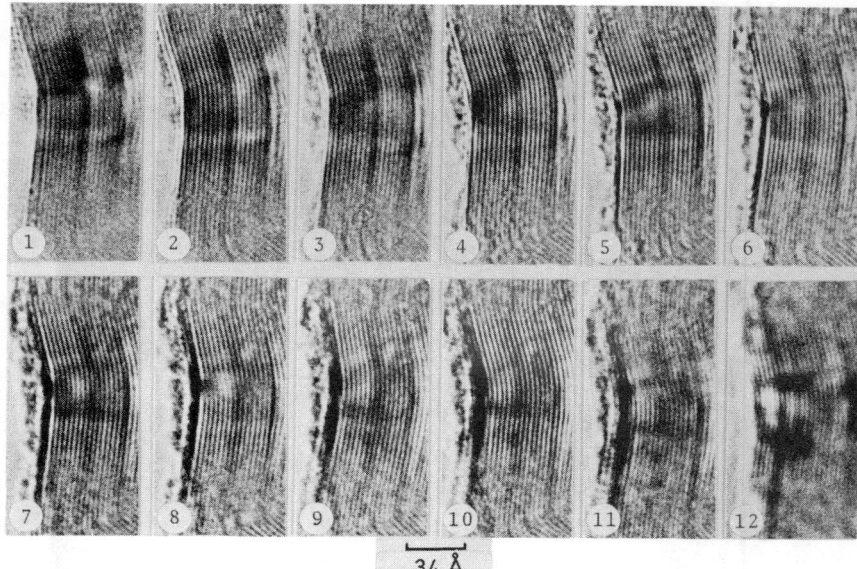

Figure 17 Focusing series with $\Delta L = 125$ Å increments from 1—11 showing the contrast variation in a segment of a heat-treated ISAF. Image 1 is $\Delta L \approx -125$ Å away from Gaussian focus; image 2 is at $\Delta L \approx 0$; image 12 is at $\Delta L \approx 2000$ Å. Image 8 approximates the correct defocus at $\Delta L \approx 750$ Å according to equation (5)

value with OD, even if the focus is varied (Figure 17). The 3.44 Å value was also obtained for randomly distributed straight segments which showed lattice images of varied contrast. This contrast variation appears to be attributable to the change in angle of orientation to the electron beam, since the thickness seemed to be approximately $t_0/4$ for all segments. One can therefore conclude that the angle of alignment does not significantly affect the interlayer spacings in the images, *i.e.* within the limits of ± 0.02 Å accuracy for the OD method.

Carbon Films. The microstructure of evaporated thin carbon films, produced by the method of Bradley,[54] has been studied by many workers, most of whom have utilized electron diffraction. A very detailed study was carried out by Kakinoki *et al.*,[55] who reported two different carbon bond lengths in these films: 1.41 Å which was assumed to be graphite-like, and 1.55 Å which is diamond-like. However, Dove[56] found only the graphitic type of

---

[54] D. E. Bradley, *Brit. J. Appl. Phys.*, 1954, **5**, 65.
[55] J. Kakinoki, K. Katada, T. Hanawa, and T. Ino, *Acta Cryst.*, 1960, **13**, 171.
[56] D. B. Dove, Proc. 26th Annual EMSA, ed. C. J. Arceneaux, Claitor's Pub. Div., Baton Rouge, 1968, p. 396.

Direct Study of Structural Imperfections

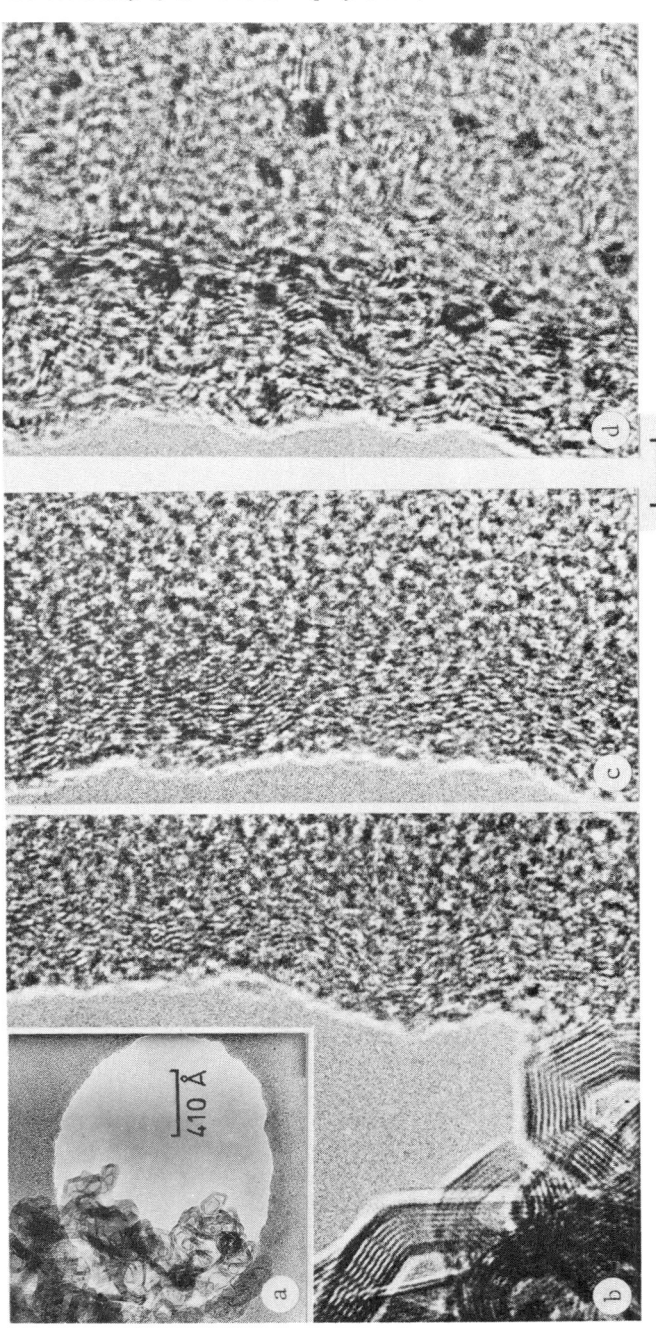

Figure 18 Carbon film microstructure: Image (a) shows a hole in a carbon film with the graphitized ISAF bridged across it. Images (b) and (c) show graphitic layer type detail at the edge of the hole. The graphitized ISAF with $d_{002} = 3.44$ Å is also shown in (b). (d) 1100 °C heat-treated carbon film showing greater development of layer structure. The dark spots are clusters of tungsten atoms that migrated from the tungsten grid which supported the film during the heat treatment

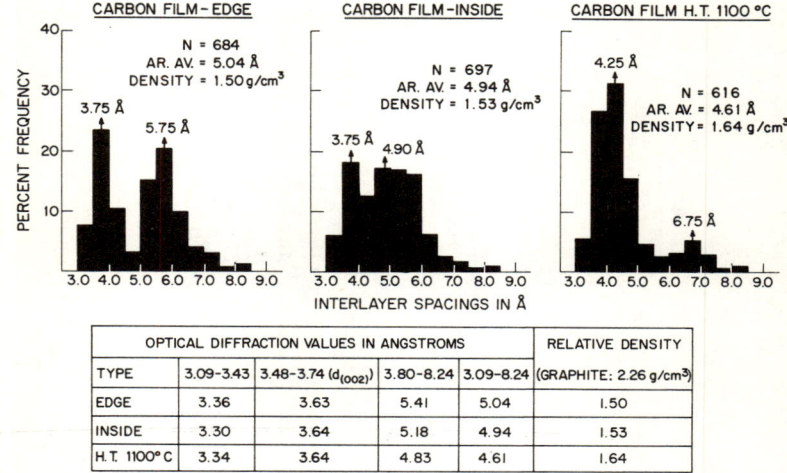

Figure 19 *Histogram of interlayer spacings from OD patterns on carbon films. The statistical parameters and relative densities are also included*

structure. He employed an improved electron diffraction camera which records only the elastically scattered electrons for the atomic radial distribution study.

High-resolution phase contrast electron microscopy has also indicated a graphitic structure [44,57] which is illustrated in Figures 18 and 19. The carbon films were deposited by vacuum evaporation onto perforated Formvar films (polyvinyl formal resin).

The Formvar films are subsequently dissolved away by washing them with ethylene dichloride. It is generally assumed that, if a material is composed of graphitic layer type short-range order, then the layers will primarily be aligned parallel to the surface. Hence, at the edge of a hole in a carbon film they would tend to be parallel to the electron beam. At these points, the (00.2) interlayer diffraction will occur quite strongly. Away from the hole, the bulk of the layers are almost perpendicular to the electron beam and, with this orientation, will produce the (hk) reflections. There is, however, always a small percentage of layers or groups of layers that are positioned for the (00.2) reflection. A typical carbon film (also containing the partially graphitized ISAF) is illustrated at low and high magnification in Figure 18. A well-defined layer structure is apparent at the edge, while a more random pattern is apparent away from the edge which is composed of the regular defocusing phase structure and the (00.2) type layer structure. The high-resolution images do not indicate a separation into two distinct zones of graphite-like

[57] L. L. Ban and W. M. Hess, 10th Biennial Conf. on Carbon, Columbus, Ohio, Battelle's Laboratories, Defense Ceramic Inf. Centre, 1971, p. 161.

and diamond-like structure as proposed by Kakinoki et al.[55] However, if there are diamond-like bonds, they could possibly occur in narrow single channels scattered throughout the film, in a manner similar to that suggested for carbon blacks.

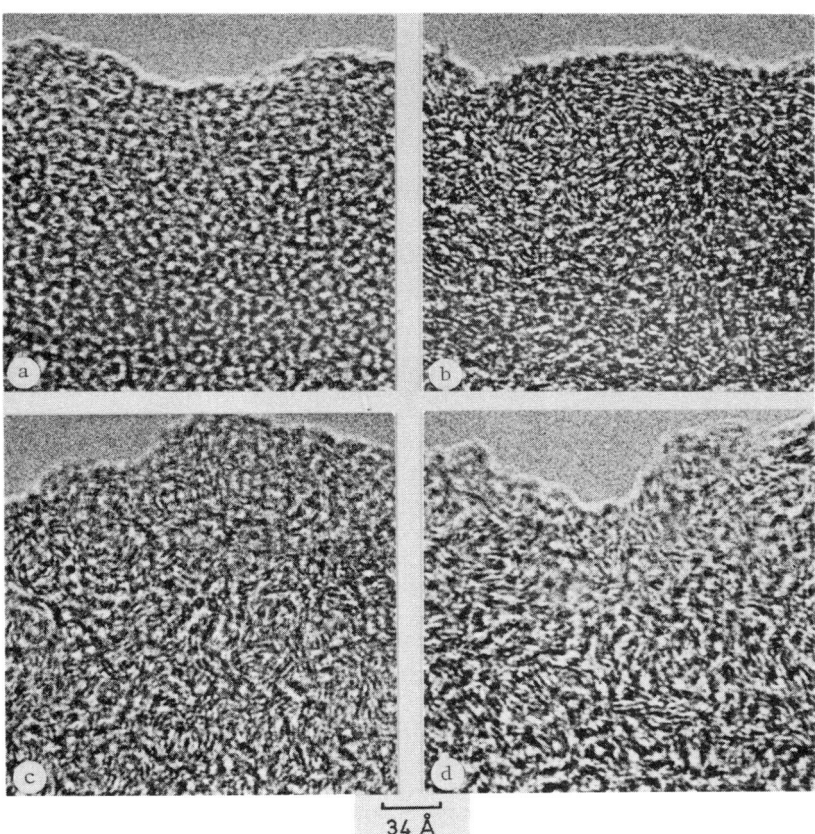

Figure 20 *Carbonized polymers: heat treatment at* 873 K: (a) *cellulose*, (b) *polyethylene showing somewhat more short-range order than cellulose; heat treatment at* 1173 K: (c) *cellulose*, (d) *polyethylene, both showing definite graphitic layer type development. The polyethylene carbon indicates somewhat larger layers than the cellulose carbon*

The low activation energy (2 eV) for carbon films at the early stages of graphitization reported by Presland and White,[58] appears to indicate the annealing, or straightening of the layers. This is indicated in Figure 18(d) for a film heated to 1100 °C.

[58] A. E. B. Presland and J. R. White, *Micron*, 1970, **2**, 73.

The distribution of interlayer spacings in carbon films were measured by OD using the method employed for carbon blacks (Figure 19). The relative densities (to that of graphite at 2.26 g cm^{-3}) calculated for the edge and inside of the films and for the 1373 K heated film are also indicated in Figure 19. The mean inner potentials calculated from the arithmetic averages of the interlayer spacings are 7.6 V for the edge and 7.8 V for the inside. These values are in good agreement with a value of 7.8 ± 0.6 V obtained by electron interference microscopy [59] (the mean inner potential of diamond is 20.8 volts).

Pyrolysed Polymers – Glassy Carbon. Carbons prepared from pyrolysed polymers heated to 873 K show the incipient stages of short-range layer-type order (Figure 20a and b).[57] The images indicate short-layer segments with a mixture of straight and distorted groupings. Subsequent heating of these carbons to 1173 K produced a pronounced layer-type ordering, but without much preferred orientation (Figure 20c and d). There appears to be grouping of two or more layers, but rarely more than five. The lengths appearing in the images are not necessarily the true layer sizes. At the present time, only apparent layer lengths can be measured because, as the layers bend, they often deviate too far from the necessary alignment for diffraction, thus causing gaps in the images. It appears that the carbonized polyethylene is more ordered at this temperature than the cellulose carbon.

Carbons formed from phenolic resins (called glassy or vitreous carbons) also resemble the pyrolysed polymers examined above. The sample heated to 773 K (Figure 21a) shows a random pattern with very little layer type development. The image structure is similar to the evaporated carbon film a short distance away from the edge of a hole. At 1173 K the layer type short-range order is apparent (Figure 21b). In this sample, there appears to be a parallel packing of short-layer segments, both distorted and straight. These show an average stacking of 2—4 layers which would possibly give $L_c \approx 10$ Å and L_a of similar dimension. In the carbon heated to 2973 K (Figure 21c) the layers remain random with bent and twisted layer groups containing about 2—15 layer packets, e.g. $L_c \approx 30$ Å. The size of the layers appears to be larger in comparison to the samples treated at lower temperatures. This indicates both straightening and the subsequent growth of the layers in all directions. The size of the layers perpendicular to and at different angles to the layer images can be as large as some of the projected lengths shown. This explanation is not in line with the model proposed by Jenkins and Kawamura [60] who proposed a uniform layer thickness interwoven throughout the material. Their model could, however, possibly be applicable to the images observed for the 1173 K heated carbon. There is also a possibility for internal cavities to be formed which are enclosed by the layer packets in a manner similar to carbon blacks. However, there are no distinct growth

[59] A. Buhl, *Z. Physik*, 1959, **155**, 395.
[60] G. M. Jenkins and K. Kawamura, *Nature*, 1971, **231**, 175.

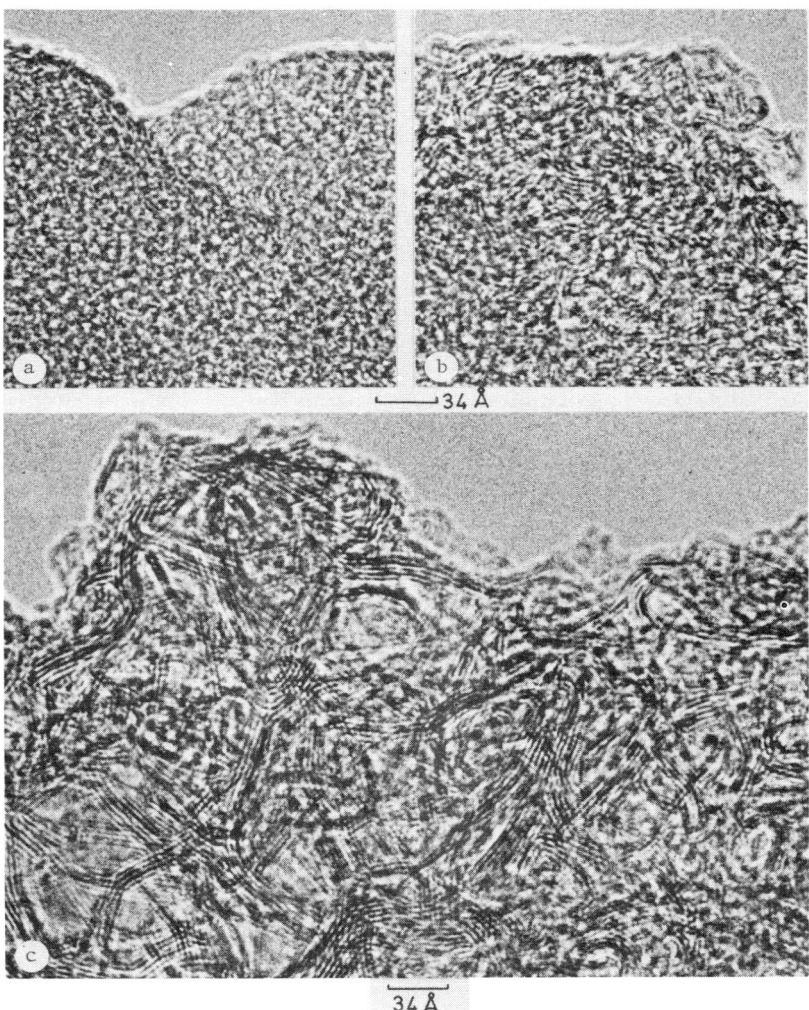

Figure 21 *Glassy carbon:* (a) *heat-treated at* 773 K *showing random detail without the appearance of layer-type short-range order;* (b) *heat-treated at* 1173 K *indicating graphitic layer-type development;* (c) *heat-treated to* 2973 K *showing extensive graphitic layer development*

centres (which are clearly seen in carbon blacks); areas around which the layers would grow tangentially. The absence of growth centres also indicates a different formation mechanism from that of carbon blacks, which seem to form from droplets of liquid crystalline structures prior to the completion of carbonization.

This glassy carbon fits the category of non-graphitizing carbons. However, the nature of the disorganized phase as proposed by Franklin can now be debated. The high-resolution images suggest that the completely disorganized phase, indicated by the X-ray radial distribution studies of Franklin, may possibly be an artifact resulting from the scattering by the highly distorted layers. It is most likely that the interatomic distances in the distorted layers vary to such an extent that the perfect hexagonal packing of atoms is interrupted. However, the extent of such imperfections and the actual structure of the layers cannot be discerned from high-resolution images at this time.

Coal and Char Samples. The microstructure of one sample of each of these two types of carbons was studied. It should be pointed out, however, that the results presented here should not be taken as general information on all coals and chars. Specific information on the many varieties of these materials necessitates a detailed study of each type of material.

The coal sample, when heated to 1273 K (Figure 22a), exhibits a structure similar to that of the polymer carbons. The same interpretation regarding the size and the distortion of the layers is also applicable. In this instance, however, there appear to be larger gaps between adjoining layer groups indicating a lower density. Occasionally, particulate material similar to carbon blacks can be found (Figure 21b), which indicates a possible liquid aromatic crystalline development before completion of carbonization. At higher temperatures there is a considerable rearrangement of the graphitic layers. Large flakes are formed in samples heated above 2273 K along with kink bands and polygonization of layer segments around growth centres (Figure 22c).

The char material (coconut) heated to 1273 K also shows microstructural detail similar to the carbonized polymers (Figure 23a). At temperatures above 2273 K, however, there are significant differences in the form of thin flakes, fibres, and narrow random layer groupings (Figure 23b). The images of the ends of the fibres reveal a folded type growth pattern found also in lamellar polymer crystals (Figure 23c). It appears that the connecting layers occur only at the broken ends of the fibre and are most likely formed during the carbonization process. The folding of layers is not apparent deep inside the fibres nor on the sides. This growth pattern has a very important relationship to the microstructure of carbon fibres [61] which are formed by the carbonization and subsequent graphitization of polymer fibres. During the carbonization and graphitization process, the original polymer orientation along the fibre axis is maintained. The layers also appear to be tangential to the surface of the fibre. The splitting of layer groups can also be noticed at the ends of the fibre, but not as extensively internally. The optical Moiré

---

[61] A. Fourdeaux, C. Herickx, R. Perret, and W. O. Ruland, *Compt. rend.*, 1969, **269**, C 1597; J. A. Hugo, V. A. Phillips, and B. W. Roberts, *Nature*, 1970, **226**, 144; D. J Johnson, *Nature*, 1970, **226**, 750.

Direct Study of Structural Imperfections

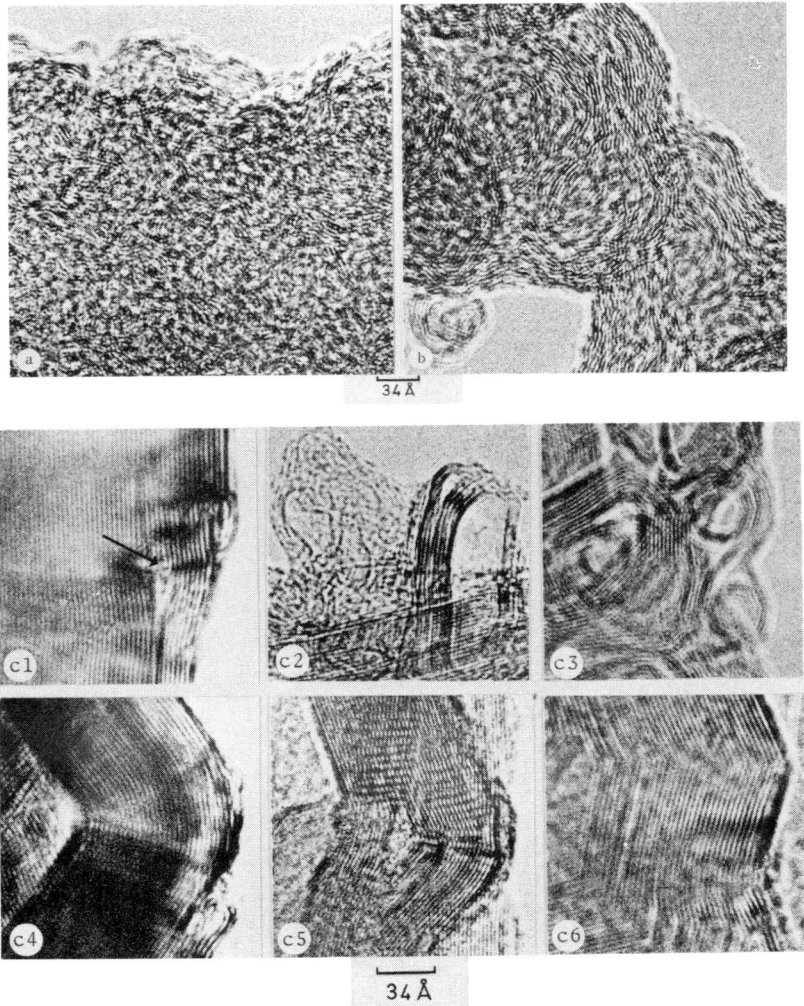

Figure 22 Coal microstructure: images (a) and (b) show the structure after 1273 K heat treatment: with random layer development in (a) and concentric orientation of layers around growth centres in (b). The images in (c) show the microstructure after heat treatment to 2573 K: c1 and c2 are segments from flaky areas (in c1 the arrow indicates a loop formation), while c3 is a segment from a random network of layers, and c4—6 are segments from particulate areas indicating bending and polygonization of the layer groups

patterns observed in the images also indicate that layer groups are crossing on top of one another.

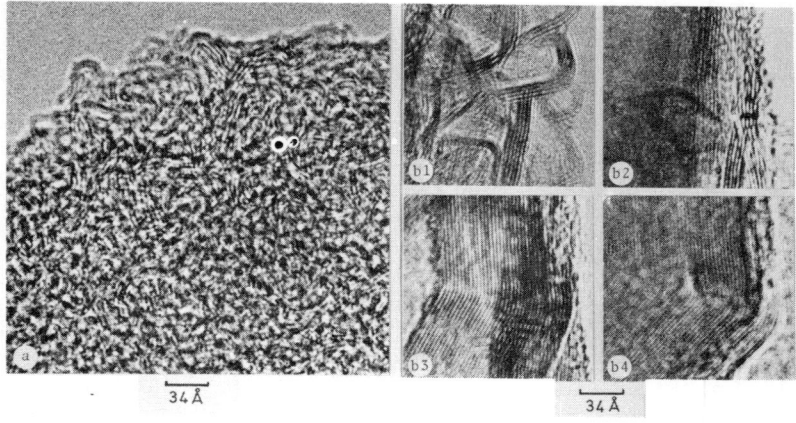
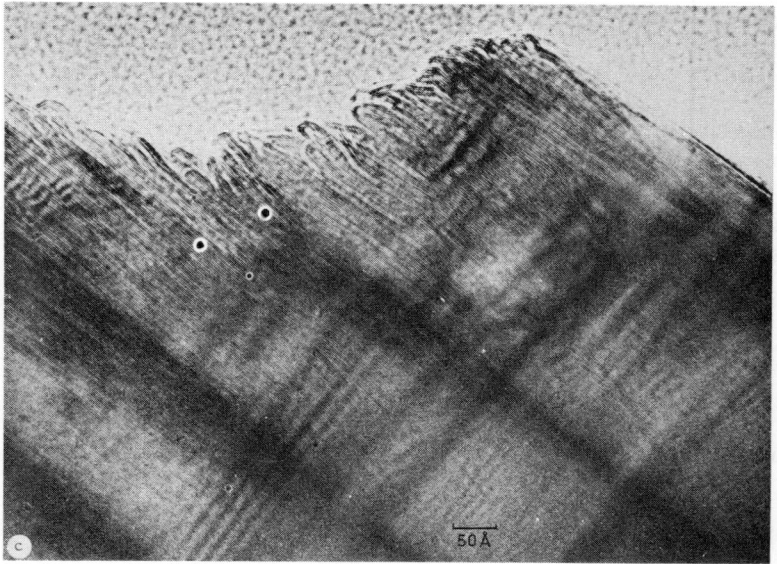

Figure 23 Char (coconut) microstructure: (a) heated to 1273 K showing random graphitic layer development; (b) heated to 2273 K: b1 is a segment of a random network of layers, b2 is a segment of a flake, and b3 and b4 are parts of particulate areas showing the bending and polygonization of the graphitic layers; (c) segment of a fibre showing the broken tip where the graphitic layers formed loops, some of which appear to be twisted. Optical Moiré's indicate the overlapping of layer segments

Direct Study of Structural Imperfections

Polymers.—Studies on the microstructure of organic polymers with the electron microscope are considerably more difficult than those on the carbons discussed in the previous sections. In interpreting the structure of a polymer network, either crystalline or amorphous, one must always be aware of possible damage caused by the electron beam in conjunction with the temperature rise of the sample. Nevertheless, the electron microscope has been quite useful in elucidating the gross morphological details of polymers, and electron diffraction has aided in determining the structure of small single polymer crystals.[62]

Amorphous Polymers. The microstructure of two rubber polymers (elastomers) will be considered in this section. These are *cis*-1,4- and *trans*-1,4-polyisoprene (natural rubber and gutta-percha). The *cis* form is generally considered to be amorphous at room temperature, but crystallizes upon stretching and cooling.[63]

The purpose of this preliminary study of the rubbery polymers by high-resolution microscopy is an attempt to resolve short-range order within the randomly coiled molecules. If there is some short-range order of the molecular chains, one ought to be able to detect it by applying OD to the high-resolution images. The problem with obtaining meaningful information on the packing of molecules is similar to that of obtaining information on the structure of an evaporated carbon film. However, in the case of amorphous polymers one can expect less preferred order relative to a surface, since the molecules are not platelet-like as are the rather large aromatic layers in carbons. Also, the polymer molecules are considerably more flexible.

Thin films of *cis*-1,4- and *trans*-1,4-polyisoprene, 50—100 Å in thickness, were prepared from benzene solution ($\approx 0.05\%$) on a water surface. The films were picked up on perforated carbon films which were coated in vacuum with gold particles to serve as magnification calibration with their resolved (111) planes. The high-resolution phase contrast images of both polymers (Figure 24a and b) are similar to those of carbon films. This is not surprising, since the superposition of molecules can give the random phase in the elastically scattered electron waves which is necessary for such an image. The main reason for using the two different isomers of polyisoprene was that the gutta-percha rubber should be crystalline at room temperature. The α crystal, which exists in the tree, is formed if it is cooled very slowly from its melting point of 338 K. The β crystal is formed if it is cooled rapidly, becoming amorphous at 329 K. It appears that the above-mentioned melting points were exceeded since no crystalline structure was observed in the images. The OD patterns of areas selected at random indicate some type of preferred orientation with a range of 4—5 Å for both isomers (Figure 25a

---

[62] P. H. Geil, 'Polymer Single Crystals', Interscience, New York, 1963; A. Keller, *Rep. Progr. Phys.*, 1968, **31**, Part 2, 623.

[63] G. W. Bunn, *Proc. Roy. Soc.*, 1942, **A180**, 40, 82; S. C. Nyburg, *Acta Cryst.*, 1954, **7**, 385.

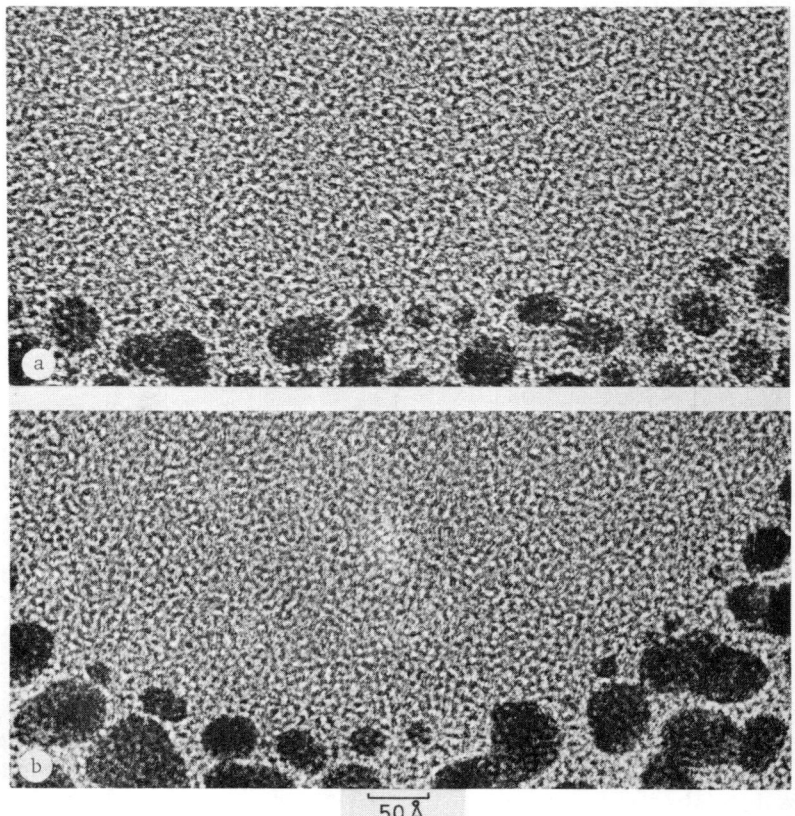

Figure 24 *Polymer microstructure:* (a) cis-1,4-*polyisoprene* (b) trans-1,4-*polyisoprene (gutta-percha)*. *Both images show short-range order, but the* trans-*isomer is more ordered than the* cis-*isomer*

and b). The gutta-percha film showed the most orientation. Using smaller and smaller mask openings, and a very high resolution dispersive diffraction in the optical diffractometer, the short-range order became more obvious.

At this stage of experimentation, it is premature to reach conclusions on the type of order that is being detected in these polymers. However, it is likely that some type of paracrystalline order is being imaged in both isomers. Further high-resolution studies at lower temperatures, along with the use of image intensification, will probably be necessary for further progress on samples of this type.

Macromolecules – DNA. The majority of electron microscope studies on the structure of macromolecules have been carried out by means of the staining

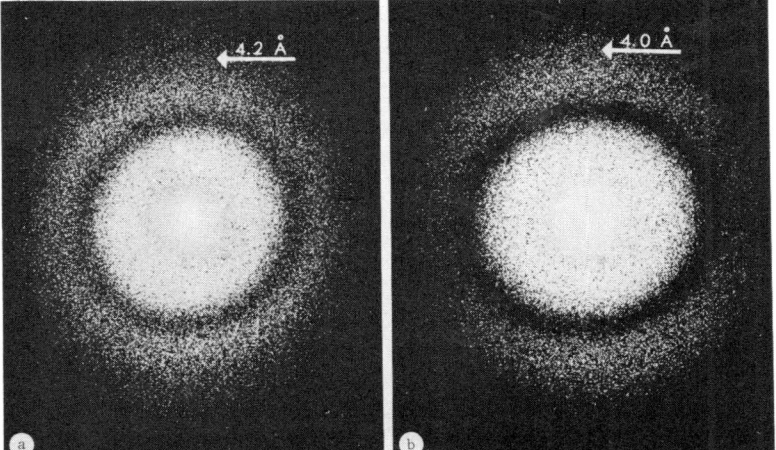

Figure 25 OD patterns of the images shown in Figure 24: (a) cis, (b) trans. The OD pattern of the trans-*isomer* indicates more order than that of the cis-*isomer*. The continuous rings are diffraction from the random phase contrast detail

and shadowing methods.[64] These methods are extremely useful but do restrict the level of resolution at which the molecular structure can be ascertained.

Utilizing the present understanding of the phase contrast properties of the objective lens, along with a rather simple specimen preparation technique, high-resolution imaging of DNA has been achieved,[65] but only with limited information on the molecular structure. The DNA preparations were of salmon testes and calf thymus. The preparations were in a buffer solution [66] of 0.01—0.005% concentration. To prepare a specimen for electron microscope observation, one drop was applied to a perforated carbon substrate on a specimen grid also containing graphitized ISAF carbon black. The preparation is carried out on filter paper. The carbon black serves a dual purpose, first in forming relatively small gaps to support short molecule segments (*e.g.* 1000 Å or shorter) and secondly for magnification calibration (Figure 26). The molecules were examined without shadowing or staining using an electron current density of about 0.12 A cm^{-2}. It was not possible to determine exactly the extent of radiation damage on the molecules. However, despite this relatively large electron dosage, it appears that the double-stranded molecule did not completely separate. The images were formed by

[64] D. H. Kay in, 'Techniques for Electron Microscopy', ed. D. H. Kay, Philadelphia, 1965.
[65] L. L. Ban, 'High Resolution Electron Microscopy in the Molecular Domain', Ninth Symposium on Polymer and Fibre Microscopy, Sept. 12–13, 1968, Text. Res. Inst. Princeton, N.J., unpublished.
[66] C. E. Hall, *J. Biophys. and Biochem. Cytol.*, 1956, **2**, 625; 1958, **4**, 1.

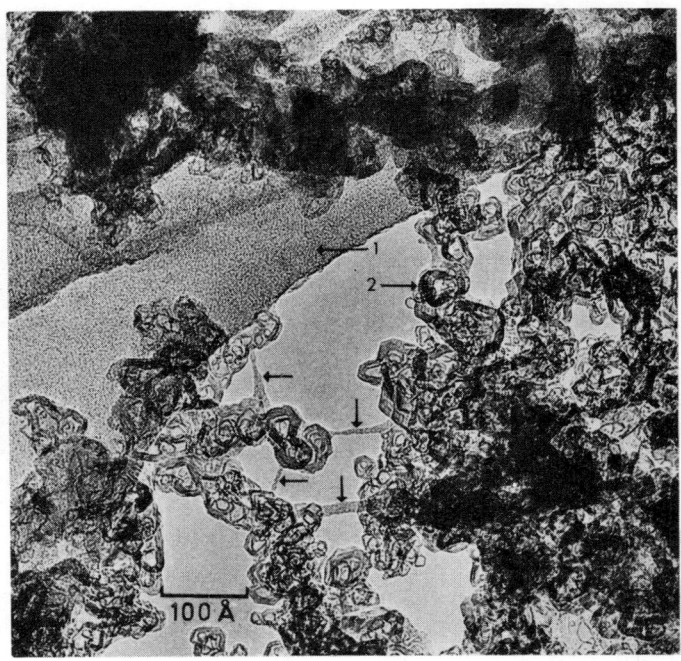

Figure 26 *DNA preparation; the area marked 1 is the carbon film, 2 is the graphitized ISAF carbon black, and the arrows (without numbers) indicate DNA segments. The narrower segments represent the double-stranded molecules*

defocusing the objective lens, using equation (5), for 10 Å detail, which is the radius of the 20 Å double-stranded molecule, *i.e.* 10 Å is approximately the diameter of one of the strands with its bases.[67] Only those strands were selected for recording which were about 20 Å in diameter. There were many DNA strands (practically 100 or more) which were recorded throughout this study and most of them showed some form of double sinuous coiling in the images. Some of these are illustrated in Figure 27. These micrographs were obtained from different preparations, all having approximately the same concentration. If the concentration was not kept low enough, thicker strands were observed. At times, continuous films about 50—100 Å thick were also formed. The detail in the images of these films was similar to that in carbon films, and information on the molecular structure could not be determined.

[67] J. D. Watson and F. H. C. Crick, *Nature*, 1953, **171**, 737; M. H. F. Wilkins, A. R. Stokes, and H. R. Wilson, *ibid.*, p. 738; R. E. Franklin and R. G. Goshing, *ibid.*, p. 740.

Direct Study of Structural Imperfections

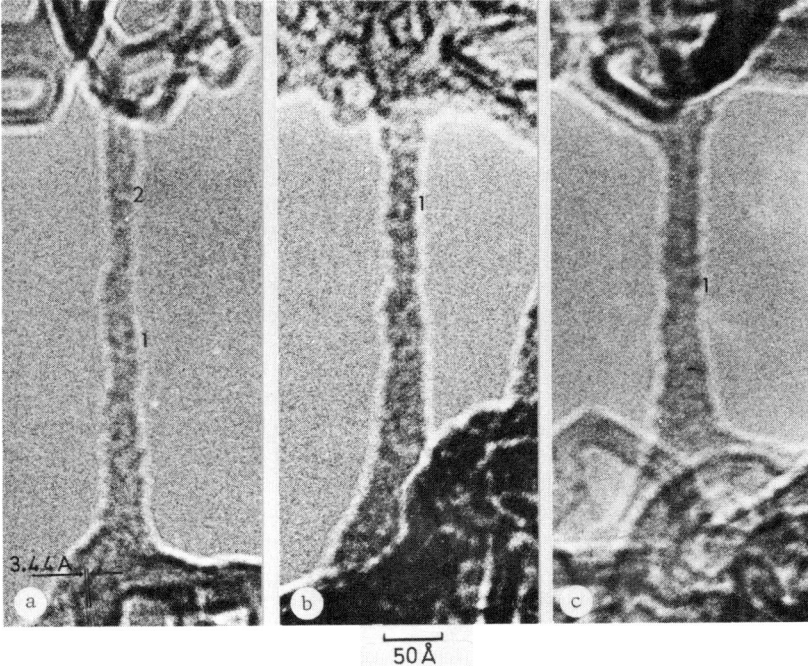

Figure 27 *DNA molecules shown at high magnification: areas marked* 1 *indicate double stranded forms. In* (a) *the area marked* 2 *appears to represent single stranded coiling. The* 3.44 Å *graphitic layer spacing is visible in the same image*

In the DNA images shown in Figure 27, the direction of coiling has been reversed by the printing process. In printing, the negative emulsion faces the photographic print, and is thus rotated 180° around the axis of the molecule. When corrected for printing, the DNA images (assuming that they represent the two strands of the molecule), will become right-handed, which is the correct rotation of the molecule according to the X-ray diffraction studies. It is difficult to ascertain the value for the pitch of the double helix from these images. This varies from approximately 34 Å (10 nucleotide bases per helix turn), for the B configuration of the molecule at high humidity, to 28 Å for the A configuration at approximately 30% humidity, and 31 Å for the C form which exists when the DNA is partially dried.[68] One can expect the images to be most representative of the C form. The 3.4 Å base sequence has not yet been resolved in the DNA images. This suggests a definite disruption in the molecular structure which may occur either in the prepara-

---

[68] M. H. F. Wilkins, *Science*, 1963, **140**, 941.

tion of the samples or is caused by radiation damage. The latter problem may eventually be eliminated by means of image intensification, thus aiding in decoding the molecule to its individual base-pairs.[69]

5 Summary and Conclusions

Although many problems still exist, it is possible to obtain meaningful information on the solid state of matter by electron microscope imaging in the crystal lattice domain. Instrumentation and interpretation of the phase contrast images are both important factors to be considered in the useful application of this technique. For many technologically important carbons, lattice imaging has provided important microstructural information which was not attainable by diffraction methods.

The information obtained on polymer microstructure, although not as definitive, is encouraging at this time in that the samples were able to sustain relatively large electron dosages without severe deterioration. The future use of image intensification and low-temperature stages should bring about further improvements in this very important area.

Optical diffraction analysis of high-resolution images of carbon black was found to be an excellent method for studying the range of interlayer spacings, including distortions up to about 8 Å. It is hoped that this technique will also provide useful information on the degree of paracrystalline order in amorphous polymers.

High-resolution transmission electron microscopy is also capable of providing useful information in the study of radiation damage in different materials, as well as in classifying the microstructure of small particle size colloids, both organic and inorganic. One particularly important application will be in the study of dislocations with improved resolution. Such a study will also aid in refining the dynamical theory of electron diffraction.

The author would like to acknowledge the encouragement and assistance of William M. Hess. I am also indebted to Miss Margaret Karvas for her excellent photographic work, to Paul C. Vegvari for carrying out the numerous optical diffraction studies, to Mrs. Sandra Litz for typing the manuscript and to the Cities Service Company for permission to publish this work.

[69] M. Beer and E. N. Moudrianakis, *Proc. Nat. Acad. Sci. U.S.A.*, 1962, **48**, 409; E. N. Moudrianakis and M. Beer, *ibid.*, 1965, **53**, 564.

3
The Rôle of Defects in Solid-phase Polymerization

BY C. H. BAMFORD AND G. C. EASTMOND

1 Introduction

It is difficult to write a definitive account of the mechanism of solid-phase polymerization. Although reports of work in the field of solid-phase polymerization constitute a sizeable fraction of the total literature of organic solid-phase reactions, relatively little is known about the detailed reaction mechanisms. The reasons for this are evident. Many monomers have been subjected to a rather perfunctory examination to determine whether their solid-phase polymerization will produce polymers different from those obtained by more conventional techniques, with the result that a large number of cursory reports, presenting few kinetic data, have appeared. More detailed investigations are fraught with difficulties, usually arising from a lack of knowledge of the organic solid state. Solid-phase polymerizations have been studied for fifteen years but, at present, crystal structures of few monomers are known and there appears to have been no direct examination of the types and properties of the defects present in any crystalline monomer. Consequently, the rôle of imperfections in these reactions must be inferred from an accumulation of pieces of evidence obtained from studies of different aspects of related systems.

From the outset two doctrines have developed. Although early experiments on the solid-phase polymerization of acrylamide yielded only amorphous polymer,[1] one school of thought held that the crystal lattice was all-important in determining the course of reaction, while a second maintained that crystal defects played a critical part. This polarization of opinion was strengthened when Okamura et al.,[2] reported that the γ-induced solid-phase polymerization of trioxan produced a crystalline polymer, with the c-axis of the polymer aligned along the c-axis of the monomer crystal.[3] Gradually, opinions have been modified and at the present time many workers adopt the hypothesis that the formation of amorphous polymer signifies polymerization in crystal defects while the appearance of crystalline polymer implies control

---

[1] R. B. Mesrobian, P. Ander, D. S. Ballantine, and G. J. Dienes, *J. Chem. Phys.*, 1954, **22**, 565.
[2] S. Okamura, K. Hayashi, and Y. Nakamura, *Isotopes and Radiation*, 1960, **3**, 416.
[3] K. Hayashi, Y. Kitanishi, M. Nishii, and S. Okamura, *Makromol. Chem.*, 1961, **47**, 237.

of the reaction by the crystal lattice. As we shall see, there are reasons for questioning the validity of the latter half of this hypothesis, but it is also clear that the structures of defects invoked in the former part must be controlled by the structure of the lattice. It is probable that in all cases the truth lies between the two extremes.

In earlier reviews,[4—7] the solid-phase polymerizations of unsaturated and cyclic monomers have usually been considered separately and we shall follow this pattern here; however, it will become apparent that the distinction may not be so sharp as hitherto assumed. In attempting to demonstrate the rôle of imperfections in all these reactions we shall, where appropriate, offer explanations of experimental data different from those of the original workers. We shall restrict ourselves here mainly to aspects which have a direct relevance to the title; we have recently reviewed other aspects elsewhere.[6, 7]

2 Some General Considerations

It is not possible to introduce chemical initiators homogeneously into crystalline monomers. Consequently, most workers have used either high-energy or u.v. radiation to initiate polymerization. A few studies have involved crystallization of monomers in the presence of chemical initiators or chemical initiation at crystal surfaces, and a number of more unusual techniques of initiation have been described.[8] Two basic techniques have been adopted for studying radiation-induced polymerization: the sample may be irradiated continuously at the reaction temperature, with continuous monitoring of the extent of reaction ('in-source' studies) or, alternatively, in 'post-irradiation' or 'after-effect' investigations, irradiation is interrupted, and the ensuing 'dark' reactions are followed. In the latter case, irradiation may be carried out at temperatures too low for significant polymerization to occur, the course of reaction arising from the active centres produced and trapped during irradiation being observed at high temperatures. This approach has the advantage of minimizing irradiation damage to the products. High-energy radiation can produce reactive centres throughout a large sample, while u.v. irradiation generally leads to reaction in a thin layer of crystals due to the high extinction coefficients of many monomers. U.v. initiation offers interesting possibilities for varying the energy input by variation of wavelengths.

When high-energy radiation is used to initiate polymerization the nature of the active propagating species may be in doubt; the available energy is sufficient to generate both radical and ionic species. Normal chemical methods for determining the nature of chain-carriers cannot be used in the solid phase, since conventional free-radical retarders, such as $\alpha\alpha$-diphenyl-β-

---

[4] H. Morawetz, in 'Physics and Chemistry of the Organic Solid State', ed. D. Fox, M. M. Labes, and A. Weissberger, Interscience, New York, Vol. I, 1963, Vol. II, 1965.
[5] A. Charlesby, *Reports Progr. Phys.* , 1965, **28**, 463.
[6] G. C. Eastmond, in 'Progress in Polymer Science,' ed. A. D. Jenkins, Pergamon Press Oxford, Vol. 2, 1970.
[7] C. H. Bamford and G. C. Eastmond, *Quart. Rev.*, 1969, **23**, 271.
[8] Various references cited in refs. 6 and 7.

The Rôle of Defects in Solid-phase Polymerization

picrylhydrazyl, are generally unable to diffuse into the crystal. For example, molecular oxygen has little influence on the polymerization of pure crystalline acrylamide but it is a strong retarder for the polymerization of acrylamide in solid solutions with propionamide.[9] This situation arises because oxygen is unable to diffuse through the acrylamide lattice, except along imperfection lines, while it can diffuse through the propionamide lattice.

It is sometimes possible to make a decision about the nature of the propagating species from chemical evidence. Trioxan, for example, is known to polymerize in solution only by ionic mechanisms, while acrylic acid polymerizes only by a radical mechanism. It is probably safe to assume that the same holds true in the solid phase. With photo-initiation there is little doubt that in almost all cases the active species will be radicals. Similarly, when hydrogen atoms are introduced into crystalline vinyl monomers only radicals will be produced. In many cases, therefore, e.s.r. investigations can usefully supplement information from polymerization studies. However, the nature of the chain-carrier is not of primary interest to us at present, since we are more concerned with physical than with chemical phenomena.

3 Vinyl Polymerization

First we consider the solid-phase polymerization of vinyl monomers since it is comparatively easy to understand the necessity for invoking the influence of defects in these systems, without recourse to any experimental evidence. In each addition step the configurations of two carbon atoms are changed from sp^2 to sp^3 hybridization [reaction (1)].

$$\begin{array}{c} H \\ \end{array}\!\!\!\!C = C\!\!\!\!\begin{array}{c} X \\ \end{array} \longrightarrow \begin{array}{c} H \\ \end{array}\!\!\!\!\begin{array}{c} H \\ | \\ C \end{array}\!\!\!\!-C\!\!\!\!\begin{array}{c} \\ | \\ \end{array} \qquad (1)$$
$$ H Y X Y$$

All the bonds attached to the carbon atoms concerned lie in a single plane in the monomer molecule. Polymerization therefore involves changes in the separation of the carbon atoms and in the disposition of groups X and Y, and the hydrogen atoms, relative to the carbon atoms and to each other. A general consideration of organic crystal structures indicates that the molecules pack together in the lattice with very little free space. It is difficult to envisage a crystal structure of a vinyl monomer which would allow sufficiently extensive movements of the types described to permit zipping-up of monomer molecules with formation of a polymer containing, say, one thousand monomer units.

As examples of the crystal structures of vinyl monomers we may quote those of acrylic acid (Figure 1), determined by Chatani *et al.*,[10] and of acrylamide.[11] It is apparent from the crystal structure of acrylic acid (especially

---

[9] G. Adler, *J. Polymer Sci. (C)*, 1965, **16**, 1211.
[10] Y. Chatani, Y. Sakata, and I. Nitta, *J. Polymer Sci. (B)*, 1963, **1**, 419.
[11] G. Adler and W. Reams, *J. Polymer Sci. (A)*, 1964, **2**, 2617.

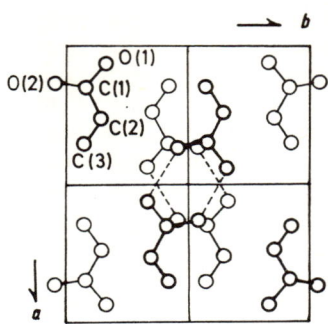

Figure 1 *Crystal structure of acrylic acid. Molecules with thick and thin lines are on planes* z = 0, z = 1/2, *respectively. Broken lines represent hydrogen-bonds*

from a model) that polymerization in the perfect lattice is impossible. It has been suggested that, in acrylamide, propagation may occur along a screw axis as far as the dimer, or possibly the trimer, stage, at which point the active radical at the chain-end would be too far out of step with the lattice to allow further propagation.[11] The necessity for invoking the involvement of defects, to provide space and freedom for the necessary movement, is therefore obvious.

Perhaps, with a little hindsight, it is even possible to predict the type of defect which will allow polymerization to occur. Let us consider the effectiveness of point defects, such as interstitial molecules or vacancies. An interstitial molecule may place vinyl groups in sufficient proximity and suitably oriented for reaction while a vacancy may allow sufficient space for some reorientation of the substituent groups in the monomer molecules on reaction. In either case the total distortion in the lattice is very limited and it is difficult to see how such a defect could support sufficient reaction to produce high polymer, even though it may provide a suitable site for the formation of an active propagating centre, or allow limited reaction. Clearly, we require a more extended defect; suitable candidates for reaction sites would, therefore, appear to be line defects, *i.e.* dislocations, or sub-grain boundaries, which are themselves a series of dislocations.

The mere existence of a dislocation in a crystal cannot be a sufficient condition for reaction; it is also necessary that molecules, at least in the vicinity of the dislocation, are able to undergo suitable molecular motions. Consequently, we may expect to find that the mobility of monomer molecules in the crystal is an important factor in these reactions. We shall continue to develop our theme on the basic assumptions that imperfections and molecular mobilities are two major factors controlling solid-phase polymerization.

To simplify the development of these ideas we shall now consider experimental evidence obtained with monomers carrying small substituent groups, *e.g.* acrylamide and acrylic acid, to avoid complications arising from the presence of large, bulky substituents. Polymerization of these monomers invariably produces amorphous polymers.

Reaction Kinetics.—Kinetic studies of solid-phase polymerization have been carried out on a variety of vinyl monomers, for both in-source and post-irradiation reactions. Care must be taken in determining conversion–time curves to prevent further reaction arising from the release of active species during isolation of the reaction products. Post-irradiation reactions rely on the presence of trapped active species; for example, large concentrations of free radicals are produced in crystalline methacrylic acid by u.v. irradiation and when the radiation is cut off these give rise to continued polymerization.[12] Under similar conditions it is difficult to detect trapped radicals in acrylic acid and no post-irradiation reaction is observed; however, the existence of some active species is demonstrated by the rapid reaction which occurs when such samples are melted. Physical techniques, such as measurements of birefringence and X-ray diffraction, are also employed to determine residual crystallinity. Their use is increasing; for example, it has been suggested that a correlation of kinetic data with rates of formation and lifetimes of positronium can provide a useful insight into the processes involved in these reactions.[13]

Post-irradiation reactions are usually characterized by a fairly rapid initial reaction, the rate of which decreases steadily with time, tending to a limiting conversion, although reaction can proceed very slowly for long periods of time. In many cases, *e.g.*, after γ-irradiation of acrylamide, the shape of the conversion–time curve is consistent with bimolecular termination of radicals.[14–16] However, the radical concentration under such conditions remains essentially constant,[15–17] demonstrating that normal bimolecular radical termination, involving diffusion of reactive species, does not occur. The decrease in rate with time is a result of physical factors which limit the growth of polymer chains.

There are numerous reports in the literature of in-source polymerizations, *e.g.* high-energy-induced polymerizations of acrylamide[17] and substituted acrylamides,[18] N-vinylcarbazole,[19] tributylvinylphosphonium bromide,[20] and methyl methacrylate,[21] and u.v.-initiated polymerizations of acrylic and methacrylic acids,[12] which exhibit autocatalytic conversion–time curves. These curves are very similar in shape to those displayed by many other

[12] C. H. Bamford, G. C. Eastmond, and J. C. Ward, *Proc. Roy. Soc.*, 1963, **A271**, 357.
[13] Y. Tabata, Y. Ito, and K. Oshima, *Mol. Cryst. Liquid Cryst.*, 1969, **9**, 417.
[14] T. A. Fadner, I. Rubin, and H. Morawetz, *J. Polymer Sci.*, 1959, **37**, 549.
[15] H. Morawetz and T. A. Fadner, *Makromol. Chem.*, 1959, **34**, 162.
[16] T. A. Fadner and H. Morawetz, *J. Polymer Sci.*, 1960, **45**, 475.
[17] B. Baysal, G. Adler, D. Ballantine, and P. Colombo, *J. Polymer Sci.*, 1960, **44**, 117.
[18] J. Zurakowska-Orszagh, *J. Polymer Sci. (C)*, 1968, **16**, 3291.
[19] A. Chapiro and G. Hardy, *J. Chim. phys.*, 1962, **59**, 993.
[20] C. S. H. Chen and D. G. Grabar, *J. Polymer Sci. (C)*, 1964, **4**, 849.
[21] T. Lipscomb and E. C. Weber, *J. Polymer Sci. (A-1)*, 1967, **5(1)**, 779.

D*

solid-phase reactions in which reaction is considered to be initiated at specific sites (imperfections) in the crystals, and which involve nucleation of a separate product phase in the host crystal. Such conversion–time curves can be explained mathematically either by postulating reaction at the interface between the expanding product particles and the lattice, or by assuming increasing numbers of product particles which need only develop to a limited size. To attempt a detailed analysis of the kinetics is obviously unrealistic without further experimental data and a more complete understanding of the nature of the reactions.

Nature of Reaction Sites.—The essential correctness of the view that polymerization is accompanied by the formation of a separate polymer phase was clearly demonstrated by X-ray diffraction studies of the in-source polymerization of acrylamide. It was observed [22] that the intensities of the diffraction spots of the crystalline monomer decayed uniformly on polymerization while a diffuse halo arising from amorphous polymer gradually developed. Polymer could be detected at conversions as low as 4% and the undistorted monomer pattern was still observable at 90% conversion. These results demonstrated the isotopic nature of the polymerization and the lack of distortion in the monomer lattice produced by growing polymer particles.

If our view that polymerization occurs in dislocations is correct, rates of polymerization cannot be uniform throughout the crystal. Non-uniform behaviour has been observed experimentally in the case of acrylamide. Optical examination of acrylamide crystals after irradiation has demonstrated the development of relatively large polymer particles (observable at magnifications of 500 ×) during the post-irradiation reaction.[23] Such particles appear to be predominantly associated with the 100 slip-plane and individual particles appear to grow at different rates. It was observed that the particles lie along lines or in loops, as would be expected for reaction in dislocations. Evidence was also obtained indicating that polymerization occurs preferentially at mosaic boundaries. Some similar evidence was found for polymerization in potassium acrylate crystals.[23] It is not certain whether the particles observed in this work are single particles or whether they arise from the coalescence of smaller units.

Sella and co-workers [24] studied the (in-source) polymerization of acrylamide by electron microscopy, and observed, at low conversion, the development of small polymer particles on the monomer crystal surfaces. The particles were about 2 nm in diameter, and were aligned along preferred crystallographic directions. As reaction proceeded the particles grew to diameters of 30—40 nm, eventually coalescing to give the appearance of polymer fibrils oriented along crystallographic directions. Ultimately, the monomer crystals were converted to a mass of globular particles.

[22] G. Adler and W. Reams, *J. Chem. Phys.*, 1960, **32**, 1698.
[23] G. Adler, D. Ballantine, and B. Baysal, *J. Polymer Sci.*, 1960, **48**, 195.
[24] C. Sella and J. J. Trillat, *Compt. rend.*, 1961, **253**, 1511; C. Sella and R. Bensasson, *J. Polymer Sci.*, 1962, **56**, S1.

The observations described above are very reminiscent of the results of dislocation-decoration studies and were taken to indicate that reaction is associated with dislocations. The limited experimental evidence available does not permit us to say much about the growth of the polymer nuclei, although there are indications in the photographs of Sella et al.[24] that the particles do not have a constant rate of radial growth, but grow rapidly to a certain size. Continued polymerization is associated with the generation of new particles. Such behaviour would be consistent with the report of Adler et al.[23] that reaction shows periods of acceleration and deceleration in different parts of the crystal at different times. On the basis of this observation it was suggested that polymerization generates stresses within the crystal which inhibit further reaction. Eventual release of stress allows rapid polymerization to recommence. Microscopic studies of acrylic acid polymerization[12] did not reveal such marked fluctuations in rate, possibly because the reaction temperature was much closer to the monomer melting point than with acrylamide and strain-releasing processes may have occurred more rapidly. We shall return to this point subsequently.

Although our understanding of the structures and properties of dislocations in organic crystals is still limited, it becomes increasingly apparent that the dislocations show many similarities with those in metallic crystals. Bamford, Eastmond, and Ward[12] anticipated the behaviour of dislocations in organic crystals, and presented evidence of a different type, pointing to the importance of these defects as reaction sites. Formation of amorphous polymer by solid-phase polymerization of thin (10—20 μm) single crystals of acrylic acid, under continuous u.v. irradiation at 277 K, was followed by observing the optical retardation of the crystals, which decreased with time in a sigmoid manner. Application of a mechanical stress ($\sim$10 atm) perpendicular to the large faces of the crystals, at about 50% loss of optical retardation, caused the reaction to cease immediately. No further change was observed until the stress was released, when reaction was resumed after a short induction period. Similar behaviour was observed with methacrylic acid, except that in this case the reaction under continuous irradiation did not stop completely and resumed its full rate immediately on release of the stress. The post-irradiation reaction which occurs with the latter monomer was completely suppressed by the same stress but immediately reappeared when the stress was removed. These observations are summarized in Figure 2. Unlike solution polymerizations, where very high pressures are needed to produce any significant effect on reaction rates, these reactions are clearly very sensitive to small stresses.

The experimental observations described above indicate that the solid-phase reaction proceeds in a stress-sensitive environment. Edge-dislocations are stress-sensitive regions in metal crystals,[25] and it was assumed that dislocations are the reaction sites in the solid-phase polymerizations of acrylic

---

[25] A. H. Cottrell, 'Dislocations and Plastic Flow in Crystals', Oxford University Press 1953.

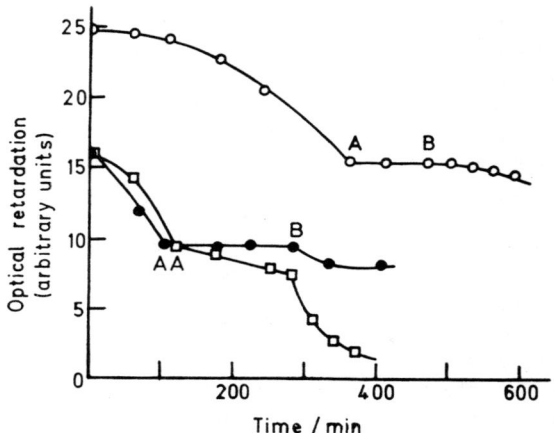

Figure 2 *Effect of stress on the polymerization of acrylic and methacrylic acids initiated by u.v. irradiation. Stress applied at point A and removed at point B in each case:* ○ *Acrylic acid at* 277 K; □ *methacrylic acid at* 280 K; ● *methacrylic acid at* 280 K, *irradiation stopped at A. This curve illustrates the suppression of the after-effect by stress* (ref. 7)

and methacrylic acids.[12] Further, it was assumed that when a radical species responsible for polymerization is produced in a dislocation, application of stress causes the dislocation to move away from the radical, leaving the latter in an imperfection analogous to a point defect. In such an environment reaction may not occur, or may occur only to a limited extent. To explain the cessation of reaction in acrylic acid under continued irradiation it was proposed[12] that the dislocations move to obstacles such as grain boundaries, probably forming piled-up groups,[25] where reaction cannot take place or is not observed under the low magnifications employed. When stresses applied to metallic crystals are removed, the dislocations in piled-up groups repel each other and become redistributed throughout the lattice.[25] The reappearance of the post-irradiation reaction in methacrylic acid indicates that the dislocations become re-localized at sites containing radicals.[12] The detailed results are extremely sensitive to both temperature and the presence of impurities, and are discussed in more detail later.[26]

Molecular Mobility.—The results described in the preceding section provide strong evidence that dislocations act as reaction sites in solid-phase polymerizations of vinyl monomers. We have already stated our belief that molecular mobility is another factor of major importance. It is, of course, the molecular mobility in the dislocations which is significant initially. On one side of an edge-dislocation line there is a region of tensile stress in which freedom of movement is greater than elsewhere in the crystal. Molecular

[26] C. H. Bamford, A. Bibby, and G. C. Eastmond, *Polymer*, 1968, **9**, 653.

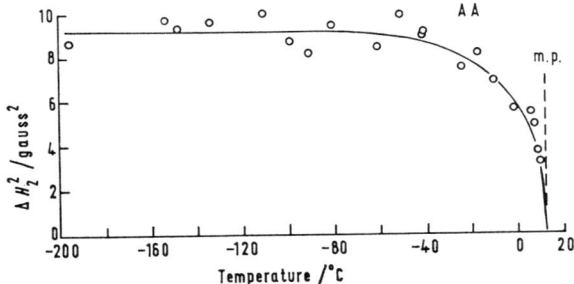

Figure 3 *Variation of the second moment of the broad-line n.m.r. spectrum of acrylic acid with temperature, from Arnold and Eastmond (ref. 27)*

motions in these regions, therefore, will probably be more extensive than in the remainder of the crystal, at a given temperature. Since the distortion associated with a dislocation, within one molecular plane, is limited, and the same types of intermolecular force will prevail as in the rest of the crystal, although at a reduced level, the variations in molecular mobility with temperature within dislocations may well follow the same pattern as within the remainder of the lattice.

We may continue the development of our basic ideas by reference to molecular mobilities in acrylic and methacrylic acids, for which Arnold and Eastmond [27] have investigated the temperature variation of the second moment (ΔH_2^2) of the broad-line n.m.r. spectrum. Figure 3 shows the variation in ΔH_2^2 with temperature for acrylic acid. Consideration of these data, together with the crystal structure of acrylic acid (Figure 1), established that the perfect lattice is rigid at temperatures below about $-80\,°C$; at this temperature molecular motions become apparent and steadily increase in magnitude with increasing temperature to the melting point. The reductions in ΔH_2^2 were attributed to rotational oscillation of the hydrogen-bonded dimers about the crystal a-axis (Figure 1), the amplitude of oscillation increasing with temperature. The experimental data for methacrylic acid were of exactly the same form as those in Figure 3, and very similar to the earlier data of Odajima *et al.*[28] No evidence was found for the anomalous transition between a low-temperature phase of high molecular mobility and a high-temperature phase of low molecular mobility, reported by Sakai and Iwasaki.[29] The crystal structure of methacrylic acid is unknown, but there is little doubt, from the value of ΔH_2^2, that the methyl group is rotating at all temperatures above 77 K and that as the temperature increases the same molecular motions as occur in acrylic acid become operative.

As we shall see later, a limited amount of reaction can be observed in these monomers at temperatures at which the perfect lattice is effectively rigid;

[27] B. Arnold and G. C. Eastmond, *Trans. Faraday Soc.*, 1971, **67**, 772.
[28] A. Odajima, A. E. Woodward, and J. A. Sauer, *J. Polymer Sci.*, 1961, **55**, 181.
[29] Y. Sakai and M. Iwasaki, *J. Polymer Sci.*, (A-1), 1969, **7**, 3143.

as the temperature increases the extent of reaction increases. However, it is not until temperatures of about 263 K are reached that polymerization in acrylic acid [12, 30] and methacrylic acid,[12, 31, 32] initiated either by u.v. or high-energy radiation, is at all evident, and polymerization is rapid only above 273 K. Examination of the n.m.r. lineshapes shows that under conditions of rapid polymerization a small fraction of the monomer molecules are undergoing rapid and extensive reorientation, as evidenced by the appearance of a small, extremely narrow line which can be resolved into the resonances of the individual proton types.[27] It is believed that these highly mobile molecules are situated in lattice defects and are preferentially involved in polymerization. In this event, it might be anticipated that, if polymerization is restricted to the molecules within the cores of the dislocations, there would be rapid reaction to a very low conversion. When the mobile molecules are converted to polymer, the character of the polymer–monomer interface will approach that of the interface between polymer and the perfect lattice. Under such conditions reaction would be expected to slow down, or even stop. This type of behaviour may indeed be inferred from the results of Sella et al.[24] on the polymerization of acrylamide, which indicate the limited growth of the individual polymer nuclei. However, we know that under continued irradiation the polymerizations of all these monomers proceed autocatalytically to high conversion, and the electron microscope data suggest that this probably occurs through the appearance of new reaction sites.

The formation of nuclei of amorphous polymer inside the monomer crystal undoubtedly produces stresses within the crystal. Such stresses could be relieved by the generation of a series of new dislocations centred around the original. The formation of new dislocations would produce a new generation of mobile monomer molecules in the dislocation cores and thus increase the number of available reaction sites. Local fluctuations in the rate of polymerization in acrylamide, observed by Adler et al.[23] and attributed to a stress-releasing process, may arise from local fluctuations in the rate of dislocation multiplication. This type of behaviour appears to have been observed directly in the thermal decomposition of sucrose,[33] and its occurrence in the solid-phase polymerization reactions may be responsible for the observed autoacceleration.[34]

Although there is a good correlation between molecular mobility and polymerizability for acrylic and methacrylic acids, with other monomers the parallellism is less striking. Crystal lattices of barium methacrylate hydrate (originally reported to be a dihydrate [35] but now known to be a monohydrate [36]) and the anhydrous salt are probably rigid, apart from methyl

[30] Y. Shioji, S. I. Ohnishi, and I. Nitta, *J. Polymer Sci.*, (A), 1963, **1**, 3373.
[31] Y. Sakai, *J. Polymer Sci.* (A–1), 1969, **7**, 3191.
[32] Y. Sakai, *J. Polymer Sci.* (A–1), 1969, **7**, 3177.
[33] J. M. Thomas and J. O. Williams, *Trans. Faraday Soc.*, 1967, **63**, 1922.
[34] G. C. Eastmond, E. Haigh, and B. Taylor, *Trans. Faraday Soc.*, 1969, **65**, 2497.
[35] J. B. Lando and H. Morawetz, *J. Polymer Sci.* (C), 1964, **4**, 789.
[36] M. J. Bowden and J. H. O'Donnell, *J. Phys. Chem.*, 1969, **73**, 2871.

group and water molecule rotation at temperatures in the range 173 to 313 K.[27] Above the latter temperature molecular motions in both forms become operative and gradually increase with increasing temperature. It is known that barium methacrylate hydrate starts to lose water near 313 K. Between 313 K and 343 K there is rapid partial loss of water and above 353 K complete dehydration is rapid. The kinetics of dehydration have been studied by Bowden and O'Donnell.[36] Around 343 K the second moments of the broad-line n.m.r. spectra of the two forms become identical,[27] due to loss of water from the hydrate and the modification of the crystal structure on dehydration.[36] The second moment of the spectrum of the anhydrous salt continues to decrease as the temperature is increased and molecular motions become more extensive. N.m.r. data show that lithium methacrylate (anhydrous) behaves in a similar way to anhydrous barium methacrylate,[27] molecular motions appearing at 223 K and steadily increasing with temperature to at least 423 K.

Hydrated barium methacrylate, irradiated at low temperature, undergoes post-irradiation polymerization at about 303 K,[35] and at 330 K conversion to polymer reaches 58% in 162 h. At higher temperatures rates of post-irradiation polymerization are higher but lower limiting conversions are reached (44% at 374 K). Both anhydrous barium and lithium methacrylates are reported to be unreactive in the solid phase[35] (irradiated barium salt gave 1% conversion in prolonged heating at 373 K and the lithium salt 0% at 433 K), although the mobilities of the methacrylate groups are higher than in barium methacrylate hydrate. It has been suggested that polymerization occurs in the hydrate prior to dehydration or during the rearrangement of the molecules during the transformation from the crystal structure of the hydrate to that of the anhydrous salt.[27] According to Costachuk et al.,[37] solid-phase polymerization of calcium acrylate is most rapid in the interphase-boundary between crystalline dihydrate and partially crystalline anhydrate.

U.v. Initiation. —As a result of e.s.r. studies on various monomers there is little doubt that in most cases initiation involves addition of radiolytic hydrogen atoms to monomer molecules. This could, in principle, occur anywhere in the lattice, but there is little detailed information available on the nature of the energy transfer processes following the interaction of radiation with the monomer. We shall, therefore, continue our theme by considering the mechanism of radical formation in the u.v.-initiated polymerization of acrylic acid.

According to the picture which has emerged so far, reaction is essentially restricted to rapidly reorientating molecules in dislocations. Consequently, we should not expect anisotropic behaviour to be observed in these reactions. In fact, when polymerization is initiated with polarized u.v. radiation there is a marked dependence of rate, in the initial stages, on the relative orientations

[37] F. M. Costachuk, D. F. R. Gilson, and L. E. St. Pierre, *Macromolecules*, 1970, **3**, 393.

of the crystal and the plane of polarization of the initiating radiation.[34] The nature of this dependence strongly indicates that the bulk of the radiation ultimately responsible for radical formation is absorbed, not by molecules in defects, but by molecules in the perfect lattice. It has been suggested [34] that absorption of radiation by molecules in the perfect lattice produces excitons which are free to travel through the lattice until trapped and localized in defects. Subsequent decay of the excitons in suitable defects (dislocations associated with sufficient molecular mobility) could result in radical formation, or an enhancement of local thermal motions which could also assist reaction.

Little can be said concerning the detailed chemical processes accompanying exciton decay, except that there is e.s.r. evidence suggesting that at least four monomer molecules must have sufficient mobility and freedom of movement to be involved in reaction (see next section).

E.S.R. Studies.—Apart from the observations that high concentrations of free radicals are stable in many monomers for prolonged periods at polymerization temperatures, demonstrating that normal bimolecular termination through radical diffusion is improbable in the solid state, the most useful information obtained from e.s.r. studies has come from investigations at temperatures too low to permit polymer formation. The existence of radical reactions in many solid monomers under these conditions has been demonstrated. Such low-temperature reactions are very limited in extent and it is probably reasonable to assume that they are confined to defects initially present in the crystalline monomers.

High concentrations of free radicals are produced in a number of monomers, including acrylamide,[38] methacrylamide,[39] acrylic acid,[30] methyl acrylate,[40] methacrylic acid,[41] methyl methacrylate,[40] barium methacrylate,[42–44] itaconic acid,[45] and dimethyl itaconate,[46] when subjected to high-energy radiation at 77 K. Except in the case of methacrylamide, the only primary radicals observed are those resulting from the addition of a hydrogen atom to a monomer molecule, *i.e.* the radicals MeĊXY, where X,Y, are the substituent groups in the monomer molecules. Additional lines in the e.s.r. spectrum of irradiated methacrylamide [39] have been attributed to the presence of the radicals $H_2N\dot{C}=O$ and $RCH_2\dot{C}MeCONH_2$, where R may be $CH_2=C$ $(CONH_2)$ CH_2 or $CH_2=CMe$. Irradiation of some materials is known to produce pairs of radicals in close proximity,[47] exhibiting additional e.s.r.

[38] G. Adler and J. H. Petropoulos, *J. Phys. Chem.*, 1965, **69**, 3712.
[39] H. Ueda, *J. Polymer Sci. (A)*, 1964, **2**, 2207.
[40] T. Gillbro, P. O. Kinell, and A. Lund, *J. Polymer Sci.*, (A-2), 1971, **8**, 1495.
[41] Y. Sakai and M. Iwasaki, *J. Polymer Sci. (A–1)*, 1969, **7**, 1749.
[42] J. H. O'Donnell, B. McGarvey, and H. Morawetz, *J. Amer. Chem. Soc.*, 1964, **86**, 2322.
[43] M. J. Bowden and J. H. O'Donnell, *Macromolecules*, 1968, **1**, 499.
[44] M. J. Bowden and J. H. O'Donnell, *J. Phys. Chem.*, 1968, **72**, 1577.
[45] M. Fujimoto, *J. Chem. Phys.*, 1963, **39**, 846.
[46] Y. Tabata, T. Miyairi, S. Katsura, Y. Ito, and K. Oshima, 'Large Radiation Sources for Industrial Processes', Internat. Atomic Energy Agency, Vienna, 1969, p. 233.
[47] Y. Kurita, *J. Chem. Phys.*, 1964, **41**, 3926.

transitions, and such pair formation has been observed in irradiated methyl acrylate and methyl methacrylate.[40] Direct addition of H and D atoms (produced in a microwave discharge) to methacrylic acid,[48] methyl methacrylate,[49] barium methacrylate,[50] and other salts of methacrylic acid,[51] acrylonitrile,[49] and styrene [49] also produces MeĊXY or CH_2DĊXY radicals.

Single-crystal studies of the primary radicals in acrylamide [38] and acrylic acid [30] have shown that they are orientated along preferred directions in the crystal lattices which correspond to the directions of the unreacted vinyl groups. Anisotropy in the e.s.r. spectra have also been reported for dimethyl itaconate,[46] methyl acrylate,[40] and methyl methacrylate.[40] In view of the small size of the hydrogen atom it is conceivable that such primary radicals could be produced anywhere in the crystal, even in the perfect lattice. However, on warming the crystals many of the radicals undergo reactions which cannot occur in the perfect lattice and these radicals are, of course, ultimately responsible for the post-irradiation polymerization. It is probably safe to assume, therefore, that the majority of such radicals are situated in lattice defects. Even so, it is not surprising that they exhibit strong orientational dependence. At temperatures near 77 K the crystal lattices of most monomers are rigid, the lattice distortions in, say, dislocations are limited, and the free volumes in the vicinity of the defects have a shape determined by the lattice structure. The radicals themselves have definite shapes and in many cases are hydrogen-bonded to neighbouring monomer molecules, so that the orientations available to them in such situations are limited.

The major features of the processes which take place on warming the samples after irradiation at 77 K are the same for all monomers. In general, the radical concentration, stable at any one temperature, decreases with increasing temperature and the primary radicals are gradually transformed into normal propagation radicals. Adler and Petropoulos[38] studied the behaviour of radicals during the warming of pre-irradiated acrylamide (m.p. 356.8 K). The first major change was observed to occur at temperatures above 148 K (*e.g.* 193 K) when the five-line spectrum of the primary radicals changed to an anisotropic triplet. This change was attributed to formation of a dimeric species orientated in the lattice. Crystallographic considerations suggested that the primary radical may have added to a nearest neighbour monomer in the direction of the two-fold screw axis of the crystal. Further reaction, which would probably have to occur in some other direction (with consequent loss of anisotropy) is prevented through lack of mobility and lack of register with the crystal lattice. Warming to 233 K brought about a slight change in the e.s.r. spectrum and some decrease in radical concentration. Further warming to temperatures in the range 243 to 253 K produced a complete loss of anisotropy. Presumably, it is in this temperature range

---

[48] B. Arnold and G. C. Eastmond, unpublished results.
[49] C. Chachaty and M. C. Schmidt, *J. Chim. phys.*, 1964, **62**, 527.
[50] H. C. Heller, S. Schlick, H. C. Yao, and T. Cole, *Mol. Cryst. Liquid Cryst.*, 1969, **9**, 401.
[51] C. H. Bamford, G. C. Eastmond, and D. Thomas, unpublished results.

that local molecular mobilities in the vicinities of the radicals become sufficiently great to allow further propagation and the development of a separate polymer phase.[22, 38] It is only at temperatures above 253 K that post-polymerization is observable.

Primary radicals in acrylic acid,[30] orientated in two directions, decay slightly on warning to 173 K. Subsequent warming to 193 K produces a further decrease in radical concentration and a simultaneous collapse of the anisotropic double quartet of the primary radicals to the isotropic triplet of randomly orientated propagating radicals.[30] It is known from n.m.r. studies that at 193 K the perfect lattice is almost completely rigid.[27] At 223 K the radical concentration decreases to an undetectably low level which, however, is sufficient to produce polymerization above about 248 K,[30] at which temperature molecular motions are becoming more extensive.[27] It is also difficult to detect free radicals in this monomer after u.v. irradiation.

The characteristic feature of these radical transformations is that they do not take place at specific temperatures but occur gradually over a wide temperature range. On storage at any given temperature a fraction of the primary radicals will be transformed while the rest remain unchanged; the higher the temperature the more extensive is the transformation. Had the primary radicals been predominantly situated in the perfect lattice the spectra of most radicals would change simultaneously. Observations of this type led Adler and Petropoulos to conclude [38] that the radicals are situated in a variety of defect structures associated with different molecular mobilities and that any specific radical undergoes a transformation at some temperature where the local mobility is sufficiently great. In contrast, high concentrations of radicals stable up to the melting point of the monomer can be generated in methacrylic acid (m.p. 289 K). γ-Irradiation of this monomer,[41] or addition of hydrogen atoms,[48] at 77 K produces Me_2CCOOH radicals which exhibit a seven-line e.s.r. spectrum, showing that both methyl groups are rotating. Storage at 77 K causes a slight change in the e.s.r. spectrum without radical decay.[48] Subsequent thermal treatment of samples produced in both studies produces almost identical changes in e.s.r. spectra.[41, 48] The spectral change observed on storage at 77 K becomes more pronounced when the storage temperature is raised to about 173 K and corresponds to the transformation of the primary radicals to a species [designated N(1) radicals] exhibiting a nine-line spectrum in which line intensities increase towards the centre of the spectrum; the radical concentration decreases during this change. Continued warming to about 223 K produces a further decrease in concentration and a gradual change of N(1) radicals to radicals characterized by a 13-line spectrum (T radicals). On warming to 253 K or 273 K the radical concentration continues to decrease and the 13-line spectrum changes steadily to a nine-line spectrum with line intensities alternately strong and weak, arising from N(2) radicals. During the later stages of heating the radical concentrations become very low. However, high concentrations of T and N(2) radicals can be obtained conveniently at the higher

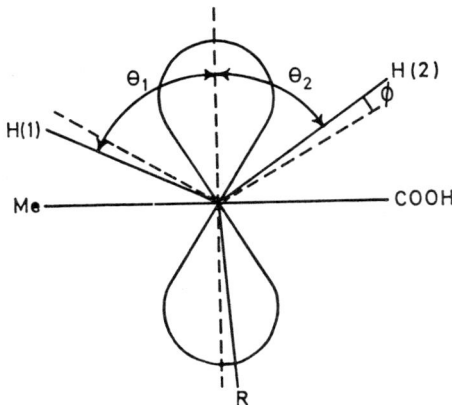

Figure 4 *Conformation of the radical* $\sim C^\alpha H_2 - C^\beta MeCOOH$. *Projection along the* $C^\alpha - C^\beta$ *bond. Dotted lines represent symmetrical conformation of H atoms, with respect to axis of half-filled p-orbital. R represents the remainder of the chain*

temperatures by u.v. irradiation of the monomer above about 323 K.[12, 52, 53]

The spectra of N(1), T, and N(2) radicals are all interpretable in terms of conformations of the propagating radical $\sim C^\beta H_2 \dot{C}^\alpha Me$ COOH, resulting from hindered rotation about the $C^\alpha - C^\beta$ bond. These conformations can be understood by reference to Figure 4 which is a projection along the $C^\alpha - C^\beta$ bond; H(1), H(2) are the methylene hydrogens, R is the remainder of the polymer chain, and $\theta_1 + \theta_2 = 120°$. In the symmetrical situation $\theta_1 = \theta_2 = 60°$, $\varphi = 0$. The N(2) spectrum can result from the co-existence of two preferred conformations[54] with different values of φ (15° and 0°, approximately) or from a single preferred conformation with a low value of φ (say 5°) with a distribution of conformations about this value.[43] Whatever may be the detailed interpretation of the N(2) spectrum, it is characteristic of propagating radicals in an amorphous matrix. The T radical spectrum, first observed after u.v. irradiation of the monomer,[52] was attributed to a single preferred conformation of the radical with $\varphi = 5°$, approximately. The N(1) spectrum is produced by radicals in the conformation $\varphi = 2°$, approximately,[41, 48] The exact conformations of N(1) and T radicals are not known since the spectral changes occur gradually and probably in no case has a pure spectrum from a single conformation been observed. In spite of this there is no evidence for the existence of intermediate species.

Detailed studies of radical conformations and concentrations during u.v. irradiation under various conditions have allowed the process T→N(2) to be investigated and certain conclusions to be drawn concerning the nature of

[52] C. H. Bamford, G. C. Eastmond, and Y. Sakai, *Nature*, 1963, **200**, 1284.
[53] C. H. Bamford, A. Bibby, and G. C. Eastmond, *Polymer*, 1968, **9**, 629.
[54] M. C. R. Symons, *J. Chem. Soc.*, 1963, 1186.

the radical sites.[53] Irradiation at temperatures above about 273 K, where polymerization is rapid, yields only N(2) radicals, indicating that the radicals are in an amorphous polymer phase and are not influenced by the surrounding crystal. At lower temperatures, 223 K to 253 K, irradiation for short times produces mainly T radicals. On continued irradiation the total radical concentration increases and the proportion of T radicals decreases, while that of N(2) radicals increases. Eventually, the radical concentration and the proportion of T to N(2) radicals approach limiting values. These limits are a function of temperature; as the irradiation temperature increases the radical concentration increases and the proportion of T radicals decreases. That is, the higher the temperature, and the higher the molecular mobility, the greater is the proportion of radicals existing in a separate polymer phase.

The limiting values mentioned were taken to correspond to the establishment of a stationary state and, since bimolecular termination by long-range diffusion is negligible, it was proposed that reaction is limited to specific defects and that initiation and termination in these sites are both photo-induced in nature. Since exciton decay must produce radicals in pairs, and the T radicals are at least dimeric, at least four monomer molecules must be involved in movements for each act of radical generation. It was postulated that reaction is limited to those defects in which there is adequate molecular mobility and the limiting radical concentration was taken to correspond to the situation in which all such defects contain radicals. These ideas, and particularly the way in which the limiting proportions of T and N(2) radicals arise, can be understood in terms of the following simple model, intended to describe the sequence of reactions in a single site.

$$\begin{array}{ccc} T & \rightarrow & N(2) \\ \uparrow & & \downarrow h\nu \\ N(2)^{N(2)}_{T} & \leftarrow & N(2)^{T}_{T} \end{array}$$

At the top left-hand corner we have a site containing a T radical which relaxes (thermally) to N(2), the rate of relaxation depending on local molecular mobility. This process might occur through detachment of the T radical from the lattice, arising, for example, by rupture of hydrogen bonds. Further initiation in the vicinity of this radical generates two T radicals, one of which similarly relaxes to N(2). The two N(2) radicals, free from the lattice and in close proximity, would be able to terminate, leaving a site containing a T radical. The temperature variation of the limiting proportions of T and N(2) radicals can be understood in terms of the rates of the various processes; a relatively high mobility corresponds to fast relaxation and a low proportion of T radicals in the stationary state.

The dependence of the limiting radical concentration on temperature can be explained in terms of a variety of defect structures associated with different local molecular mobilities. The existence in many organic crystals of a series of dislocations with such properties is readily understandable. Edge-dislocations with various structures can be produced through the use of different slip-planes in the crystal. In the unit cell of acrylic acid, and probably of methacrylic acid, there are eight monomer molecules (four hydrogen-bonded dimer units) and to generate at least some of the dislocations with unit Burgers vector it is necessary to remove, or introduce, more than one half-plane of monomer molecules. Such half-planes may be separated by various distances. Thus, it is possible to produce a distribution of structures with different local molecular mobilities. In only a small fraction of such sites will the mobility be sufficiently high to support reaction at low temperatures, but with increasing temperature and general level of molecular mobility the number of 'active' sites, and hence the limiting radical concentration, will increase. The e.s.r. data indicate that the distribution of mobilities in the sites is influenced by the conditions of crystallization [53] and by the presence of traces of impurities,[55] rapid crystallization increasing the total number of potential sites but decreasing, on average, the associated mobilities.

The presence of impurities (e.g. 0.02% isobutyric acid) has a marked influence on reactions in methacrylic acid crystals,[55] increasing to some extent the density of reaction sites, but also greatly increasing the local molecular mobilities. Thus, the proportion of T radicals is drastically reduced at short reaction times, but at long irradiation times it is identical with that obtained in the absence of impurity; i.e. the effect of the impurity is lost as reaction proceeds. It is visualized that the impurity concentrates around the dislocation cores in the form of Cottrell clouds,[25] effectively influencing only local properties. In the early stages reaction is therefore greatly influenced by the impurities, but as reaction proceeds towards the more perfect regions of the lattice the influence is lost. Thus, the nature of the reaction site changes us reaction proceeds.

Individual methacrylic acid molecules are undoubtedly hydrogen-bonded to neighbouring molecules in both the perfect lattice and in imperfections. The T radicals, considered to be dimeric or trimeric, are probably formed when a primary radical has sufficient mobility to add to a neighbour with retention of the hydrogen-bonds. The disposition of the hydrogen-bonds connecting the radical to the remainder of the crystal probably determines the conformation under conditions of low molecular mobility.[53]

It has been demonstrated that N(1) and T radicals are chemically identical and that the change N(1) →T can be reversed if samples are cooled sufficiently slowly.[41] Sakai and Iwasaki [41] suggested that the N(1) spectrum arises from a collapse of certain lines in the T radical spectrum in a low-temperature phase of relatively high mobility. However, Arnold and Eastmond [48] found no evidence for the existence of such a phase and interpreted the conformational change in terms of a change in defect structure and disposition of hydrogen-bonds.

Studies on γ-irradiated barium methacrylate hydrate have also shown the

[55] C. H. Bamford, A. Bibby, and G. C. Eastmond, *Polymer*, 1968, **9**, 645.

existence of different radical conformations. Raising the temperature of the primary radicals from 77 K to 193 K produces propagating radicals with $\varphi = 2°$, said to be dimer species in a monomer environment.[44] Warming to 323 K produces an effect which is only detected by subsequent cooling of the specimens, when additional conformations ($\varphi = 3$ or $4°$) become apparent; this change is attributed to an increase in chain length.[44] The radicals are orientated in the lattice,[42] and are considered to be polymer radicals in a monomer environment. Storage at 323 K produces a five-line spectrum of the propagating radical (with $\varphi = 7°$ or greater), and polymerization proceeds. From the kinetics of the conformational change at low temperatures it was concluded that a range of radical-trapping sites exists in the monomer crystals and that reaction occurs preferentially in some sites.[43] No decrease in radical concentration accompanies spectral changes below 313 K; above this temperature post-polymerization occurs and radical concentrations decay rapidly at first and then more slowly. It was proposed that termination occurs preferentially between short radicals in high local concentration.[43]

Comparison of low-temperature e.s.r. and n.m.r. data for both methacrylic acid and its barium salt demonstrates quite clearly that at temperatures at which the lattice is effectively rigid (apart from small-group rotations), reactions take place which require molecular movements. The occurrence of such processes supports the idea that radicals are located in defects where small-scale motions assist reaction.

Further evidence for a multiplicity of reaction sites, and for preferential termination of closely situated radicals, has arisen from studies of γ-irradiated methyl methacrylate, which show that one type of radical pair disappears at 133 K while other pairs, in which the separation is presumably greater, do not decay until higher temperatures are reached.[40]

Obviously, e.s.r. studies of the type outlined above are capable of providing much information on the nature and properties of radical sites, and on the mechanisms of solid-phase polymerization. By correlating such work with crystallographic data it may be possible eventually to obtain quantitative information on the structures of imperfections and the distribution of their properties.

Polymerization at Low Temperatures.—So far we have not discussed the polymerization of acrylonitrile, although this has received considerable attention. Acrylonitrile exhibits a first-order solid-phase transition at 160 K with high- and low-temperature phases stable above and below this temperature, respectively.[56] Bensasson *et al.*[56] have shown that both forms can be prepared at 77 K by suitable thermal treatment, and that the high-temperature phase reverts to the stable low-temperature phase on warming to 143 K.

X-Irradiation at 77 K produces a small amount of polymerization in

[56] R. Bensasson, A. Dworkin, and R. Marx, *J. Polymer Sci.* (C), 1964, **4**, 881.

both phases, with limiting conversions of about 1% in the low-temperature phase and 4% in the high-temperature phase.[56] Similarly, low conversions at this temperature have been obtained using γ-irradiation,[57] high-energy electrons [58] and protons,[59] reactor radiation,[60] and recoil particles from the ^{10}B (n, α)^{7}Li reaction.[61] The details of the reactions, rates, limiting conversions, and polymer structures are reported to vary with the type of radiation used. The mechanisms of these reactions are not clear at present and the reader is referred to the original data for details. Further information on the mechanisms of the energy transfer processes and their influence on the reactions would obviously be of interest.

Studies of the in-source polymerization have been extended down to 4 K; under such conditions it is difficult to distinguish between reaction at these low temperatures and reaction during warming and isolation of the polymer. However, various workers have concluded from their data that there is a true reaction at low temperatures.[62—64] There appears to be general agreement that the rates and limiting conversions increase slowly (with almost zero activation energy) with increasing temperature from 4 K to the first-order transition, and then increase more rapidly in the high-temperature phase to just below the melting point of the monomer, when the rate decreases rapidly; very low rates of radiation-induced polymerization are observed at low temperatures in the liquid phase.

Post-irradiation polymerization does not occur in either monomer phase below 133 K.[56, 58] After X-irradiation of the low-temperature form at 77 K post-polymerization does not occur until a temperature of 160 K is reached, when 4% conversion accompanies the first-order transition to the high-temperature phase.[56] Slow warming of the irradiated metastable high-temperature phase produces 1% conversion to polymer at 138 K, when the monomer recrystallizes to the stable low-temperature form. On the other hand, rapid warming of the latter samples from 77 K to temperatures above 160 K, *i.e.* to temperatures where the high-temperature form is stable, produces about 10% polymer formation.

Various explanations have been advanced to explain the existence of low-temperature polymerizations with almost zero activation energy and the sudden variations in rate with temperature. These explanations, which we have reviewed elsewhere,[6, 7] usually involve some form of concerted mechanism. However, they run into difficulties; for example, they do not explain the

[57] Y. Tabata, S. Shu, and K. Oshima, *Mol. Cryst. Liquid Cryst.*, 1969, **9**, 436.
[58] I. M. Barkalov, V. I. Gol'danskii, N. S. Enikolopyan, S. F. Terekhova, and G. M. Trofimova, *J. Polymer Sci.*, (C) 1964, **4**, 897.
[59] Y. Tabata, *J. Macromol. Sci.* (A), 1967, **1**, 1407.
[60] Y. Tabata, *J. Macromol. Sci.* (A), 1968, **2**, 919.
[61] Y. Tabata, *J. Macromol. Sci.* (A), 1968, **2**, 931.
[62] I. M. Barkalov, V. I. Gol'danskii, and V. G. Rapaport, *Doklady Akad. Nauk, S.S.S.R.*, 1965, **161**, 1268.
[63] M. A. Bruk, V. F. Gromov, I. V. Chernyak, P. M. Khomikovskii, and A. D. Abkin, *Vysokomol. Soedineniya*, 1966, **8**, 961.
[64] J. Danon, A. Dworkin, and M. Roussel, *J. Chim. phys.*, 1969, **66**, 879.

low limiting conversions. Nor do they explain the observations of Barkalov et al.,[65] that polymerization of vinyl acetate at 77 K, which is said to resemble that of acrylonitrile, is more rapid in the glass than in the crystalline state, suggesting that disorder favours polymerization. There appears to be no good reason for adopting a more sophisticated approach to these reactions than that developed for other vinyl monomers. The increases in rate with increasing temperature can probably be correlated with variations in molecular mobility. Indeed, Bensasson et al.[56] suggested that post-irradiation polymerization is restricted to the high-temperature phase in which there is probably more freedom for molecular movement, or is associated with the movements of molecules during a phase transformation.

The low limiting conversions observed for the in-source polymerization of acrylonitrile at very low temperatures, and similar observations on vinyl acetate,[65] hexamethylcyclotrisiloxane,[66] and tetrafluoroethylene,[63] are understandable if reaction occurs in a limited number of defects, which may be sufficiently extensive to contain mobile molecules at these temperatures or may be thermally activated by exciton decay. Barkalov et al.[67] have considered the 'loosening' of molecules during energy transfer processes. Continued polymerization to high conversion may be prevented at low temperatures if the crystal lattices are so hard that strains produced during polymerization cannot be relieved by processes such as dislocation multiplication; in these circumstances reaction is restricted to the defects initially present.

The use of the term 'low temperatures' thus becomes purely relative since the remarks made above are equally applicable to many solid-phase polymerizations which proceed only to extremely low conversions at a given temperature.

Topotactic Polymerization.—Probably the most useful definition of topotaxy in this context is that of Glassler et al.[68] which states that 'for true topotaxy there must be some three-dimensional correspondence between the structures of the product and its host'. The formation of a topotactic product does not necessarily imply the occurrence of a topochemical process, which would require reaction to be controlled by the perfect lattice. In view of the preceding discussion we cannot expect the chain polymerization of any simple vinyl monomer to produce a crystalline polymer with an exact correspondence between the crystal structures of the polymer and monomer. However, it is always possible, given suitable intra- or inter-molecular interactions, for the polymerization of such a monomer to yield a polymer with some degree of crystallinity or orientation.

[65] I. M. Barkalov, V. I. Gol'danskii, N. S. Yenikolopyan, S. F. Terekhova, and G. M. Trofimova, *Vysokomol. Soedineniya*, 1964, **6**, 98.
[66] I. M. Barkalov, D. A. Gareyeva, V. I. Gol'danskii, N. S. Yenikolopyan, and A. A. Berlin, *Vysokomol. Soedineniya*, 1966, **8**, 1140.
[67] I. M. Barkalov, V. I. Gol'danskii, N. S. Yenikolopyan, S. F. Terekhova, and G. M. Trofimova, *Vysokomol. Soedineniya*, 1964, **6**, 92.
[68] L. S. D. Glassler, F. P. Glassler, and H. F. W. Taylor, *Quart. Rev.*, 1962, **16**, 343.

Polymerization of crystalline vinyl stearate produces a partially crystalline polymer.[69] Monomeric vinyl stearate crystals are monoclinic with the vinyl groups situated in well-separated layers; the paraffin chains are in an orthorhombic sub-cell. The orientation of the stearate residues is retained during polymerization but their packing changes from orthorhombic to hexagonal. (This change in structure often accompanies an increase in molecular mobility in long-chain hydrocarbons as the melting point is approached,[70] and involves an overall expansion in the sub-cell area perpendicular to the long axis of the paraffin chains.) At 30% conversion the presence of the monomer lattice can still be detected, but at 40% conversion all stearate residues are packed in the hexagonal arrangement. The polymer is not stereoregular,[69] so that the backbones are unable to crystallize and crystallinity must be restricted to the side-chains; it is known that non-stereoregular vinyl stearate will undergo side-chain crystallization.[71] No evidence of topochemical control was obtained[69] and it has been suggested that the backbone of the polymer and terminal units of the stearate residues are in a disordered phase.[7]

p-Benzamidostyrene, which polymerizes thermally in the crystalline state,[72] also has large strongly-interacting substituent groups. When polymerized below the glass transition temperature of the polymer (393 K) the i.r. dichroism of the N—H stretching vibration does not change. Jakabhazy *et al.*[72] have remarked that some molecular motion must occur during reaction to accommodate the change in configuration of the carbon atoms of the vinyl group with retention of the orientation of the N—H bonds. No information on the direction of polymer growth has been obtained but it has been suggested that the vinyl groups are situated in well-separated layers in the monomer crystal and that polymerization is essentially a two-dimensional process.

The polymerizations of vinyl stearate and *p*-benzamidostyrene are two cases in which the polymerization of a vinyl monomer leads to a product with some degree of crystallinity or orientation. In neither case is there any evidence of topochemical control of polymerization. Electron microscopy has demonstrated that during the polymerization of the latter monomer protuberances develop on the surfaces of the crystals, and these may indicate the points of emergence of dislocations in the monomer crystals.

Interesting examples of topotactic polymerizations which have been investigated recently are those of 2,5-distyrylpyrazine (DSP) and related compounds.[73–76] DSP (1) and *trans,trans*-1,4-bis-(β-pyridyl-2-vinyl)benzene (2)

[69] N. Morosoff, H. Morawetz, and B. Post, *J. Amer. Chem. Soc.*, 1965, **87**, 3035.
[70] E. R. Andrew, *J. Chem. Phys.*, 1950, **18**, 607.
[71] D. A. Lutz and L. P. Witnauer, *J. Polymer Sci. (B)*, 1964, **2**, 31.
[72] S. Z. Jakabhazy, H. Morawetz, and N. Morosoff, *J. Polymer Sci. (C)*, 1964, **4**, 805.
[73] M. Hasegawa and Y. Suzuki, *J. Polymer Sci. (B)*, 1967, **5**, 815.
[74] M. Iguchi, H. Nakanishi, and M. Hasegawa, *J. Polymer Sci. (A-1)*, 1968, **6**, 1055.
[75] M. Hasegawa, Y. Suzuki, F. Suzuki, and H. Nakanishi, *J. Polymer Sci.*, (A-1) 1969, **7**, 743.
[76] F. Suzuki, Y. Suzuki, H. Nakanishi, and M. Hasegawa, *J. Polymer Sci.* (A-1), 1969, **7**, 2319.

$$\text{Ph-CH=CH-(2-pyrazinyl)-CH=CH-Ph} \quad (1)$$

$$\text{(2-Py)-CH=CH-(1,4-C}_6\text{H}_4\text{)-CH=CH-(2-Py)} \quad (2)$$

have the same crystal structure and undergo step-wise photopolymerization in the solid phase to produce very crystalline high polymers with cyclobutane rings in the main chain.[74,75] These reactions are analogous to the photo-dimerizations of cinnamic acid derivatives.[77] 1,4-Bis-(β-pyridyl-3-vinyl)-benzene and 1,4-bis-(β-pyridyl-4-vinyl)benzene have different crystal structures and do not polymerize. DSP has been investigated in most detail.[73—76, 78] Photolysis of crystalline DSP ($\lambda < 400$ nm) produces a crystalline oligomer which on irradiation with shorter wavelengths yields a crystalline polymer.[79] If the oligomer is recrystallized, presumably with formation of some other crystal structure, it will not yield high polymer. Similarly, photo-irradiation of amorphous oligomer does not lead to high polymer.[79] Crystals of monomeric DSP consist [80] of columns of monomer molecules arranged parallel to the crystal c-axis, with their long axes tilted at an angle to the c-axis. The polymer chains grow parallel to the c-axis.[74] Since the centres of gravity of the monomer molecules remain almost stationary ($< 2\%$ contraction along the c-axis), polymerization apparently only requires the molecules to tilt further about their short axes (parallel to the b-axis) producing an 11% contraction along the a-axis.[81] To allow for the change in configuration of the carbon atoms on polymerization there is a 12% expansion along the b-axis. On the evidence available this reaction appears to fulfil the requirements for a topochemical process more closely than any other solid-state polymerization. The data are also consistent with reaction in dislocations since the predominant dislocation lines would run parallel to the c-axis, but there is no evidence that such an interpretation is necessary, unless it is found that the tilting of the monomer molecules and the expansion in the b-axis direction cannot take place in the perfect lattice with suitable co-operative motions.

The elegant work of Wegner and his colleagues [81a] over the last few years has demonstrated the topotactic photo- and thermal-polymerization of crystalline monomers containing conjugated triple bonds, notably those of types (3) and (4).

[77] M. D. Cohen and G. M. J. Schmidt, *J. Chem. Soc.*, 1964, 1996.
[78] H. Nakanishi, Y. Suzuki, F. Suzuki, and M. Hasegawa, *J. Polymer Sci.* (*A*-1), 1969, **7**, 753.
[79] M. Hasegawa, Y. Suzuki and T. Tamaki, *Bull. Chem. Soc. Japan*, 1970, **43**, 3020.
[80] Y. Sasada, H. Shimanouchi, H. Nakanishi, and M. Hasegawa, *Bull. Chem. Soc. Japan*, in the press.
[81] M. Hasegawa, private communication.
[81a] G. Wegner, *J. Polymer Sci.* (*B*), 1971, **9**, 133. This paper gives earlier references.

R—C≡C—C≡C—R (3)

(4) [aromatic ring with R¹]—C≡C—C≡C—[aromatic ring with R¹]

These compounds are only polymerizable in the solid phase if the R or R^1 groups interact strongly by hydrogen-bond formation or dipolar forces, and if the aromatic rings in (4) are *o*- or *m*-disubstituted. The polymers formed are crystalline and fully conjugated and therefore highly coloured. Individual monomer molecules are arranged in a ladder-like fashion in the lattice, the conjugated triple-bond systems being the (tilted) rungs which are held together by the interacting R or R^1 groups. On polymerization, the angle of tilt of the rungs increases, *i.e.* the ladder is sheared. This type of mechanism is consistent with the structural requirements mentioned [81a]. These polymerizations are probably free-radical in character, and proceed by 1,4-additions.

Studies of the structure of poly[2,4-hexadiynylenebis(phenylurethane)] (3; R = CH$_2$O·CO·NHPh) prepared by solid-phase polymerization have been reported by Hädicke *et al.*[81b] The polymer contains alternate double and triple bonds in conjugation, with an all-*trans* structure. No essential change in the crystal lattice occurs on polymerization.

Evidently, the polymerizations of DSP and derivatives, and of the diacetylenes, are sterically much more favourable for topochemical reaction than the polymerizations of simple olefins (See Vinyl Polymerization).

Polymerizations through sequences of monomer molecules in urea- and thiourea-canal complexes have received some attention.[82-87] Restrictions on the orientation of monomer molecules often control the polymer structures. Thus, 2,3-dimethyl-1,3-butadiene, 2,3-dichloro-1,3-butadiene, and 1,3-cyclohexadiene undergo γ-initiated polymerization in their thiourea clathrates to produce almost exclusively *trans*-1,4-polybutadiene derivatives.[82] Stephan *et al.*[84] found some evidence for the presence of monocyclic structures in the polymer from 2,3-dimethyl-1,3-butadiene. The polymers are crystalline and can be isolated as needles, with the polymer molecules maintaining their original orientation along the needle axis. Smaller monomers, 1,3-butadiene and vinylacetylene, do not give stereoregular polymers on reaction in the thiourea clathrates,[82] but by the use of urea clathrates it is possible to

[81b] E. Hädicke, E. C. Mez, C. H. Krauch, G. Wegner, and J. Kaiser, *Angew. Chem. Internat. Edn.*, 1971, **10**, 266.
[82] I. F. Brown and D. M. White, *J. Amer. Chem. Soc.*, 1960, **82**, 5671.
[83] D. M. White, *J. Amer. Chem. Soc.*, 1960, **82**, 5678.
[84] V. Stephan, J. Vodhenal, I. Kössler, and N. G. Gaylord, *J. Polymer Sci.* (*A*-1), 1967, **5**, 503.
[85] L. Kiss, paper presented at I.U.P.A.C. Symposium, Prague, 1965.
[86] V. S. Ivanov, T. A. Sukhikh, Yu. V. Medvedev, A. Kh. Berger, V. B. Osipov, and V. A. Gol'din, *Vysokomol. Soedineniya*, 1964, **6**, 782 (*Polymer Sci., U.S.S.R.*, 1964, **6**, 856).
[87] Y. Chatani, S. Nakatani, and H. Tadokoro, *Macromolecules*, 1970, **3**, 481.

prepare the all-*trans* polymer from 1,3-butadiene and crystalline syndiotactic polyvinyl chloride and crystalline polyacrylonitrile.[83] Little is known about the mechanisms of these polymerizations, but investigations have been started by Chatani *et al.*[87] who have carried out an X-ray crystallographic study of the 2,3-dichloro-1,3-butadiene–thiourea system. These workers have shown that the thiourea matrix is maintained during polymerization and that the length of canal occupied per monomer unit decreases from 6.3 Å to 4.8 Å, showing that polymerization involves movement of monomer molecules along the canals. Conformations of the polymer in the clathrate and the polymer crystal are identical.

The Influence of Additives.—We have considered elsewhere (see E.S.R. Studies) some influences of additives at impurity levels of concentration. Here we shall consider two systems in which high concentrations of additives are present.

Acrylamide and propionamide form solid solutions in all proportions.[16] In solid-phase polymerizations at 298 K, the rates of fractional conversion in solid solutions containing up to 10% propionamide are identical with those of pure monomer, and for conversion below 10% are identical for solid solutions containing 50% propionamide.[16] However, the molecular weights of the polymers formed in solid solutions were found to be lower. A tentative explanation for this has been presented elsewhere.[6, 7] Adler and Reams [11] obtained 100% conversion of acrylamide in solid solutions containing as little as 10% acrylamide, and 60% conversion in systems containing only 4% acrylamide; in all cases the products were high polymers. These workers suggested that a radical can react with any one of a number of near-neighbour molecules. As polymerization proceeds, recrystallization occurs at the polymer–monomer interface and is accompanied by diffusion of acrylamide; thus, more of the monomer becomes available for reaction, and the high conversions are understandable. It has also been suggested that at very low acrylamide contents there may be some tendency for acrylamide to concentrate around dislocation lines in the propionamide lattice, thus minimizing molecular motions necessary for reaction.[55] If, under these conditions, there is little recrystallization, some acrylamide in the more ordered regions of the lattice may be unable to react, so that there is a limiting conversion.

Methacrylamide and isobutyramide do not form solid solutions over the whole range of composition[88] but produce an intermediate phase with a congruent melting point at about 65% methacrylamide, and two eutectics (m.p. ~366 K). The intermediate phase can accommodate either component in the form of solid-solutions over a range of compositions. X-Ray diffraction studies show that the d spacings vary with the composition of the reaction mixture.[88] On continuous irradiation of the solid solutions at 298 K the methacrylamide polymerizes and the X-ray diffraction patterns change. An amorphous halo produced by the polymer builds up, while the

[88] A. Faucitano and G. Adler, *J. Macromol. Sci.* (*A*), 1970, **4**, 261.

intensity of the solid-solution pattern decreases. Simultaneously, the d spacings from the solid solutions change, as anticipated from the depletion of monomer in the lattice, and under suitable conditions phase changes have been observed. The changes in the diffraction patterns have been studied in detail and they demonstrate quite clearly that considerable recrystallization and annealing accompanies solid-phase polymerization in the solid solutions, relieving the stresses produced in the crystals as a result of polymerization.[88]

Effects of Pressure.—A number of workers [57, 89-96] have studied the effects of high pressures (300—40 000 atm) on solid-phase polymerizations of both vinyl and cyclic monomers, usually in the polycrystalline state. Although this work has produced much interesting information, we do not discuss it in detail here since interpretation of the results and their relevance to the rôle of defects in solid-phase polymerization are unclear. We have summarized some of the findings elsewhere.[6, 7]

We have discussed (see Nature of Reaction Sites) the sensitivity of the solid-phase polymerizations of acrylic and methacrylic acids to small stresses (~10 atm) as part of the basic evidence for the occurrence of polymerization in dislocations. The remarkable reductions in rate under continuous irradiation which occur when stress is applied at about 50% conversion [12] require the stress to be efficiently transmitted throughout the monomer phase. This may necessitate some form of recrystallization during polymerization to maintain a coherent monomer phase in which dislocations can exhibit their normal structures and properties. Alternatively, stress may be transferred through the original monomer crystal containing small inclusions of polymer. Although the effects of these stresses are large in both acrylic and methacrylic acids at about 277 K, they diminish at higher and lower temperatures.[26] The stress-effect in acrylic acid completely disappears at 280 K, at which temperature the monomer is in the pre-melting region (as evidenced by a rapid decrease in monomer birefringence with increasing temperature[12]); and under these conditions there is probably such high mobility and local disorder that the discrete structures associated with dislocations at lower temperatures are lost and the defects cannot respond to the stress. The stress-effect in methacrylic acid also disappears when the reaction temperature is decreased to 259 K. Under these conditions reaction is very slow, and it is supposed that the crystals are so hard that small stresses are unable to move the dislocations; this phenomenon may be related

[89] Y. Tabata and T. Suzuki, *Makromol. Chem.*, 1965, **81**, 223.
[90] P. J. Fydelor and A. Charlesby, *J. Polymer Sci.* (C), 1969, **16**, 4493.
[91] Y. Tabata, T. Miyairi, S. Katsura, Y. Ito, and K. Oshima, in 'Large Radiation Sources for Industrial Processes', Internat. Atomic Energy Agency, Vienna, 1969, p. 233.
[92] M. Prince and J. Hornyak, *J. Polymer Sci.* (B), 1966, **4**, 493.
[93] M. Prince and J. Hornyak, *J. Polymer Sci.* (A-1), 1967, **5**, 531.
[94] T. Suzuki, Y. Tabata, and K. Oshima, *J. Polymer Sci.* (C), 1967, **16**, 1921.
[95] H. Rao and D. S. Ballantine, *J. Polymer Sci.* (A), 1965, **3**, 2579.
[96] T. Suzuki, Y. Tabata, and K. Oshima, *Makromol. Chem.*, 1967, **104**, 236.

to the low rate of reaction resulting from a low rate of dislocation multiplication.

Concentrations of isobutyric acid in methacrylic acid as low as 0.1% cause a marked reduction in the stress-effect at 277 K and 0.3% isobutyric acid almost eliminates it.[26] At 259 K, addition of 0.1% isobutyric acid causes a reappearance of the effect, but further increase to 0.5% reduces the sensitivity to stress. All these data may be interpreted in terms of concentration of the impurity around the defects and an increase in local molecular mobility, as deduced from the results of e.s.r. studies (see relevant section earlier). Addition of the impurity is equivalent to a local increase in temperature in these respects. The failure of the impurity cloud to pin down the dislocations may arise from the proximity of the reaction temperatures to the melting points of the crystals. A significant difference between the behaviour of impurities in experiments on the effect of stress and in the low-temperature e.s.r. studies, is that, in the latter, the influence of the impurity is lost at very low extents of reaction while in the former it is maintained to high conversions. This difference is understandable if some recrystallization accompanies polymerization at the higher temperatures but is impossible at the lower temperatures where molecular mobility is lower.

4 Polymerization of Cyclic Monomers

Solid-phase polymerization of cyclic monomers has been known since the 1930's, when Kohlschütter and Sprengel [97] observed that, in the presence of formaldehyde vapour, a crystalline polymer formed on the surfaces of trioxan crystals and that the polymer was orientated along the c-axis of the trioxan crystals. This early work was 'lost' and not until some 30 years later, when Okamura *et al.* observed that the radiation-induced solid-phase polymerizations of trioxan,[2] β-propiolactone,[98] diketen,[99] and 3,3-bis-(chloromethyl)oxetan [100] yielded crystalline polymers, was an active interest shown in this type of reaction. These observations, especially the finding that the polymers were highly oriented in the monomer crystals, immediately lent strong support to the idea of lattice control in solid-phase polymerizations. The c-axis of the crystalline polyoxymethylene formed by polymerization in a single crystal of trioxan was found to be parallel to the c-axis of the original monomer crystal.[3] Similarly, the crystalline polymer fibres produced by the polymerizations of crystalline β-propiolactone and 3,3-bis-(chloromethyl)oxetan are aligned at 90° and 45° to the growth directions of the monomer crystals, respectively.[3]

Lattice control of polymerization also received support from early kinetic

[97] H. W. Kohlschütter, *Annalen*, 1930, **482**, 75; H. W. Kohlschütter and L. Sprengel, *Z. phys. Chem. (B)*, 1932, **16**, 284.
[98] S. Okamura, K. Hayashi, Y. Kitanishi, and M. Nishii, *Isotopes and Radiation*, 1960, **3**, 510.
[99] S. Okamura, K. Hayashi, and Y. Kitanishi, *Isotopes and Radiation*, 1960, **3**, 346.
[100] S. Okamura, K. Hayashi, and H. Watanabe, *Isotopes and Radiation*, 1961, **4**, 73.

observations. It was reported that radiation-induced polymerizations of the cyclic monomers would not proceed in the liquid phase,[101, 102] the rates of polymerization in the solid state increasing with temperature to a maximum just below the monomer melting point then decreasing rapidly to virtually zero in the liquid. Maximum rates were observed at 203 K in β-propiolactone (m.p. 240 K) and diketen (m.p. 208 K) and near 281 K in 3,3-bis(chloromethyl)oxetan (m.p. 291 K); the reported temperatures varied somewhat with radiation dosage. Similar kinetic results have also been obtained with hexamethylcyclotrisiloxane.[103] Hence, it was suggested that the monomer crystal structure plays a dominant rôle and that polymerization occurs only in the perfect lattice. Support for this view was adduced from reports that both in-source and post-irradiation polymerizations of these monomers proceed to limiting conversions of less than 100%.[95, 101,102, 104—109] It was supposed that aggregation of the individual polymer molecules into bundles caused exclusion and disorientation of residual monomer which, as a result of its disordered nature, was unable to polymerize.[104] Attempts were made to analyse polymerization kinetics on this basis.[104, 110, 111] Although such mechanisms could predict the general shape of the kinetic curves they were unsatisfactory in some respects; in the polymerization of diketen at 195 K, for example, it was necessary to assume the disorientation of 11.4 monomer molecules for every act of propagation.[111]

Since these early reports, considerable interest has been shown in the solid-phase polymerization of cyclic monomers. Much of the effort has been devoted to investigations of relationships between the crystal structures of the monomers and the directions of growth and crystal structures of the polymers. Some of this work provides a measure of support for propagation through the lattice, while the remainder raises doubts of the validity of this simple approach. The significance of the limiting conversions observed is debatable. Under apparently identical conditions, Hayashi et al.[102] reported maximum conversions in the polymerization of trioxan of 17% at 323 K and 55% at 328 K, while Sakamoto et al.[109] obtained maximum conversion of 80% over the temperature range 318—328 K. Some authors[101, 106] have reported

[101] S. Okamura and K. Hayashi, *J. Chim. phys.*, 1962, **59**, 429.
[102] K. Hayashi, H. Ochi, and S. Okamura, *J. Polymer Sci. (A)*, 1964, **2**, 2929.
[103] E. J. Lawton, W. T. Grubb, and J. S. Balwit, *J. Polymer Sci.*, 1956, **19**, 455; W. Burlant and C. Taylor, *J. Polymer Sci.*, 1959, **41**, 547.
[104] S. Okamura, K. Hayashi, and Y. Kitanishi, *J. Polymer Sci.*, 1962, **58**, 925.
[105] A. Chapiro and S. Penczek, *J. Chim. phys.*, 1962, **59**, 696.
[106] S. Nashakio, M. Kondo, H. Tsuchita, and M. Yamada, *Makromol. Chem.*, 1962, **52**, 79.
[107] C. David, J. Van der Paaren, F. Provcost, and A. Ligotti, *Polymer*, 1963, **4**, 341.
[108] N. S. Marans and F. A. Wessells, *J. Appl. Polymer Sci.*, 1965, **9**, 3681.
[109] M. Sakamoto, I. Ishigaki, A. Shimizu, M. Kunakura, M. Nishii, H. Yamashino, T. Iwai, and A. Ito, paper presented at I.U.P.A.C. Macromolecular Symposium, Japan 1966.
[110] K. Hayashi and S. Okamura, *Makromol. Chem.* 1961, **47**, 230.
[111] A. Charlesby, *Reports Progr. Phys.*, 1965, **28**, 463.

limiting conversions of approximately 10% with 3,3-bis(chloromethyl)-oxetan; however, Kagiya et al.[112] found no evidence of a limiting conversion with carefully purified monomer. Addition of water or *sym*-dichloroacetone gave rise to this phenomenon, and conversely, addition of drying agent to the unpurified monomer enhanced the maximum yield,[113] presumably by removal of water. Clearly, adventitious impurities may play an important part in these reactions. The reported failure of cyclic monomers to polymerize in the liquid phase may be the result of inhibition by impurities; more recently it has been found that trioxan and 3,3-bis(chloromethyl)oxetan when pure will undergo radiation-induced polymerization in the liquid phase,[114] but the rates are very sensitive to traces of impurities. It seems that impurities have been present in most of the early investigations. As would be anticipated, observed rates and conversions have varied with the conditions of crystallization of the monomers. Rao and Ballantine took considerable care to standardize their experimental conditions, but were unable to obtain reproducible kinetic data.[95]

The effects of adding impurities, such as water and methylene chloride, to trioxan have been summarized by Ito and Hayashi,[115] who have also summarized information on polymer molecular weights. The nature of the atmosphere has a profound effect on the post-irradiation polymerization of trioxan,[112] much greater conversions being achieved if irradiation and polymerization are carried out in air rather than in vacuum. Rao and Ballantine[95] state that only a trace of polymer is obtained if both irradiation and polymerization are carried out in vacuum. Such results indicate that an atmospheric constituent is involved in initiation, probably entering and leaving the crystal by diffusion along imperfections. Radiolysis of 3,3-bis(chloromethyl)oxetan produces carbon dioxide, and additives which remove this product are reported to accelerate polymerization.[113] Herz and Stannett[116] state that trithian crystallized from chloroform solution will undergo radiation-induced solid-phase polymerization, while pure trithian and samples recrystallized from dimethyl formamide will not polymerize. These various data therefore suggest that impurities and imperfections may have a large rôle to play in solid-phase polymerization of cyclic monomers.

In addition to the monomers already mentioned, trithian,[117, 118] triselenan,[119] tetroxocan,[120, 121] and pentoxecan[122] have all been reported to undergo

[112] T. Kagiya, M. Izu, and K. Fukui, *J. Polymer Sci.* (*B*), 1964, **2**, 93.
[113] T. Kagiya, M. Izu, and K. Fukui, *Makromol. Chem.*, 1967, **102**, 39.
[114] K. Ueno, H. Tsukamoto, K. Hayashi, and S. Okamura, *J. Polymer Sci.* (*B*), 1967, **5**, 395.
[115] A. Ito and K. Hayashi, *Hydrocarbon Processing*, 1968, **47**, 197.
[116] J. E. Herz and V. Stannett, *J. Polymer Sci.* (*B*), 1966, **4**, 995.
[117] J. B. Lando and V. Stannett, *J. Polymer Sci.* (*A*), 1965, **3**, 2369.
[118] G. Carazollo, M. Mammi, and G. Valle, *Makromol. Chem.*, 1967, **100**, 295.
[119] G. Carazollo and M. Mammi, *Makromol. Chem.*, 1967, **100**, 28.
[120] K. Hayashi, M. Nishii, and S. Okamura, *J. Polymer Sci.* (*C*), 1963, **4**, 839.
[121] Y. Chatani, T. Uchida, H. Tadokoro, K. Hayashi, and S. Okamura, *J. Macromol. Sci.* (*B*), 1970, **2**, 567.
[122] Y. Chatani, K. Kitahama, H. Tadokoro, T. Yamauchi, and Y. Miyake, *J. Macromol. Sci.* (*B*), 1970, **4**, 61.

solid-phase polymerization with formation of crystalline polymers, the direction of growth of the polymer chains lying along preferred crystallographic directions in the monomer lattices. Such evidence would appear to support the contention that the monomer lattices control the polymerizations. The crystallographic data themselves raise certain problems in this connection. Growth of crystalline polymer does not necessarily occur along a single direction in the monomer lattice. Polyoxymethylene derived from trioxan,[120, 123] tetroxocan,[120, 121] and pentoxecan,[122] and polythiomethylene derived from trithian [117, 118] have 'twinned' structures, with preferred directions of polymer growth and formation of small amounts of polymer corresponding to growth in other specific directions. Thus, polymerization is not restricted to a unique arrangement of monomer molecules in their crystallographic lattices. There is also the difficulty that the crystal structures of the monomers and polymers are not compatible. Polyoxymethylene produced by the polymerizations of trioxan, tetroxocan, and pentoxecan crystallizes in the form of a 9/5 helix,[121, 122] or the virtually identical 29/16 helix,[123] in which the distance along the c-axis is 1.93 Å per —CH_2—O— repeat unit. Formation of crystalline polyoxymethylene usually involves a contraction perpendicular to the c-axis of the polymer but an expansion along this axis. Chatani et al.[121, 122] have summarized the intermolecular distances in the crystal structures of the three monomers. For trioxan the available distance per —CH_2—O— unit in the direction of major polymer growth (along the monomer c-axis) is 1.39 Å, while in tetroxocan the corresponding distance (along the monomer b-axis) is only 1.05 Å, and in pentoxecan (along the diagonals of the ab-plane) it is 1.35 Å. Clearly, even the addition of a single monomer unit (*i.e.* three —CH_2—O— units in the case of trioxan and four for tetroxocan) cannot possibly occur in the main polymer growth directions in the perfect lattices without considerable disruption of the lattices.

To overcome these difficulties Chatani et al.[121] considered the possibility that polymerization of tetroxocan may occur randomly with subsequent crystallization of the polymer, but they could find no reason why the polymer should then show such a high degree of orientation. It was suggested that in trioxan the formation of sub-crystals in the subsidiary growth direction, where the available distance per —CH_2—O— unit is 1.87 Å, could help to relieve the stress produced by main crystal formation. In the case of tetroxocan the same workers considered various molecular motions which might occur in the monomer lattice to facilitate polymer growth. With pentoxecan both main and sub-crystal formation occurs along the directions of closest approach of monomer molecules. In the polymerization of trioxan the amount of sub-crystal formation relative to main crystal formation decreases as the polymerization temperature is raised, and it has been suggested[121] that the stress responsible for sub-crystal formation could be relieved at the higher temperatures by thermal motions and lattice distortion.

Since polymerization in these crystals obviously involves considerable

[123] G. Carazzolo, S. Leghissa, and M. Mammi, *Makromol. Chem.*, 1963, **60**, 171.

molecular reorientation, it is relevant to consider the relationship between molecular mobility and the polymerization reaction. At room temperature trioxan is a classic example of a plastic crystal. Its rate of polymerization, low below 293 K, increases rapidly with increasing temperature above 293 K, and still more rapidly above about 313 K.[101] Broad-line n.m.r. studies show the existence of two transitions associated with increases in molecular motions at 293 K and 313 K.[124, 125] There is, therefore, a strong correlation between molecular mobility and ease of polymerization; above 313 K a small narrow line in the centre of the n.m.r. spectrum indicates the presence of extensive molecular motions of some of the molecules. Most investigations of the polymerization of single crystals have been carried out near 323 K. Although facile molecular motions may assist the reorientations required for polymerization, they can hardly be reconciled with lattice propagation.

Adler has obtained interesting information from experiments with polycrystalline samples of trioxan in the form of a thin layer of single crystals.[126] Polymerization at 297 K gave a small amount of crystalline fibrous polymer on the crystal surfaces; the fibres exhibited no preferred direction and even changed direction. At 319 K extensive post-irradiation polymerization occurred, producing bundles of crystalline fibrous polymer. The fibres grew across intercrystalline boundaries and intervening space with little or no change in direction, although the crystals through which they grew had different orientations. Apparently, the propagation process is uninfluenced by the orientation of the monomer crystals and of the monomer molecules. Polymerization at 327 K gave polymer fibres aligned almost exclusively along the c-axes of the residual monomer crystals; this phenomenon, however, arose from extensive recrystallization of the monomer around the polymer. Similar results have also been obtained in the polymerization of trithian.[126]

Adler [126] provided a partial reconciliation of these observations with those obtained from single-crystal studies by suggesting that polymerization starts along a preferred crystallographic direction, while subsequent growth proceeds by the addition of monomer to a growing chain which is already in the form of a helix, irrespective of the orientation of the monomer molecules. The high level of molecular mobility no doubt allows reorganization of the monomer around the growing ends of the polymer chains. On this view it is the crystallizability of the polymer, rather than the crystal structure of the monomer, which controls the nature of the product. It is notable that all the polymerizations of cyclic monomers we have so far discussed yield polymers which have no asymmetric carbon atoms and which are readily crystallizable. The polymer of hexamethylcyclotrisiloxane crystallizes much less readily and solid-phase polymerization of this monomer gives only an amorphous product.[103]

[124] K. Hayashi and S. Okamura, U.S. Atomic Energy Comm., TID 7643, 1962, p. 150.
[125] A. Komaki and T. Matsumoto, *J. Polymer Sci.* (*B*), 1963, **1**, 671.
[126] G. Adler, *J. Polymer Sci.* (*A*–1), 1966, **4**, 2883.

Adler's view that polymerization commences in a certain direction while propagation proceeds independently of crystal orientation implies that reaction is initiated at a specific type of site, presumably an imperfection, and that it is the nature of this site which controls the direction in which polymer chains are initially orientated. Bassett [127] has produced photomicrographic evidence showing that the {001} sub-grain boundaries can be specific sites for initiating polymerization, and Cannavo et al.[128] have presented photomicrographic evidence that the polymerization of trioxan does not proceed uniformly throughout the crystals but is associated with defects.

Eastmond [129] has suggested that further reconciliation of the various data can be achieved by considering dislocations as preferred reaction sites. Comparison of the crystal structures of trioxan, tetroxocan, and pentoxecan with the directions of polymer growth in single crystals of the monomers (Figure 5) shows that the growth directions of both the main and sub-crystals of the polyoxymethylene are parallel to natural slip-planes in the monomer crystals. Examination of the crystal structures reveals that the c-axes of polyoxymethylene are also located in planes suitable for the introduction, or removal, of a half-plane of monomer molecules, as required for the formation of an edge-dislocation. Consequently, the c-axes of the polymer crystals are situated along probable directions of dislocation lines in the monomer crystals. The generation of edge-dislocations with unit Burgers vector in crystals of these monomers requires the introduction, or removal, of more than one half-plane of monomer molecules. In a real crystal it is not necessary for the half-planes to be adjacent to each other, and the crystals may contain a series of dislocation structures continuing for considerable distances through the monomer crystals in the direction of polymer growth. As in all other cases, dislocations are energetically favourable reaction sites. Thus, if the energy transfer processes occur predominantly in dislocations, as in the u.v.-initiated polymerizations of acrylic acid,[34] we can understand the various experimental data. Propagation would be expected to proceed more easily along the dislocation line since the molecules in these situations have higher mobility and the generation of strain in the surrounding lattice is minimal. The polymers tend to crystallize in the direction of the dislocation lines.

The relative amounts of main- and sub-crystal formation may reflect the relative densities of the various dislocation structures or the relative ease of reaction in these structures. The temperature variation of the relative amounts of these species may arise from changes in the relative populations of different structures suitable for initiation. We should expect that, under conditions of high molecular mobility, polymer growth which started in a certain direction would proceed linearly by addition of monomer to the crystalline polymer, thus producing results of the type observed by Adler.[126] The irreproducibility of the kinetic data and the sensitivity of the

---

[127] D. C. Bassett, *Nature*, 1967, **215**, 731.
[128] C. Cannavo, J. Deschamps, K. Hayashi, and C. Sella, *Compt. rend.*, 1968, **266**, C, 777.
[129] G. C. Eastmond, *Polymer J.*, in the press.

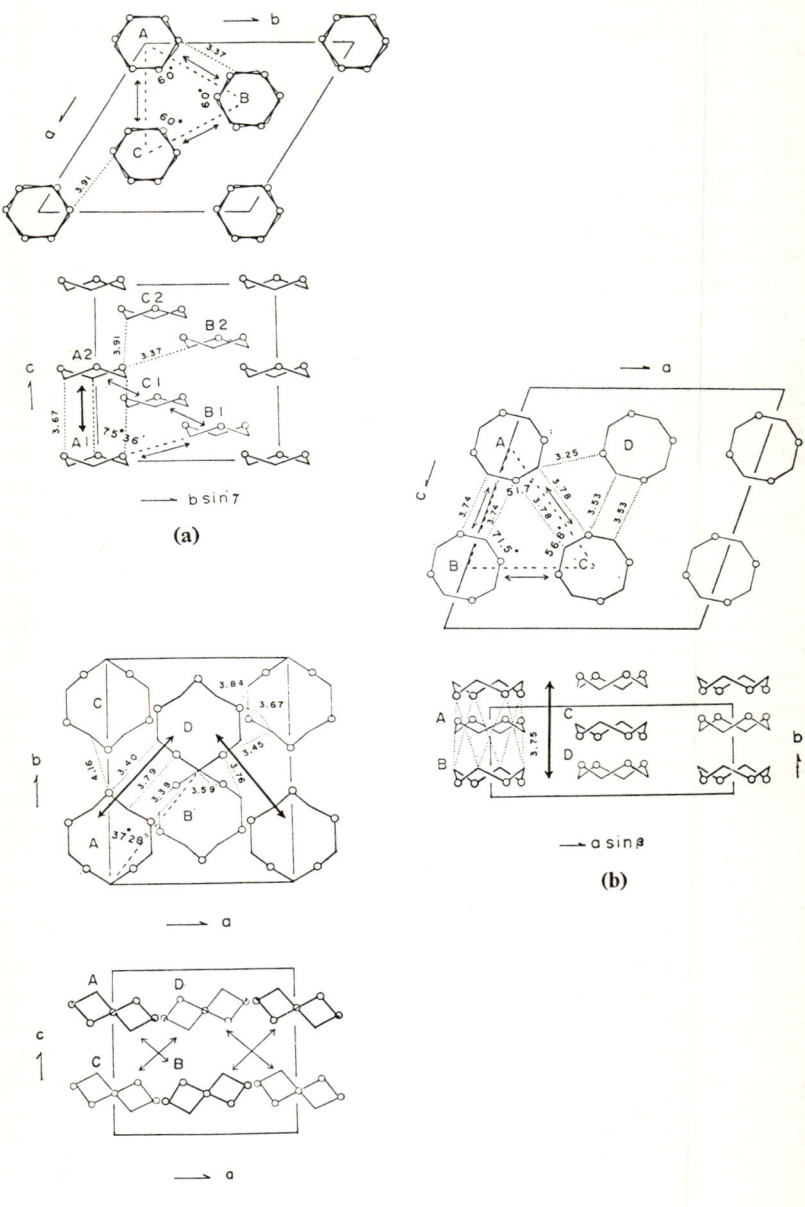

Figure 5 *Crystal structures of* (a) *trioxan,* (b) *tetroxocan, and* (c) *pentoxecan. Thick and thin arrows indicate the main and subsidiary directions of polymer growth, respectively*
[Reproduced by permission from *J. Macromol. Sci.* (*B*), 1970, **4**, 61]

reactions to impurities are readily reconciled on this model. A detailed crystallographic study of these monomers, with the object of determining the structures and populations of the various dislocations, would probably be most rewarding.

5 Co-polymerization

A number of solid-phase co-polymerizations have been studied with a variety of types of monomer pairs, and the experimental data have recently been summarized in a review by Herz and Stannett.[130] We shall not discuss this aspect in detail since many of the data are too fragmentary to allow any definite conclusions to be drawn, and we shall simply mention a few general features. In all cases the results of co-polymerization experiments are dominated by physical rather than chemical factors. Thus, in the co-polymerizations of oxetans, the co-polymers have the same compositions as the initial monomer mixtures, leading to apparent reactivity ratios of unity, quite different from those observed in the liquid state.[131] This result is a consequence of a lack of diffusive processes in the solid phase.

Maleic anhydride and acenaphthylene produce an alternating (1 : 1) co-polymer in the liquid state. These monomers crystallize separately and form a eutectic; polymerization produces a mixture of polyacenaphthylene and the alternating co-polymer.[131] The co-polymer is considered to form at intercrystalline boundaries in the initial stages and throughout the system in the later stages when considerable disorder has been produced.

Rates of reaction are often maximal at eutectic compositions, as in the acrylamide–acrylic acid system which forms a 1 : 2 molecular compound and two eutectics. At fairly low temperatures the main products are co-polymers with compositions corresponding to the eutectics and the molecular compound.[132]

In many cases it appears that co-polymerization occurs most effectively in disordered regions associated with intercrystalline boundaries and is often more rapid than homopolymerization of the component monomers.

6 Concluding Remarks

In this report we have summarized much of the detailed information available on solid-phase polymerizations and have demonstrated that in virtually all the systems studied the experimental results are consistent with reaction in defects, especially dislocations. Indeed, in most cases such defects are essential for reaction to proceed. Consideration of dislocations as preferred sites for the formation of active species and regions of high local free volume and molecular mobility, which facilitate reaction, reveals a simple

---

[130] J. E. Herz and V. Stannett, *Macromol. Rev.*, 1968, **3**, 1.
[131] K. Hayashi, M. Nishii, A. Shimizu, and S. Okamura, *Polymer Preprints*, 1964, **5**, 951.
[132] Gy. Hardy and L. Nagy, *J. Polymer Sci. (C)*, 1967, **16**, 2667.

general pattern that accommodates a wide variety of reactions. It is a valid criticism that almost any experimental evidence can be reconciled with reaction in defects. Proof (or otherwise) of some of the ideas we have expressed requires additional experimental information, in particular, data on the structures and properties of dislocations in monomer crystals.

4
Structural Imperfections in Organic Molecular Crystals

BY J. M. THOMAS AND J. O. WILLIAMS

Over the past decade wide-ranging studies have underlined both the occurrence and the chemical importance of a variety of structural imperfections which may occur in organic molecular solids. Thus, measurement of self-diffusion coefficients,[1–3] of exciton lifetimes and pathlengths,[4,5] of the trapping of charge carriers,[6–8] of fluorescence characteristics,[9] of optical reflectivity,[10] as well as studies of X-ray diffuse scattering,[11–13] mechanical deformation,[14–17] and the occurrence of localized regions of enhanced reactivity[18,19] within the bulk or at the surfaces of certain crystals, all reflect the existence of distinct classes of imperfections. These imperfections, as illustrated by Bamford and Eastmond's account,[20] can play as important a rôle in the chemistry of organic solids as do grain-boundaries in the mechanical behaviour of metals or point-defects in the electronic properties of inorganic semi-conductors.

Taking the broad view, it is necessary to consider at least four major classes of structural abnormality. We shall briefly discuss all classes: point defects, line defects (dislocations), planar defects (or stacking faults which may arise from dissociated dislocations), and long-range disorder of the type

[1] J. N. Sherwood and S. J. Thomson, *Trans. Faraday Soc.*, 1960, **56**, 1443.
[2] J. N. Sherwood, *Molecular Crystals and Liquid Crystals*, 1966, **1**, 97.
[3] P. J. Reucroft, H. K. Kevorkian, and M. M. Labes, *J. Chem. Phys.*, 1966, **44**, 4416.
[4] P. J. Reucroft, P. L. Kronick, A. R. McGhie, and M. M. Labes, in 'Crystal Growth' ed. H. S. Perser, Pergamon Press, Oxford, 1967, p. 105.
[5] M. Pope, Preprints of the 5th International Conf. on Molecular Crystals, Philadelphia, 1970.
[6] J. M. Thomas, J. O. Williams, and L. M. Turton, *Trans. Faraday Soc.*, 1968, **64**, 2505.
[7] J. M. Thomas and J. O. Williams, *Molecular Crystals and Liquid Crystals*, 1969, **9**, 59.
[8] J. Sworakowski, *Molecular Crystals and Liquid Crystals*, 1970, **11**, 1.
[9] P. Sarti-Fantoni and R. Teroni, *Molecular Crystals and Liquid Crystals*, 1970, **12**, 27.
[10] G. C. Morris, S. A. Rice, and A. E. Martin, *J. Chem. Phys.*, 1970, **52**, 5149.
[11] K. Lonsdale, *Reports Progr. Phys.*, 1942, **9**, 256.
[12] H. D. Flack, *Phil. Trans. Roy. Soc.*, 1970, **A266**, 561.
[13] A. M. Glazer, *Phil. Trans. Roy. Soc.*, 1970, **A266**, 593.
[14] J. O. Williams and J. M. Thomas, *Trans. Faraday Soc.*, 1967, **63**, 1720.
[15] J. M. Thomas and J. O. Williams, *Trans. Faraday Soc.*, 1967, **63**, 1922.
[16] J. N. Sherwood, *Molecular Crystals and Liquid Crystals*, 1969, **9**, 37.
[17] G. J. Ogilvie and P. M. Robinson, *Molecular Crystals and Liquid Crystals*, 1971, **12**, 379.
[18] J. M. Thomas and J. O. Williams, *Chem. Comm.*, 1967, **432**.
[19] J. M. Thomas and J. O. Williams, *Progr. Solid-State Chem.* 1971.
[20] C. H. Bamford and G. C. Eastmond, this volume, Chapter 3.

associated with diffuse *X*-ray scattering.[11–13] Several excellent articles and reviews, dealing at various levels of sophistication with all these imperfections, have appeared recently. We shall, therefore, concentrate upon the theoretical aspects and discuss the principles involved, enumerate the various solids that have been studied and, where appropriate, illustrate some of the chemical repercussions of the existence of structural defects.

1 Point Defects

Because of the large energetic requirements to form vacancies and interstitials—generally the energy of creation of a vacancy in a molecular crystal

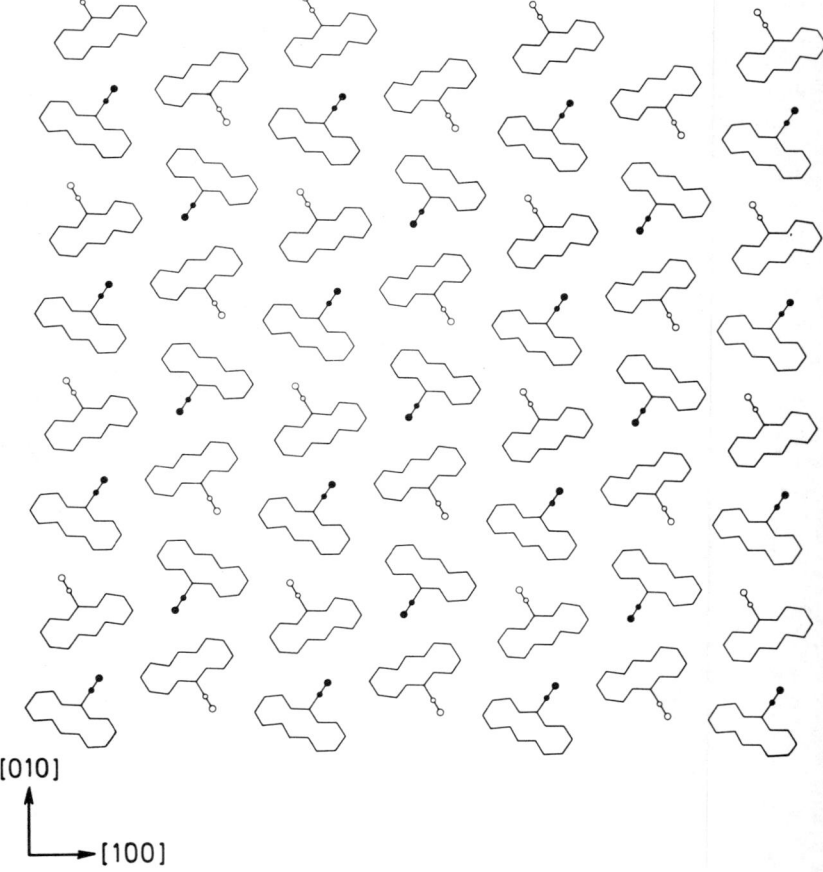

Figure 1 *Structure of 9-cyanoanthracene projected on to the (001).* **Full circles** *represent CN groups pointing up, empty circles CN pointing down*

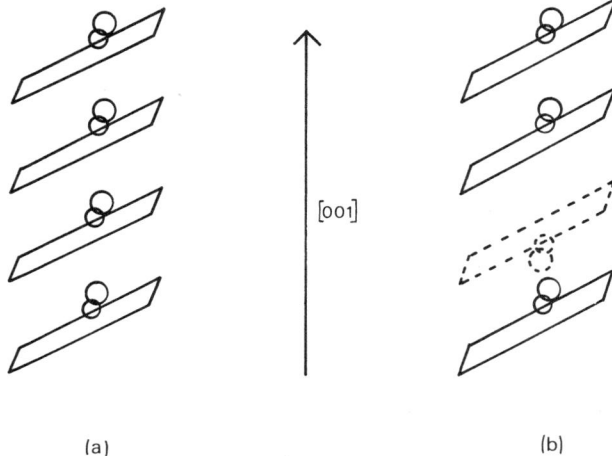

Figure 2 *Diagrammatic representation of the conversion of a perfect 9-cyanoanthracene structure (a), into one containing an orientational point defect (b). Diagram shows molecules in a stack running along [001]. Circles represent pendant carbon and nitrogen atoms of the CN group. The orientational point defect is shown dotted, and the CN group in the 'defect' molecule points downwards*

is at least as large as the enthalpy of sublimation—point defects and point defect clusters are comparatively few in number within organic molecular solids compared to their occurrence in metals, ionic salts, and many covalently bonded solids. Except for those solids which enter a rotator phase prior to melting, point defect concentrations are quite low even at temperatures approaching the melting point.

It is important to realize, however, that Schottky and Frenkel defects are not the only types of point imperfections which are possible in principle, within a structure having specifically oriented bulky molecules at each lattice point. For example, a dipolar, non-centrosymmetrical molecule such as 9-cyanoanthracene, which crystallizes with space group $P2_12_12_1$ (see Figure 1) in which all molecules in the c-direction within a given stack are similarly oriented (Figure 2a), may accommodate an orientational point defect such as that depicted in Figure 2b. Here, one of the molecules within a stack is oriented incorrectly with respect to the remainder. The same type of point defect may, of course, occur in a non-polar material (such as anthracene) but the chemical and physical repercussions of such a defect would be small and, consequently, detection of the creation of the defect (by, say, dielectric absorption) would be correspondingly more difficult. It is to be expected that the concentration of such defects would increase exponentially with increasing temperature, according to the free-energy change associated with the reorientation. Little or no experimental work has so far been carried out on

E*

such defects, largely because they still remain only a theoretical possibility. Magnetic resonance as well as dielectric measurements should, however, be capable of exploring their occurrence and properties.

Structural point defects of a basically different kind, in that they are not thermodynamically stable, could arise for at least two different reasons. First, channels of 'vacancies' or columns of 'interstitials' may be generated as a result of the annihilation of positive and negative edge dislocations and by the 'climbing' process by mechanisms which have been previously explained (see ref. 21 for diagrammatic sketches). The total number of point defects (vacancies or interstitials) present in the columns or channels could far exceed the number of isolated defects expected thermodynamically. The second mechanism of producing an excess of point defects, more especially vacancies, entails accidents of growth in which, for statistical reasons, not all lattice sites are populated. This is rather reminiscent of the type of 'vacancies' or sites left empty at metal surfaces when chemisorption of diatomic gases occurs upon them.[22]

Rather special, and as yet incompletely characterized, point defects are produced in organic molecular crystals as a result of irradiation.[23, 24] Thus, Blum et al.[23] conclude that the colour centres generated in anthracene by γ-irradiation arise from a cross-linking of neighbouring anthracene molecules.

Self-diffusion measurements along with studies of plastic deformation of organic solids, notably pivalic acid, hexamethylethane, cyclohexane, succinonitrile, camphene, and a range of aromatic hydrocarbons such as naphthalene, anthracene, and biphenyl have been studied by Sherwood et al.[25, 26] Their results shed much light on the nature of point defects in these solids.

2 Dislocations

The earliest evidence for the presence of line defects in organic solids came from deformation studies[27, 28] and observations relating to crystal growth.[29, 30] With a few notable exceptions, e.g. Dawson's ingenious replication technique of revealing edge-dislocations in β-lactoglobulin[31] and Menter's[32] use of lattice imaging as a technique for revealing defects in metal phthalocyanines, our knowledge of dislocations in a wide range of organic crystals has in the majority of instances been obtained via one or

[21] J. M. Thomas, *Adv. Catalysis*, 1969, **19**, 202.
[22] J. K. Roberts, *Proc. Roy. Soc.*, 1936, **A152**, 445.
[23] H. Blum, P. L. Mattern, R. A. Arndt, and A. C. Damask, *Molecular Crystals and Liquid Crystals*, 1961, **3**, 269.
[24] P. L. Kronick and M. M. Labes, *Molecular Crystals and Liquid Crystals*, 1961, **2**, 293.
[25] H. M. Hawthorne and J. N. Sherwood, *Trans. Faraday Soc.*, 1970, **66**, 1783, 1792, 1799.
[26] N. T. Corke and J. N. Sherwood, *J. Materials Sci.*, 1971, **6**, 68.
[27] A. Kochendorfer, *Z. Krist.*, 1937, **97**, 263.
[28] R. B. Gordon, *Acta Metallurgica*, 1965, **13**, 199.
[29] I. M. Dawson and V. Vand, *Proc. Roy. Soc.*, 1951, **A206**, 555.
[30] F. C. Frank, *Adv. Phys.*, 1952, **1**, 91.
[31] I. M. Dawson, *Nature*, 1951, **168**, 241.
[32] J. W. Menter, *Proc. Roy. Soc.*, 1956, **A236**, 119.

other, or a combination, of the following: dislocation-etching with wet reagents (or, in a few cases, photochemical 'etching'), plastic deformation in conjunction with optical microscopy, topographical studies of cleaved or as-grown crystal faces, X-ray topography using mainly the Lang approach.[33] A few relatively stable polymeric materials have been amenable to study also by transmission electron microscopy employing both Moire [34] and diffraction-contrast [35] techniques but recently, with the advent of reliable cold stages and other technical innovations, it has been possible to employ essentially the same transmission electron microscopic procedures in the study of comparatively volatile and rather soft organic materials, such as anthracene and its numerous derivatives and substituted olefins. In addition, it ought, with careful selection of materials and suitable fine adjustments, ultimately to be possible to utilize the direct lattice imaging techniques which Ban [36] has used so effectively on rather more beam-stable materials such as carbon black. These developments should herald a new era in the study of organic materials and enhance our knowledge of the defect solid state.

In enquiring about the defect structure of single-crystal organic solids chemists have been concerned first with the rather obvious, but not unimportant, quantity known as dislocation density—the number intersecting per unit area of a defined crystallographic plane or surface. This is what principally governs the number of primary nucleation centres for reactions such as thermally or photochemically stimulated [18, 37, 38] dimerization, polymerization,[20] decomposition,[39-41] etc. It is also useful in that it permits estimates of trap concentrations [5, 7, 19] (for singlet or triplet excitons, or charge carriers).

The second property relating to dislocation density is the distribution of dislocations within the bulk. For anisotropic solids (and most organic solids fall within this category) certain alignments of dislocations into various types of small-angle boundaries are more favourable than others.[42, 43]

The third and probably the most characteristic property of a dislocation is its Burgers vector [44, 45] which defines both the magnitude and direction of the slip inevitably associated with all such imperfections. We shall return to the question of the magnitude of the Burgers vector in Section 3 as it leads logically to discussions of stacking faults. It is the magnitude of the Burgers

[33] A. R. Lang, *Acta Cryst.*, 1959, **12**, 249.
[34] R. Gevers, *Phil. Mag.*, 1962, **7**, 1681.
[35] P. B. Hirsch, A. Howie, and M. J. Whelan, *Phil. Trans. Roy. Soc.*, 1960, **A252**, 499.
[36] L. L. Ban, R. D. Heidenreich, and W. W. Hess, *J. Appl. Cryst.*, 1968, **1**, 1; L. L. Ban, this volume, Chapter 2.
[37] M. D. Cohen, I. Ron, G. M. J. Schmidt, and J. M. Thomas, *Nature*, 1969.
[38] M. D. Cohen, Z. Ludmer, J. M. Thomas, and J. O. Williams, *Chem. Comm.*, 1969, 1172.
[39] J. M. Thomas, J. O. Williams, H. Marsh, and B. Rand, *Carbon*, 1966, **4**, 143.
[40] C. H. Bamford and G. C. Eastmond, *Quart. Rev.*, 1969, **23**, 271.
[41] J. M. Thomas, *Endeavour*, 1970, **29**, 149.
[42] D. Hull, in 'Introduction to Dislocations', Pergamon Press, Oxford, 1965.
[43] J. Weertman and J. R. Weertman, in 'Elementary Dislocation Theory', Macmillan, New York, 1964.
[44] J. M. Burgers, *Proc. Acad. Sci. Amsterdam*, 1939, **42**, 293, 378.
[45] A. H. Cottrell, in 'Dislocations and Plastic Flow in Crystals', Clarendon Press, Oxford, 1953.

vector which determines, in turn, (a) the severity of the distortion of the lattice at the dislocation core (hence the degree of enhanced reactivity and the tenacity or energy of trapping at the core), and (b) the total elastic strain energy of the dislocation. Of these two, by far the more important,[19, 46] chemically, is (a). The precise factors (relating to dislocations) which affect enhanced chemical reactivity have been reviewed recently both from a general viewpoint [21] and in the particular context of organic solids.[19]

The number of organic materials in which dislocations have been identified, to a greater or lesser degree, continues to grow. At present (mid. 1971), dislocations in crystals of the following have been reported: acenaphthylene;[37] anthracene [7, 14, 16, 19, 47—53] and its derivatives;[54, 55] anthranilic acid;[56] anthraquinone;[57] benzoic acid;[58] biphenyl;[58] cyclomethylenetetramine (cyclonite);[59] hexatriacontane;[29] metaldehyde;[60] β-methylnaphthalene;[61, 62] naphthalene;[27, 28, 48, 56, 63—65] nylon 66 and 610;[66, 67] oxalic acid dihydrate;[68] pentaerythritol;[69] phthalocyanines;[32] polyethylene;[70—77] polyoxymethylene and poly(4-methylpent-1-ene);[78] potassium hydrogen tartrate hemihydrate;[79] pyrene;[58] the rotator-phase solids, camphor,[80] adamantane,[81] cyclohexane,[25] hexamethylethane,[25] camphene,[25] pivalic acid,[25] succinonitrile,[25] and hexamine;[82] sucrose;[83—85] p-terphenyl;[85] tetracene;[58] and triglycine sulphate.[86, 87]

[46] J. M. Thomas, *Chem. in Britain*, 1970, **6**, 60.
[47] P. M. Robinson and H. G. Scott, *J. Cryst. Growth*, 1967, **1**, 67.
[48] P. M. Robinson and H. G. Scott, *Acta Metallurgica*, 1967, **15**, 1230.
[49] A. R. McGhie, P. J. Reucroft, and M. M. Labes, *J. Chem. Phys.*, 1966, **45**, 3163.
[50] A. R. McGhie, A. M. Voschcenkov, P. J. Reucroft, and M. M. Labes, *J. Chem. Phys.*, 1968, **48**, 186.
[51] P. M. Robinson and H. G. Scott, *Phys. Stat. Solidii*, 1967, **20**, 461.
[52] D. Michell, P. M. Robinson, and A. P. Smith, *Phys. Stat. Solidii*, 1968, **K93**, 26.
[53] J. O. Williams and J. M. Thomas, *Molecular Crystals and Liquid Crystals*, in the press.
[54] M. D. Cohen, Z. Ludmer, J. M. Thomas, and J. O. Williams, *Proc. Roy. Soc.*, 1971, **A324**, 459.
[55] J. M. Thomas, J. O. Williams, J. Bouas-Laurent, J. P. Desvergne, and G. Guarini, in preparation.
[56] J. O. Williams and J. M. Thomas, unpublished observations.
[57] J. O. Williams, I. Adams, and J. M. Thomas, *J. Materials Sci.*, 1969, **4**, 1064.
[58] J. N. Sherwood, in 'Organic Solid-State Chemistry', ed. G. Adler, Gordon and Breach, London, to be published.
[59] W. Connick and F. G. J. May, *J. Crystal Growth*, 1969, **5**, 65.
[60] P. J. Jackson, *Acta Metallurgica*, 1965, **13**, 1057.
[61] M. I. Koslovskii, *Soviet Phys. Crystal.*, 1957, **2**, 146.
[62] G. G. Lemmlein, E. D. Dukova, and A. A. Cherov, *Soviet Phys. Cryst.*, 1959, **2**, 426.
[63] N. T. Corke, A. A. Kawada, and J. N. Sherwood, *Nature*, 1967, **212**, 62.
[64] J. M. Thomas, J. O. Williams, W. C. Evans, and E. Griffiths, *Nature*, 1966, **211**, 181.
[65] G. W. Sears, *J. Chem. Phys.*, 1962, **37**, 2155.
[66] B. N. Dey, *J. Appl. Phys.*, 1967, **38**, 4144.
[67] D. A. Zaukelies, *J. Appl. Phys.*, 1962, **33**, 2797.
[68] D. Michell, A. P. Smith, and T. M. Sabine, *Acta Cryst.*, 1969, **B25**, 2458.
[69] J. O. Williams, J. M. Thomas, J. R. N. Evans, and I. Minkoff, to be published.
[70] H. D. Keith, in 'Physics and Chemistry of the Organic Solid State', Interscience, New York, 1963, Vol. 1, p. 461.
[71] A. Keller, in 'Growth and Perfection of Crystals,' eds. R. H. Doremus, B. W. Roberts, and D. Turnbull, Wiley, New York, 1958, p. 499.
[72] V. F. Holland, *J. Appl. Physics*, 1964, **35**, 1351, 3235.
[73] V. F. Holland and P. H. Lindenmeyer, *J. Appl. Phys.*, 1965, **36**, 3049, 4051.
[74] P. Predecki and W. O. Statton, *J. Appl. Phys.*, 1966, **37**, 4053.

3 Stacking Faults

X-Ray crystallographers have pictured stacking faults in a manner which sometimes does not coincide with the stacking faults generally described by electron microscopists. A classic example of the way in which the X-ray crystallographers formulate stacking faults or stacking disorder is cited by Lonsdale.[88] Some materials, such as metallic cobalt, contain stacking disorder of the following kind. The cobalt structure is mainly arranged in close-packed hexagonal layers, but changes to a close-packed cubic arrangement approximately every eleven layers or so; and then it reverts back to the hexagonal form:

ABABABABABABCBCBCBCBCACACACACACACACABCB . . . .

X-Ray crystallographers frequently state [89] that graphite behaves in a like manner, in that, interspersed amongst the hexagonal matrix are occasional slivers of the rhombohedral form. Thus, for graphite, we may picture three variants of this type of stacking fault according as to whether there are one, two or three infractions (*i.e.* rhombohedral arrangements) to the normal hexagonal stacking sequence (Figure 3).

The electron microscopist, as a result of seeing directly (by diffraction contrast) ribbons of stacking fault (Figure 4) threading through a solid at various levels within the bulk, does not picture the fault as extending throughout the entire crystallite or crystal. The faults are pictured to follow naturally from the dissociation of a unit strength dislocation into partials.[19] Consequently, the percentage of such stacking faults may be increased by increasing the dislocation content. For graphite, deformation introduces more basal-plane dislocations which dissociate into partials, and thereby increases the rhombohedral content of the hexagonal material.

As yet, hardly any studies of dissociated dislocations have been completed for organic crystals (apart from a few polymers). There are circumstances which could, undoubtedly, permit dislocations lying on certain planes in some types of organic solids to dissociate to yield stacking faults.

[75] J. M. Peterson, *J. Appl. Phys.*, 1966, **37**, 4047.
[76] P. H. Geil, 'Polymer Single Crystals', Interscience, New York, 1963.
[77] P. Predecki and W. O. Statton, *J. Appl. Phys.*, 1967, **38**, 4140.
[78] D. C. Basset, *Phil. Mag.*, 1964, **10**, 595.
[79] V. A. Meleshina, *Soviet Phys. Cryst.*, 1967, **12**, 322.
[80] W. J. Dunning, *J. Phys. and Chem. Solids*, 1961, **18**, 21.
[81] H. M. Hawthorne, Ph.D. Thesis, Strathclyde, 1969.
[82] A. R. Lang, private communication.
[83] W. J. Dunning and N. Albon, 'Growth and Perfection of Crystals', eds. R. H. Doremus, B. W. Roberts, and D. Turnbull, Wiley, New York, 1958.
[84] H. E. C. Powers, *Nature*, 1962, **196**, 58.
[85] J. O. Williams, *J. Chem. Soc. (A)*, 1970, 2939.
[86] V. P. Konstantinova, *Soviet Phys. Cryst.*, 1963, **7**, 605.
[87] V. A. Meleshina, *Soviet Phys. Cryst.*, 1964, **9**, 304.
[88] K. Lonsdale, *May and Baker Laboratory Bulletin*, 1964, **VI**, 34.
[89] J. M. Robertson, 'Organic Crystals and Molecules,' Cornell Univ. Press, New York, 1953.

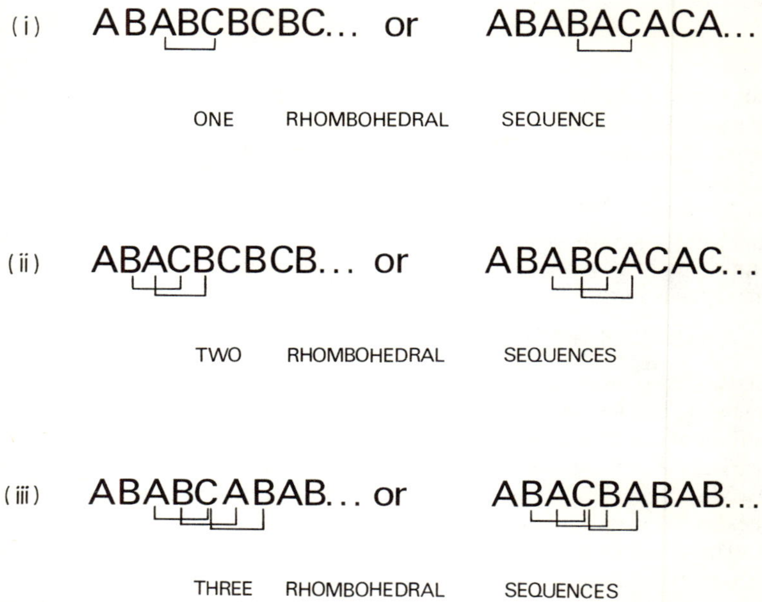

(i)   AB<u>ABC</u>BCBC... or   ABAB<u>ACA</u>CA...

ONE   RHOMBOHEDRAL   SEQUENCE

(ii)   AB<u>AC</u><u>BC</u>BCB... or   ABA<u>BC</u><u>AC</u>AC...

TWO   RHOMBOHEDRAL   SEQUENCES

(iii)   AB<u>AB</u><u>CA</u><u>BA</u>B... or   AB<u>AC</u><u>BA</u><u>BA</u>B...

THREE   RHOMBOHEDRAL   SEQUENCES

Figure 3 *Three possible types of infractions leading to stacking faults in a regular hexagonal structure*

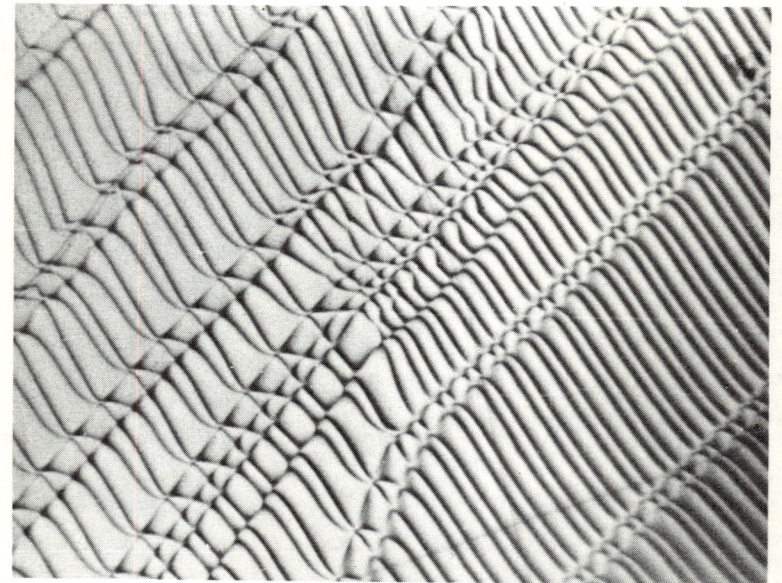

Figure 4 *Transmission electron micrograph showing stacking faults in molybdenite at dissociated basal-plane dislocations ($\times 15\,000$)*

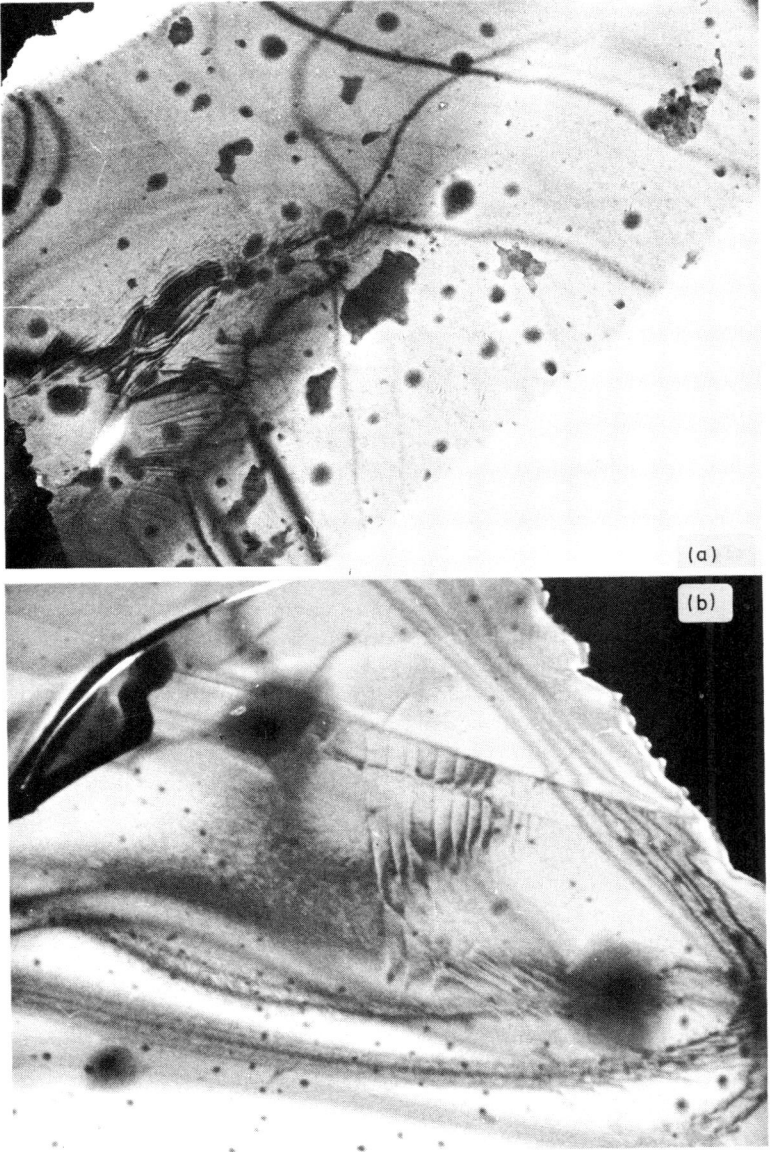

Figure 5 *Diffraction contrast arising from a dislocation network* (a) *in anthracene and* (b) *in anthraquinone. Bright field electron micrographs* [*magnifications* (a) ×6500; (b) ×10 000]

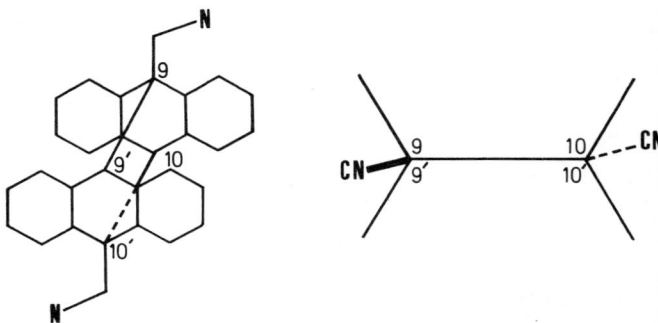

Figure 6 *Illustration of structure of the* trans-*dimer of 9-cyanoanthracene*

Prima facie evidence for such an occurrence in anthracene and anthraquinone has been obtained by Williams and Thomas [53, 90] (see Figure 5).

The major significance of stacking faults, so far as the chemical reactivity of solids is concerned, is that the unusual order associated with this type of defect can offer a favourable pathway for specific reactions. The most plausible example of this effect exists in 9-cyanoanthracene. In the solid state, this material readily dimerizes [38, 54] to yield the *trans* rather than the *cis* dimer (Figure 6). Dipole moment measurements confirm this fact.[91, 92] The question that now arises is how can the individual molecules re-orient in the solid so as to be predisposed to yield the *trans* dimer? The orientational point defect (Figure 2) must be considered as one possible answer, but it is unlikely to be significant as it is difficult to envisage how a self-sustaining (or propagative) mechanism could stem from such a fault. If, however, a stacking fault occurs as a result of one or other of the following dislocation reactions on (221) planes (see ref. 54 for fuller details):

$$[100] \rightarrow \tfrac{1}{4}[21\omega] + \tfrac{1}{4}[2\bar{1}\bar{\omega}] \tag{1}$$

and $\quad [010] \rightarrow \tfrac{1}{4}[23\omega] + \tfrac{1}{4}[\bar{2}\bar{1}\bar{\omega}] \tag{2}$

then contiguous molecules across the entire (221) plane are more suitably oriented for the ultimate production of the *trans* dimer. Indeed, as seen from Figure 7 [which schematically illustrates reaction (1) above], half the total number of molecules are already in the desired *trans* configuration, whilst the other half are partially overlapping in a favourable orientation.

It now remains to be seen whether electron microscopy can reveal the presence of stacking faults in 9-cyanoanthracene. At least for 1,8-dichloro-10-methylanthracene,[55] transmission electron microscopy does reveal direct evidence of what appears to be slip traces running in the < (104 >

---

[90] J. O. Williams and J. M. Thomas, *Molecular Crystals and Liquid Crystals*, in the press.
[91] R. Calas and R. Lalande, *Bull. Soc. chim. France*, 1959, 873.
[92] D. P. Craig and P. Sarti-Fantoni, *Chem. Comm.*, 1966, 742.

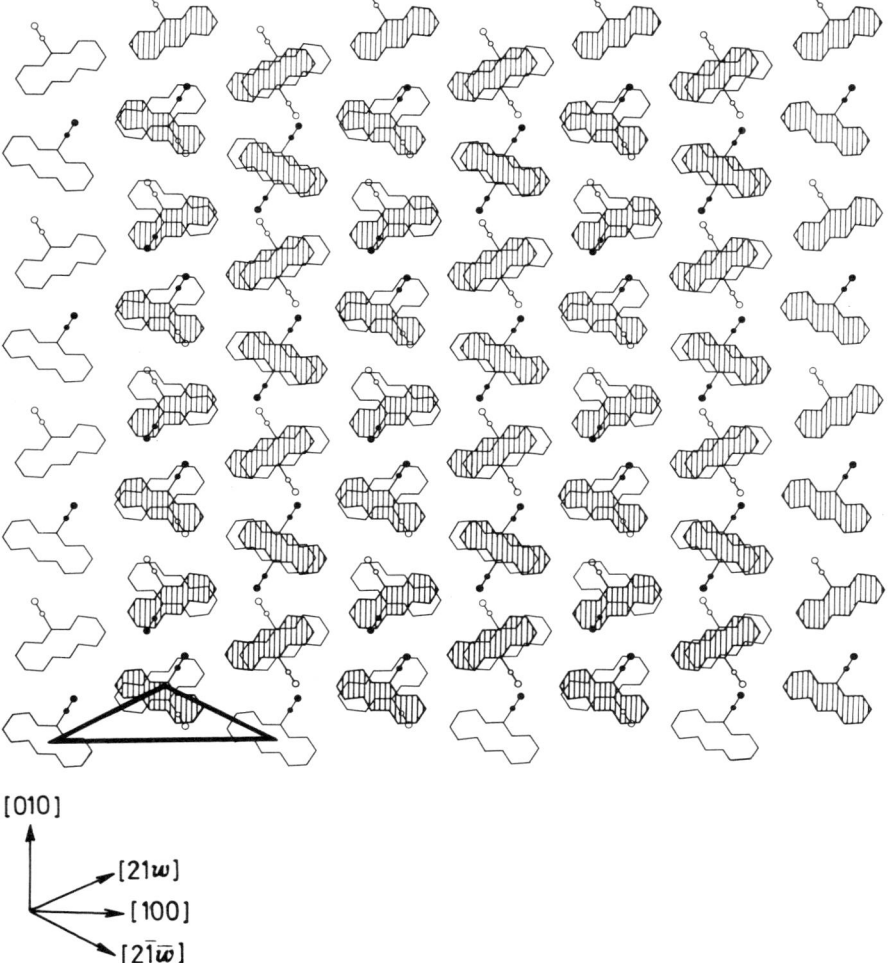

Figure 7 *Schematic illustration of nature of stacking faults that may be produced by dissociation of the type:* $[100] \rightarrow \frac{1}{4}[21\omega] + \frac{1}{4}[2\bar{1}\bar{\omega}]$, *when the dislocation glides in the* (221) *plane. It can be seen that half the total number of molecules within the stacking fault are centrically related to their nearest neighbour*

direction on the (010) face, a state of affairs which permits of a ready interpretation of *trans* dimer formation similar in many respects to that formulated for the 9-cyano-derivative.

Williams and Thomas[90] and Milledge *et al.*[93] have shown that trans-

[93] A. P. Rood, D. Emerson, and H. J. Milledge, *Proc. Roy. Soc.*, 1971, **A324**, 37.

mission electron microscopy can readily detect minute quantities of anthracene dimer forming inside the monomer matrix. The former workers have also seen, by electron microscopy, a preferential accumulation of dimer at regions reminiscent of dislocation networks.

Though transmission electron microscopy appears to be the most powerful technique for elucidating the 'ordered' structures associated with the stacking fault, particularly on a localized scale, it is often possible to derive valuable information about stacking faults from optical microscopy, and especially from measurements of diffuse reflections of X-rays. Williams [85] has outlined a possible type of stacking fault, arising from the occurrence of $(1\bar{1}0)\frac{a}{3}$ [110] partial dislocations, in p-terphenyl. Abrahams, Speakman, and co-workers [94-96] have elucidated the nature of the stacking disorder in trismethylsulphonylmethane (TMSM) (1) from a study of the diffuse X-ray

$$Me-\overset{\overset{O}{\underset{\underset{O}{\|}}{S}}}{\underset{\|}{S}}-CH\overset{\overset{O}{\underset{\underset{O}{\|}}{S}}-Me}{\underset{\overset{O}{\underset{\|}{S}}-Me}{}}$$

(1)

scattering. In this solid [97] molecules are stacked, at intervals of $c/2$, along the threefold axis and any single stack is regularly ordered. The disorder arises because the stacks do not all point in the same direction, as would be required by the space group $R3c$. If in the idealized structure all stacks point up c, or all down c (referred to respectively as A and B), then the explanation of the actual structure rests on the fact that A and B occur in the proportion of 61 : 39. The observed pattern of diffuse spectra can be accounted for by postulating a specified type of lateral packing of A and B stacks (see ref. 95 for further details). Interestingly, crystals of the closely similar compound, trisethylsulphonylmethane [96] (TESM), are free from the disorder associated with TMSM.

4 Long-range Disorder and Short-range Order

It has been known for some time that when certain molecules which lack a centre of symmetry crystallize they do so in such a way as to indicate,

---

[94] S. C. Abrahams and J. C. Speakman, *J. Chem. Soc.*, 1956, 2562.
[95] J. V. Silverton, D. T. Gibson, and S. C. Abrahams, *Acta Cryst.*, 1965, **19**, 651.
[96] D. R. McGregor and J. C. Speakman, *Acta Cryst.*, 1969, **B25**, 540.
[97] S. B. Hendricks, *Z. Krist.*, 1933, **84**, 85.

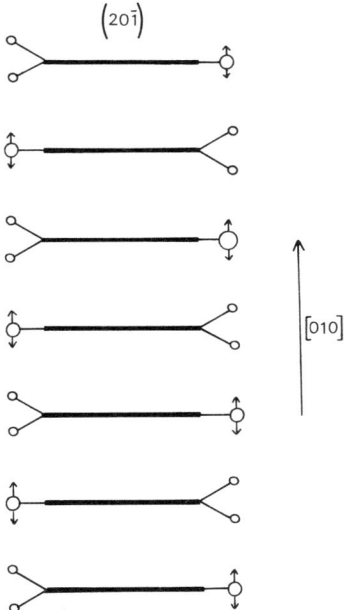

Figure 8 *Illustration of the way in which equal numbers of asymmetric (anthrone) molecules point in inverse directions in a 'disordered' domain [(201) section]—after Flack (ref. 12)*

from the space-group, that they are centrosymmetric [97–100]. Typical examples are azulene,[98] anthrone,[12, 99] 1,4-bromochlorobenzene,[97] and *N*-oxyphenazine.[100] These substances are, therefore, said to have pseudosymmetric crystal structures. It is impossible for them to be centrosymmetric, yet they behave as if they were. This must mean that they are disordered in such a way that equal numbers of asymmetric molecules are pointing in inverse directions. Figure 8 illustrates the point by reference to anthrone, which Flack[12] has investigated. (We may, in passing, note that the very large thermal expansion along [010] in anthrone arises because of the appreciable out-of-plane libration of the oxygen.)

Lonsdale, Milledge, and their co-workers [11, 12, 13, 88, 101, 102] have made detailed studies of the type of molecular packing which exists in pseudosymmetric structures and also in mixed crystals of which one component is

---

[98] J. M. Robertson, H. M. Shearer, G. A. Sim, and D. G. Watson, *Acta Cryst.*, 1962, **15**, 1.
[99] S. W. Strivastava, *Acta Cryst.*, 1964, **17**, 851.
[100] R. Curti, R. Riganti, and S. Locchi, *Acta Cryst.*, 1961, **14**, 33.
[101] H. J. Milledge, *Acta Cryst.*, 1969, **A25**, 173.
[102] H. J. Milledge and A. Graeme-Barber, *Joyce-Loebl Rev.*, 1969, **2**, 2.

pseudosymmetric. We shall now summarize their results after first outlining the general principles involved.

If we designate a molecule pointing one way R and its inverse molecule W then there are three possibilities.[103] First, there is the completely ordered arrangement:

RWRWRWRW   etc.
or   RRWWRRWW   etc.

This is equivalent to a superlattice structure. A completely random arrangement constitutes the second possibility:

RRWRRRWRRWRWRRWWWWRRWR   etc.

The third is an arrangement having regions of *short range order* which are, however, only of an approximately uniform size and which are separated by regions of random sequences:

RWRWRWRWRWWRWRWRWWRWRWRRR   etc.

The second arrangement would have a definite pseudosymmetric structure (the unit being $(R+W)/2$) yielding a diffraction pattern consisting of sharp reflexions accompanied with heavy diffuse scattering. The third arrangement would be distinguished by an appropriate kind of patterned diffuse scattering. Lonsdale, Milledge, Flack, and Glazer [12, 13] have demonstrated, using X-ray diffraction photographs of stationary single crystals taken with characteristic Cu $K\alpha$ radiation plus the continuous (white) spectrum, that the *static* effect of the short-range order gives a diffuse pattern which may be readily distinguished from the *dynamic* effect due to thermal vibration. Flack,[12] in particular, has shown how to calculate the dimensions of ordered regions in pseudosymmetric crystals from the intensity of diffuse scattering.

In anthrone [12] (space group $P2_1/a$, unit cell dimensions $a = 15.79$, $b = 3.99$, $c = 7.90$, and $\beta = 101.4°$) there are short-range order domains of average length 9 molecules (36 Å) in the [010] direction and 3 molecules (25 Å) in the [001] direction, the arrangement along [100] being random.

Now consider what happens when anthraquinone (space group $P2_1/a$, unit cell dimensions $a = 15.83$, $b = 3.97$, $c = 7.89$ Å, and $\beta = 102.5°$) is incorporated substitutionally into the anthrone lattice. It is instructive first to inquire whether the short-range order domains survive in the mixed crystals, and, second, to determine whether the theory of Kitaigorodskii,[104] which states that complete solid-state miscibility between two components is possible if the symmetry of the distribution of the molecules in the endmembers is the same, is valid. The answers, provided by Flack,[12] are clearcut. In the mixed crystals, up to 12 mole per cent of anthraquinone can

---

[103] K. Lonsdale, K. S. Krishnan Memorial Lecture, National Physical Laboratory, New Delhi, 1969.
[104] A. I. Kitaigorodskii, *Doklady Akad. Nauk S.S.S.R.*, 1957, **113**, 604.

crystallize with anthrone without disturbing the short-range order. At the other extreme of the phase diagram, no short-range order is found for up to 45 mole per cent anthrone in anthraquinone. The region 12 to 55 mole per cent of anthraquinone in anthrone seems to define a region of immiscibility in the mixed-crystal series.

Glazer [13] has provided analogous answers to the above questions posed for the phenazine and N-oxyphenazine mixed crystals. N-Oxyphenazine itself has short-range order domains [13] in (100) planes, of average size 13 Å (along b) by 50 Å (along c). Along a, the alternation of centrosymmetrically arranged molecules is random. When phenazine is incorporated into N-oxyphenazine, in the 52—100 per cent N-oxyphenazine composition, the short-range order domains are not disrupted, the phenazine being accommodated between the sheets or rods of short-range order. From 10 to 52 per cent N-oxyphenazine (in phenazine) there is a miscibility gap. Mixed crystals of low N-oxyphenazine content (*e.g.* 0 to 10 per cent N-oxyphenazine) have these molecules arranged randomly in the phenazine matrix. The results of the anthrone–anthraquinone and N-oxyphenazine–phenazine taken collectively, therefore, indicate that two components, one of which is ordered and the other showing only short-range order, cannot form a continuous mixed-crystal series.

In conclusion, it is worth emphasising that although there are obvious similarities between stacking-fault disorder associated with dislocations (and deformation) on the one hand, and stacking faults and the long-range disorder (or short-range order) associated with pseudosymmetric crystals on the other, there remains the important difference that dislocations are thermodynamically unstable, whereas the disorder present in pseudosymmetric crystals may represent a thermodynamically stable situation.

5
Surface Studies by Photoemission

BY M. W. ROBERTS

1 Introduction

Photoemission is one of the most powerful and least ambiguous techniques available for the study of the electronic band structures of solids. The advent in the sixties of photostimulated electron emission as both a valuable and fashionable approach to the determination of molecular structure has been striking. One should not lose sight, however, of the fact that the technique is in principle a resuscitation of work initiated some four decades ago. In many ways it resembles the historical pattern of development of Low Energy Electron Diffraction: discovered in the late twenties, used somewhat infrequently for some four decades, and culminating in a burst of activity (largely due to instrumental developments) during the last decade.

The current qualitative view of the photoemission process, deduced from the studies of metals and of elemental and compound semiconductors, is that it can be considered to be a three-step process:

(a) The electrons are optically excited to a characteristic depth within the material, the depth being dependent on the optical absorption coefficient of the material.
(b) A fraction of these electrons reach the surface, but on the way suffer inelastic scattering, usually through electron–electron interaction.
(c) Electrons arriving at the surface with sufficient energy escape.

Although we are for convenience considering photoemission here as a somewhat classical three-step process, from a purely quantum-mechanical viewpoint it should be considered as a whole. We will, however, retain, as is usual in such cases, the conceptual viewpoint of the whole process as being composed of three sub-processes, remembering that to understand any of these processes in detail it will be essential to consider the influence of the others. The validity of Koopmans' theorem is considered later. There are, therefore, three parameters which are of prime significance in discussing photoemission from solids: (i) the escape depth of the photoelectron, (ii) the energy-loss mechanism likely to occur in the solid, (iii) the potential barrier at the surface which excited electrons must surmount in order to escape. We will discuss these in more detail later since they are central to current thinking in the field, but first it is worthwhile reviewing very briefly how the subject has developed during the past two decades.

It was undoubtedly the very careful work of Apker, Taft, and Dickey [1] of some 20 years ago that marked the beginning of a new era in the development of photoemission techniques for the study of solids and their surfaces. In particular, they investigated the energy distribution of photoelectrons from the elemental semiconductors Se, Te, and B with a view to probing the presence of emission from surface states of the type discussed by Bardeen, Brattain, and Shockley.[2] An important conclusion of their work was that electron yield obeyed a power law near the threshold, and we will return to this point later.

Up to about 1960 the photon-stimulated electron emission from metals was assumed to result from light absorption occurring at the surface, but the work of Meesen [3a] and of Meyer and Thomas [3b] with the alkali metals indicated that photoelectrons can originate at depths of several hundred ångströms below the surface. In other words, they showed that the emission is dominated by a volume process rather than a surface process. Similar conclusions were made concerning emission from solids other than metals, and not unnaturally the question was posed: how does the space charge existing in a semiconductor influence the interpretation of the photoemission process? The work of Gobeli and Allen [4] at the Bell Telephone Research Laboratories and Van Laar and Scheer [5] at the Philips Research Laboratories, on silicon, together with that of Spicer [6] on the compound semiconductors Cs_3Sb, undoubtedly paved the way to the development of the subject in the sixties. Their results, together with the theoretical work of Kane,[7] are of much wider significance than just the understanding of photoemission from elemental and compound semiconductors, and impinge directly on the understanding of the physics and chemistry of surfaces and the mechanism of surface reactions. The rationalization of the experimental observations of gas chemisorption on oxides has relied heavily on a qualitative knowledge of the electronic band structure of solids, as also did Dowden's d-band theory [8] to account for the catalytic activity of metals.

There have, during the past decade, been extensive and important developments in the field of photoemission from solids, developments which are relevant to surface and solid-state chemistry and to catalysis. We will not in this article be concerned with electron emission stimulated by high-energy

---

[1] L. Apker, E. Taft, and J. Dickey, *Phys. Rev.*, 1948, **74**, 1462.
[2] J. Bardeen, *Phys. Rev.*, 1947, **71**, 717; W. H. Brattain and W. Shockley, *Phys. Rev.*, 1947, **72**, 345.
[3] (a) A. Meesen, *J. Phys. Radium*, 1961, **22**, 308; (b) H. Meyer and H. Thomas, *Z. Physik*, 1959, **147**, 419.
[4] G. W. Gobeli and F. G. Allen, *Phys. Rev.*, 1962, **127**, 141, 150; *J. Appl. Phys.*, 1964 **35**, 597.
[5] J. Van Laar and J. J. Scheer, *Philips Res. Reports.*, 1962, **17**, 101.
[6] W. E. Spicer, *J. Appl. Phys.*, 1960, **31**, 2077.
[7] E. O. Kane, *Phys. Rev.*, 1962, **121**, 131.
[8] D. A. Dowden, *Bull. Soc. chim. belges*, 1958, **67**, 439.

photons (*e.g.* X-rays), usually referred to as ESCA (Electron Spectroscopy for Chemical Analysis), or with secondary electrons produced by the Auger process. Both these are considered in detail in Chapter 6 by C. R. Brundle. Here we will be concerned with the more recent views on the mechanism of stimulated electron emission by photons of energy less than $\sim$10 eV, the deduction of the electronic structure of solids from energy distribution data, the nature of surfaces (including alloys), and in particular the importance of photoemission studies for investigating the very early stages of the oxidation of clean metal surfaces. If we often refer to emission from semiconductors it is because of the physicists' principal interest in these materials, and it has been the physicist who has been most involved in this field to date.

Formal Description of the Photoemission Process.—In view of the recent growth of interest by chemists in photoemission, it is worthwhile first considering here some of the physical aspects of the process. Following Apker, Taft, and Dickey [1] and Spicer,[6] if we consider a slab of material of thickness dx at a distance x below the surface (Figure 1), then the electron emission from this section of material is di ($x,h\nu$), as defined in equation (1), where $I(x,h\nu)$ is the

$$di(x,h\nu) \, d(h\nu) = \alpha_\mathrm{p}(h\nu) \, I(x,h\nu) \, P(x,h\nu) \, dx \cdot d(h\nu) \qquad (1)$$

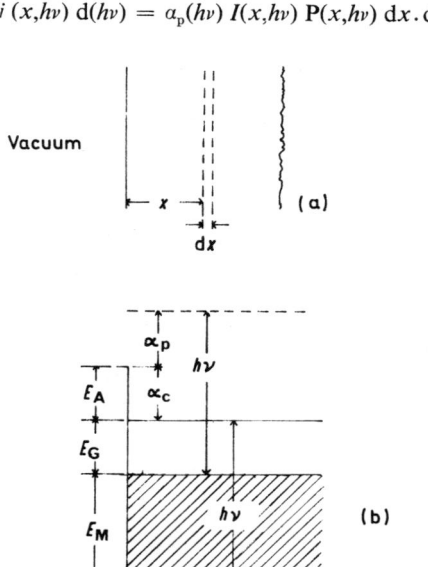

Figure 1 (a) *A section of material of thickness* dx *parallel to and at a distance* x *from the surface; light of energy* $h\nu$ *is incident on the vacuum–solid interface.* (b) *Energy band diagram for photoemitter.* E_A *is the electron affinity,* E_G *the band gap, and* $h\nu$ *is the energy of the incident photon.* E_M *is the lowest energy in the valence band from which an electron can be excited into conduction band by a photon* $h\nu$

light intensity at depth x, $P(x,h\nu)$ is the probability that an electron which has been excited at a depth x by a photon of energy $h\nu$ will escape from the material, $a_p(h\nu)$ is the absorption coefficient for excitation into states above the vacuum level, and $d(h\nu)$ is the band-pass of the monochromator used. The optical absorption can be divided into two events: the transitions into states above the vacuum level, a_p, and those into states below the vacuum level, a_c. Obviously, only transitions of type a_p will lead to photoemission. If we define a_t as the total absorption coefficient then

$$I(x,h\nu) = I_0(h\nu) \exp[-a_t(h\nu)x] \quad (2)$$

where $a_t = a_p + a_c$ and $I_0(h\nu)$ is the intensity of the incident light. Studies with Cs_3Sb suggest[9] that the probability of electron escape, P, is given by equation (3) where $B(h\nu)$ is some function of $h\nu$ and β is a constant. This leads

$$P(x,h\nu) = B(h\nu) \exp[-\beta(h\nu)x] \quad (3)$$

to equation (4) for the photoelectric yield $Y(h\nu)$ (electrons per incident photon). Equation (4) indicates the principal factors determining the photoelectric yield.

$$Y(h\nu)d(h\nu) = \frac{a_p(h\nu)}{a_p(h\nu) + \beta(h\nu)} [B(h\nu)d(h\nu)] \quad (4)$$

In the case of semiconductors the absorption process is best discussed in terms of energy *versus* the virtual crystal momentum k; the bands are assumed to be parabolic but any more generalized band structure would not alter the discussion. The general expression for the photoemission current will therefore be

$$i(h\nu)d(h\nu) = \int_{x=0}^{x=\infty} a_p(h\nu)I(x,h\nu)P(x,h\nu)dx \; d(h\nu) \quad (5)$$

where $i(h\nu)$ is the photocurrent at photon energy $h\nu$; the sample has been assumed to be sufficiently large so that the upper limit of the integral can be taken as infinity. If it is assumed that $P(x,h\nu)$ can be written in the form $B(h\nu) \exp(-\beta x)$ where $B(h\nu)$ is some function of $h\nu$ and β is a constant, equation (5) can be integrated to give the photoemission yield (equation 6).

$$Y(h\nu) = \left[\frac{a_p(h\nu)}{a_t(h\nu) + \beta}\right] B(h\nu) \quad (6)$$

Spicer[10] points out that if the escape probability P were unity the yield would be determined by a_p/a_t, but in any case where $a_t(h\nu) \gtrsim \beta$ the quotient $a_p(h\nu)/a_t(h\nu)$ will be important in determining $Y(h\nu)$. That is, if the impinging light is absorbed within a distance from the surface comparable to $1/\beta$, the reduction in yield due to electrons excited into states below the vacuum level will be very important. In the case of Cs_3Sb Burton[9] has shown that a_t and

[9] J. A. Burton, *Phys. Rev.*, 1947, **72**, 531.
[10] W. E. Spicer, *Phys. Rev.*, 1958, **112**, 114.

β are comparable, and so the reduction in yield due to excitation of electrons into states below the vacuum level will be significant. Spicer [10] has made, with some success, calculations of the theoretically predicted yield as a function of photon energy for a number of materials. He showed that $B(h\nu)$ could be taken as constant and so the yield near the threshold should be proportional to $a_p(h\nu)$. Experimentally it was shown that

$$Y(h\nu) \propto [h\nu - (E_G - E_A)]^{3/2} \tag{7}$$

where E_G is the band gap energy and E_A the electron affinity.

Photoemission near the Threshold.—The treatment of photoemission from a metallic conductor, due to Fowler,[11] indicates that when light of energy $h\nu$ impinges on the surface of a metal of work function φ at temperature T then photocurrent I at zero field near the threshold $(h\nu_0)$ is given by equation (8),

$$I = A(kT)^2 f\left(\frac{h\nu - h\nu_0}{kT}\right) \tag{8}$$

where A is taken to be constant for $h\nu \approx h\nu_0$. If $\ln(I/T^2)$ is plotted *versus* $h\nu/kT$ near the threshold, together with the function f, and the two curves superimposed, then the magnitude of the shift along the $\ln(I/T^2)$ axis gives the value of $\ln(k^2 A)$ while that along the $h\nu/kT$ axis gives $h\nu_0/kT$. This approach assumes the free-electron approximation for the energy band structure, a uniform surface with no patches and a negligible accelerating field. As in thermionic emission, a superimposed electric field will influence the emission process. We will return to consider patch and superimposed-field effects later and also the source region of the photoelectrons.

In the case of semiconductors the free-electron model is not valid and various approaches to this problem have been adopted. Following Fan's quantum-mechanical treatment [12] for the volume production of photoelectrons, Huntington and Apker [13] generalized Markinson's free-electron approach [14] to the surface production of photoelectrons by taking the band structure into account. But the paper which has had the most impact on current trends in the study of photoemission is undoubtedly that due to Kane,[7] who considered the form of the photo-yield as a function of the light energy for a general band structure and for various electron-production and electron-scattering mechanisms. It is necessary to consider first the two ways in which valence-band electrons may be excited: (i) by direct (vertical), or (ii) by indirect (photon-aided) optical transitions; these processes are illustrated in Figure 2. Only the flat band case is considered. Figure 2 (a) shows the valence- and conduction-band edges as a function of distance (z) from the

---

[11] R. H. Fowler, *Phys. Rev.*, 1932, **38**, 45.
[12] H. Y. Fan, *Phys. Rev.*, 1945, **68**, 43.
[13] H. B. Huntington and L. Apker, *Phys. Rev.*, 1953, **89**, 352.
[14] R. E. B. Markinson, *Phys. Rev.*, 1949, **75**, 1908.

Surface Studies by Photoemission

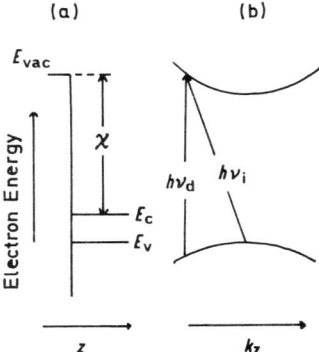

Figure 2 *Schematic representation of direct and indirect transitions in photoemission from a flat-band semiconductor*

surface and Figure 2 (b) is a plot of electron energy *versus* the z-component of the wave-vector k for the valence band and for some higher conduction band whose curve intersects the vacuum level (E_{vac}). It is assumed that the extrema of both bands lie at $k = 0$ and that the maximum valence-band energy E_v corresponds in k-space to the minimum conduction-band energy E_c. When such a coincidence occurs the semiconductor is said to have a direct intrinsic gap. The significance of the term 'direct gap' is that a photon $h\nu_d$ can excite an electron from the top of the filled band directly to one of the states at the very bottom of the conduction band, in a transition which is vertical in k-space. For indirect transitions the threshold for photoemission, E_i, corresponds to a photon energy $h\nu_i$, which is equal to the energetic separation between the top of the valence band E_v and the vacuum level E_{vac}, the energy of the phonon involved in the necessary momentum transfer being negligible. Thus $E_i = \chi + E_g$. For direct transitions, on the other hand, the wave-vector k is conserved, and as a consequence a higher-energy photon ($h\nu_d$) is necessary to effect photoemission. The direct threshold E_d may be influenced by whether or not the excited electron will undergo phonon scattering before leaving the crystal. In the absence of scattering E_d will be greater than E_i by an amount equal to the energy of the hole left behind plus the component of the kinetic energy of the emitted electron in the tangential direction, which is not affected by the emission. Scattering may reduce E_d since there is no requirement of transverse momentum conservation. There is therefore the possibility that electrons of energy too low to leave the crystal may be reoriented by scattering so as to have the threshold energy in the z-direction.

What this means is that we would expect a change, with increasing photon energy, from the indirect process to the more efficient direct mechanism after the direct threshold has been reached. Kane's theoretical approach gives the predicted form of the Y *versus* $h\nu$ relationship near the threshold for a general shape of the band structure and for a variety of excitation

Table 1 *Variation of the photoelectric yield Y with photon energy hv near the threshold E_T for different excitation and scattering processes (Kane's theory[7])*

Volume Processes

Source of electrons	Transition	Scattering processes	Threshold, E_T	n^*
Valence Band	Indirect	Unscattered	$E_T = \chi + E_c - E_v$	5/2
		Scattered	$E_T = \chi + E_c - E_v$	1
	Direct	Unscattered		
		Scattered	$E_T = \chi + E_c - E_v$	2

Surface Processes

Source of electrons			Threshold, E_T	n^*
Valence Band		Diffuse surface scattering	$E_T = \chi + E_c - E_v$	5/2
		Specular surface scattering		3/2
Surface Band States		Direct optical excitation	$E_T > E_F$	1
			$E_T = E_F$	3/2
		Indirect optical excitation	$E_T > E_F$	2
			$E_T = E_F$	5/2

* The value of n is with reference to the expression $Y \propto (E - E_T)^n$.

and scattering processes. Each of the processes involved results in a power-type law for Y (equation 8a) above the corresponding threshold E_T, where n has integral or half-integral values between 1 and 5/2 depending on the type of

$$Y \propto (h\nu - E_T)^n \qquad (8a)$$

excitation and scattering involved. It is worth recalling in the above context that for photoemission from a metal equation (8a) approximates to $Y \propto (h\nu - h\nu_0)^2$ (cf. Table 1). The value of n is therefore not truly diagnostic in distinguishing between photoemission from a metal and, say, a semiconducting oxide, a situation likely to be encountered in studies of the transformation of chemisorbed oxygen on metals to oxide, and which we consider later. Thus, it is at least in principle possible to infer production mechanisms from the photoelectric yield versus $h\nu$ relationship. The results of Kane's treatment are summarized in Table 1 and we will consider later various experimental data in the light of Kane's theory.

Escape Depth of Photoelectrons.—In the early sixties evidence became available [3] which indicated that photoelectric emission was a 'volume' process, and this meant that the position in the solid from which the photoelectrons originated was an important issue. Both the attenuation of the incident light and the range of the excited electrons must be considered for this purpose. Redfield [15] assumed an exponential decay in light intensity (I) with depth (x) from the surface so that:

$$I = I_0 \exp(-\alpha x) \qquad (9)$$

where α is the absorption coefficient. The range of the excited electrons is clearly different for different materials. If lattice scattering is the dominant scattering mechanism, then since the escape probability in this case had been shown by Hebb to be exponentially dependent on x [i.e. $\propto \exp(-\beta x)$ where $1/\beta$ is the mean escape distance], Redfield derived equation (10) for the dependence of photocurrent i, where $\gamma = \alpha + \beta$ and $1/\gamma$ is the mean depth of

$$i \propto \exp(-\gamma x) \qquad (10)$$

origin of the photoelectrons. For Cs_3Sb, β is $\sim 4 \times 10^5$ cm^{-1} and $\alpha \sim 1 \times 10^5$ cm^{-1} near the threshold, so that there is a mean depth of origin of ~ 200 Å. In the case of Ge, $\alpha \approx 1.4 \times 10^6$ cm^{-1} near the threshold, and scattering by pair production may decrease the escape depth to well below 100 Å.

We can now see how, if we assume that the density of states $n(E)$ is taken as $n(E) \propto (E - E_v)^y$, that $i \propto (E - E_v)^y$ and so the photocurrent will take the form of equation (11), which is the power-law dependence at each depth

$$i \propto (E - E_{v(x)})^y \exp(-\gamma x) \qquad (11)$$

weighted in proportion to the contribution that electrons from that depth make to the photocurrent, and which clearly depends on the functional

[15] D. Redfield, *Phys. Rev.*, 1961, **124**, 1809.

dependence of E_v on x. A point to note here is that, in a space-charge region, E_v may vary by up to ~1 eV, and therefore the value of E_v will vary by possibly ~0.2 eV over the region of origin of the photoelectrons. This can be of significance at the semiconductor–gas interface and also in situations where a metal surface has reacted with, say, oxygen to generate a metal–oxide interface.

Experimental studies of electron escape distances [16] from metals stem largely from work with the alkali metals, and the approach that has been generally used is to monitor photoemission from thin metal films as a function of film thickness (see Figure 3). In the event of a surface effect, the photocurrent at constant irradiation with monochromatic light would be expected to increase rapidly with the number of atoms evaporated on to an insulating substrate until a continuous surface had been built up. As there are two limiting surfaces (vacuum–metal and metal–substrate), then with increasing film thickness a decrease in emission should be observed, due to the decreasing number of electrons coming from the opposite interface, until the film thickness becomes larger than the diffusion range of the electrons.

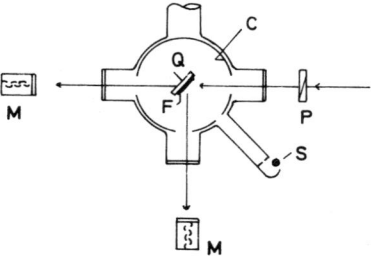

Figure 3 *Schematic representation of the apparatus used in the alkali-metal studies.[16] S is the metal source, Q is the cooled quartz substrate, P is a polarization prism, F is the metal film. C is the photocurrent cylindrical collector and M are photomultipliers for determining the reflected and transmitted light intensities*

With a volume effect, the photocurrent will increase until some value of film thickness corresponding to the diffusion range (escape depth) of the excited electrons has been reached, and should thereafter be independent of thickness. Also, the escape depth might be anticipated to depend on the light wavelength, *i.e.* on the initial energy of the excited electron, and the number of electrons excited at any distance from the surface will depend on the number of photons arriving and will thus be proportional to the light intensity. The latter is clearly not independent of film thickness but in the case of the alkali metals this dependence may be neglected as a first approximation (even to ~100 Å) because of the weak absorption. The results obtained [16] for sodium are shown in Figure 4. Neglecting the maximum at

---

[16] F. J. Pipenbring in 'Optical Properties and Electronic Structure of Metals and Alloys, Proc. Int. Colloq., Paris, 1965', ed. F. Abelès, North Holland Publishing Co., Amsterdam.

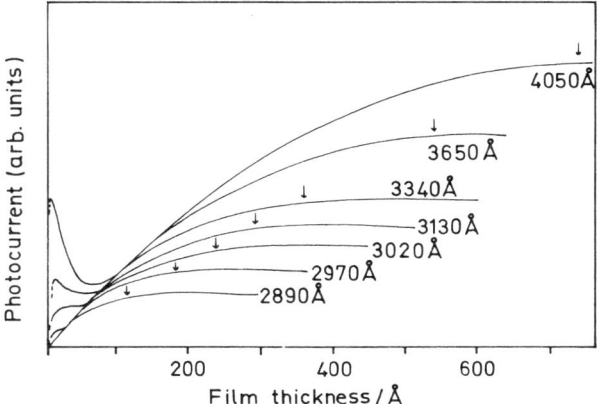

Figure 4 *Photoemission from sodium films of varying thickness by light in the wavelength range* 2890—4050 Å

about one monolayer for $hv \approx hv_0$, the photocurrent increases almost linearly with thickness until the film thickness becomes equal to or even larger than the diffusion range (escape depth) of the photoelectrons. Subsequently it is invariant with thickness. The arrows indicate the escape depths, which increase with increasing energy from $\sim$100 Å ($\lambda \approx 2890$ Å) to $\sim$700 Å ($\lambda \approx 4050$ Å). Similar results were obtained for potassium (Figure 5) but the escape depth in this case changed even more (from 10 Å to 1000 Å) with increasing wavelength than was observed with sodium. The data obtained (Figure 5) with caesium films were striking in that the photocurrent reaches saturation at a film thickness between 5 and 10 Å for all wavelengths.

These results indicate quite clearly a volume photoelectric effect with K

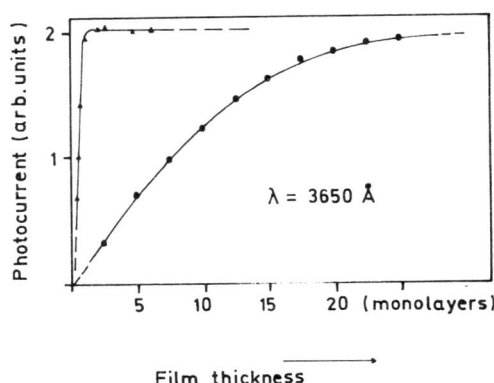

Figure 5 *Photoemission from caesium (▲) and potassium (●) films of varying thickness deposited on quartz ($\lambda = 3650$ Å). (Mayer and Wiegreffe, see ref. 16)*

and Na and also that the escape depth of the photoelectrons depends on their energy. To understand the dependence of the mean diffusion range of the photoelectrons on energy, energy-loss processes have to be considered other than collisions of electrons with ions and lattice defects, which are not likely to effect any energy loss. Loss of energy is likely to arise from electron–electron interactions either as individual collisions or by way of the excitation of plasma oscillations. Pipenbring [16] considers which of these is more likely; he argues that if $\varepsilon_i < \varepsilon_0 + h\nu_p$ where ε_i is the maximum initial excitation energy of the photoelectron, ε_0 the Fermi energy, and ν_p the plasma frequency, then loss of energy is only possible by individual electron–electron scattering processes. This is the case for Na($\varepsilon_0 \sim 3.2$ eV and $h\nu_p \sim 5.5$ eV) since the highest excitation energy was ~ 7.5 eV. The observed diffusion range (escape depth) of the photoelectrons is thus determined by the value of the mean free path for electron–electron scattering, which is defined as the product of the lifetime of the excited particle and its velocity. Calculations by Quinn [17] give results which show a very strong dependence of the mean free path and the excitation energy and which are in qualitative agreement with the observations of Pipenbring with sodium (Figure 6).

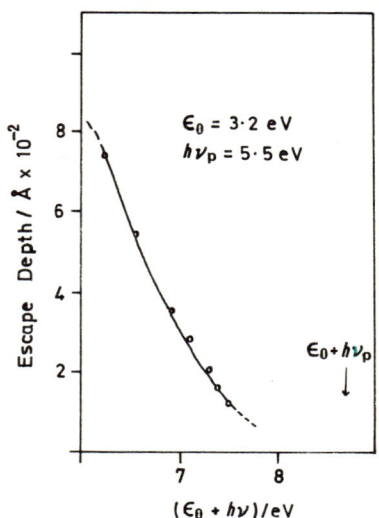

Figure 6 *Escape depth for photoexcited electrons in sodium as a function of the initial excitation energy. ε_0 is the position of the Fermi level, ν_p the plasma frequency, and ν the frequency of the impinging light*

As the initial energy ε_i increases so that $\varepsilon_i \geqslant \varepsilon_0 + h\nu_p$ the photoelectrons may lose energy by plasmon excitation. Then the mean free path for electron–plasmon interaction will determine the electron diffusion range. Assuming plasmon creation is responsible for energy loss in potassium and caesium,

[17] J. J. Quinn, *Phys. Rev.*, 1962, **126**, 1435.

Surface Studies by Photoemission

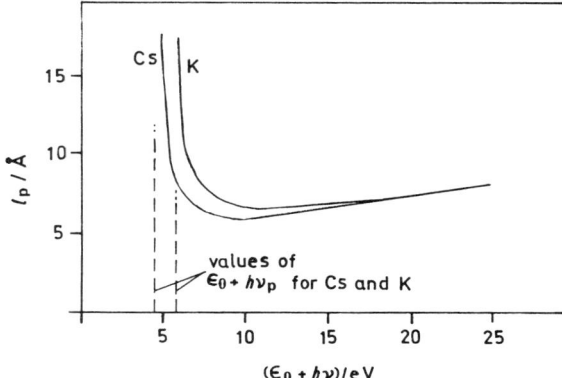

Figure 7 *Mean free path l_p for plasmon excitation plotted* [18] *as a function of the initial energy of the excited electrons for caesium and potassium*

Thomas [18] calculated the mean free path for plasmon creation as a function of the initial energy of the excited electrons (Figure 7). Above light energies of ~ 4 eV plasmon creation is feasible with potassium, so that excitation energies between 2 and 4 eV will create electrons which can only lose energy by electron–electron scattering. At lower light energies, electron–electron scattering is the only important energy-loss process. The results for potassium verify this model. The caesium case is particularly interesting ($\varepsilon_0 \sim 1.5$ eV and $h\nu_p \sim 2.8$ eV) since even light of energy near the photothreshold will excite electrons to an energy high enough for plasmon creation. Thus in this case a maximum diffusion range of only ~ 5—15 Å is anticipated, independent of light energy, and this is observed in practice (Figure 5).

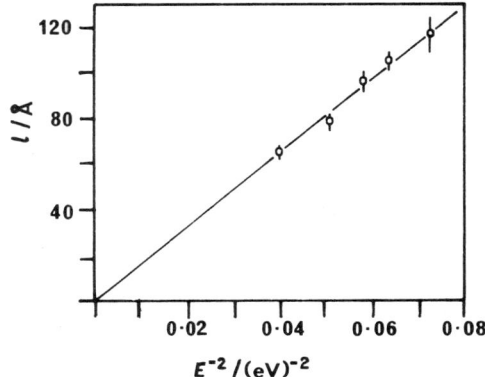

Figure 8 *Variation of the attenuation length l with electron energy* E

[18] H. Thomas, *Z. Physik*, 1957, **147**, 395.

F

More recently, Lea and Mee [19] *estimated* the attenuation length of photoelectrons in uranium films for photons in the range 3.5—5.0 eV to be between 60 and 120 Å, the length decreasing with increasing energy. Quinn,[17] and Sze, Moll, and Sugano [20] predict that the attenuation length *l* should vary approximately as the electron energy, E^{-2}, and this is shown to be the case for uranium (Figure 8). In the photon range studied by Lea and Mee they conclude that the principal energy-loss mechanism for the excited electrons is one of electron–electron scattering.

2 Experimental Approaches in Photoemission

The basic experimental set-up for determining photoelectric yield and energy distribution is relatively simple but there are a number of design prerequisites: (*a*) The collector should be so shaped that it provides a spherical equipotential for emitted electrons. In practice, a circular cylinder whose length is about twice its diameter suffices. (*b*) The vacuum system should be capable of attaining pressures of $\leqslant 10^{-9}$ mmHg (*c*) The 'window' for admitting monochromatic light must have good transmittance in the particular range of photon energy. Lithium fluoride can be used up to 11.9 eV. (*d*) The collector work-function should not change during an experiment. This is particularly crucial in photoemission studies of the interaction of gases with the emitter (*e*) Reverse current, *i.e.* photoelectrons arising from reflected light impinging on the collector, should be kept to a minimum since it will distort Energy Distribution Curves (EDC's). Provided the collector is made sufficiently positive, yield data will, however, not be affected. (*f*) Some consideration has to be given to the question of the light source not being truly monochromatic, particularly in relation to fine structure in the EDC.

The energy distribution of the photoemitted electrons is obtained by measuring $-di/dV$, the rate of decrease of photocurrent i with increasing retarding voltage V. The *actual* retarding potential for the emitted electrons is the retarding voltage plus the difference in the work function between the collector and the emitter. We will call this quantity V_R. Although it is possible to differentiate i/V_R data to give $-di/dV_R$ (*i.e.* the EDC), a very much better approach is that of Spicer and Berglund [21a] who used an a.c.-retarding-potential method. In this a small a.c. voltage is superimposed on the slowly varying (1 V min^{-1}) retarding potential. Now the small-signal a.c. conductance of the photocell is $-di/dV_R$ so that a plot of the in-phase a.c. component of *i* versus V_R is the desired EDC, since the a.c. current is proportional to the a.c. conductance. We thus have a direct measurement of EDC's. Eden [21b] has recently improved on the original Spicer approach by devising a system which has excellent noise performance, while Pierce and

---

[19] C. Lea and C. H. B. Mee, *Phys. Stat. Sol.* 1968, **25**, 613.
[20] S. M. Sze, J. L. Moll, and T. Sugano, *Solid State Electronics*, 1964, **7**, 509.
[21] (*a*) W. E. Spicer and C. N. Berglund, *Rev. Sci. Instr.*, 1964, **35**, 1665; (*b*) R. C. Eden, *ibid.*, 1970, **41**, 252.

Di Stefano [22a] have described a retarding-field energy-analyser incorporating a field-free region around the emitter. The analyser error was estimated to be $\sim 2.8\%$ of the kinetic energy.

One of the crucial points in studying photoemission from surfaces, particularly those undergoing reactions with gases, is the fact that in general it is not easy to measure photocurrents $< 10^{-14}$ A. Howling,[22b] using a Geiger-Mueller photon counter, has improved very considerably on this accuracy of measurement and has reported a sensitivity to currents as low as $\sim 10^{-21}$ A. This is of obvious advantage in studies of 'spectral tails'.

The usual method of determining the work function of a semiconductor is to measure the contact potential difference between the surface under investigation and a probe of known work function, using the capacitor method. Recently, Hughes and Dalrymple [23a] have drawn attention to an experimental approach that depends entirely on using photoelectric data. The method is claimed to be more accurate than the Kelvin technique.

3 Information from Photoemission Studies

Although the surface chemist has used photoemission in the main as a means of acquiring information concerning chemisorption on metals, it is valuable also to consider information obtained with semiconductors, since analogies with the semiconductor field will be useful when we consider the perturbation of metal surfaces during gas interaction, especially when that interaction leads to the formation of a new phase. Photoemission offers a number of advantages over the capacitor or retarding-field methods for studying changes of work-function on adsorption; in the first place, it is capable of observing surface heterogeneity. So-called patch effects have been observed [23b,c] as a consequence of the interaction of oxygen with metal surfaces (Figure 9) and have also recently been reported by Baker, Johnson, and Maire [24] for nickel; in the latter case the contribution from different crystallographic planes was recognized. Patch emission has also been observed [25] during the adsorption of CO on Mo(100); each patch has, by flash desorption, been shown to correspond to five different binding states, each with its characteristic work function. In the weakest adsorbed state the calculated surface dipole moment is close to the free-molecule value, indicating little electronic rearrangement in this state. Secondly, since the photoemission process is volume dependent, it also has a clear advantage over other methods in that it enables the sub-surface to be explored, and it is in studying new surface phases (cf. Quinn and Roberts [26]) that we believe it is having and will have the greatest impact.

---

[22] (a) D. T. Pierce and T. H. Di Stefano, *Rev. Sci. Instr.*, 1970, **41**, 1740; (b) D. H. Howling, *J. Appl. Phys.*, 1966, **37**, 1844.
[23] (a) O. H. Hughes and M. Dalrymple, *J. Phys. (C), Solid State*, 1970, **3**, L92; (b) J. S. Anderson and D. F. Klemperer, *Proc. Roy. Soc.*, 1960, **A258**, 350; (c) C. S. McKee and M. W. Roberts, *Surface Sci.*, in the press.
[24] B. G. Baker, B. B. Johnson, and G. L. C. Maire, *Surface Sci.* 1971, **24**, 572.
[25] L. D. Mathews, *Surface Sci.*, 1971, **24**, 248.
[26] C. M. Quinn and M. W. Roberts, *Trans. Faraday Soc.*, 1965, **61**, 1775.

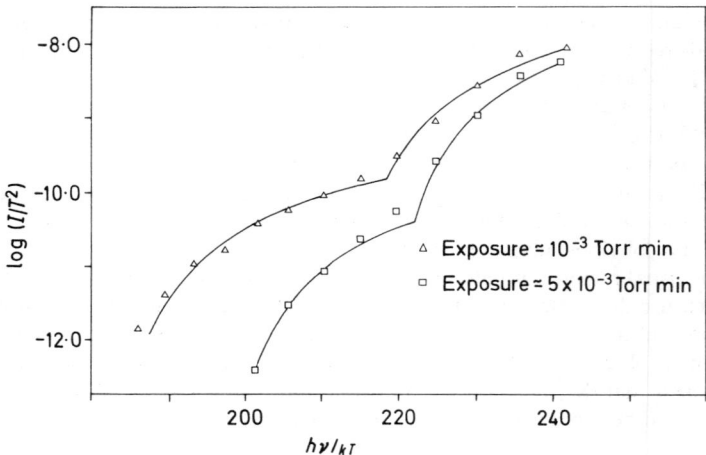

Figure 9 Split Fowler curves reflecting patch-emission after oxidation of nickel at 973 K

Alloy Surfaces.—The study of alloy surfaces is attractive to catalytic chemists in that such surfaces provide, in principle, an opportunity to vary both lattice and electronic parameters in a controlled way. In other words, the respective rô es of the geometric and electronic factors can be explored. For example, the catalytic activity of a transition metal possessing an unfilled d-band should, according to Dowden, decrease drastically on alloying it with an sp metal, which in effect fills the holes in the d-band. Recently, the Dutch School have used [27] the photoelectric work-function to characterize the surface composition of alloys. In the first instance, Sachtler, Dorgelo, and Jongepier [27a,b] showed that the surface composition of clean copper–nickel alloy films, after sintering at 473 K, is constant and independent of the bulk composition, within the limits of the miscibility gap. These results were interpreted in terms of diffusion and thermodynamic data, and the model suggested was a kernel of a nickel-rich alloy enveloped by a mantle of a copper-rich alloy. Central to this model were two facts: (a) that copper diffuses faster than nickel, and (b) the copper–nickel system possesses a miscibility gap, as suggested by free-energy calculations. Bouwman and Sachtler [27c] have, more recently, using the same approach, considered the gold–platinum system, and we will consider this in some detail.

Platinum–gold alloy films were prepared by various techniques and the surfaces formed studied by using them as cathodes in a phototube. Two parameters were used to characterize the surface: the work function (Φ) and the emission constant (M). According to the Fowler theory, M is related to

---

[27] (a) W. M. H. Sachtler and G. J. H. Dorgelo, *J. Catalysis*, 1965, **4**, 654; (b) W. M. H. Sachtler and R. Jongepier, *ibid.*, 1965, **4**, 665; (c) R. Bouwman and W. M. H. Sachtler, *ibid.*, 1970, **19**, 127.

the photoelectric yield Y and the frequency of the light ν by equation (12)

$$Y = MT^2\{(\pi^2/6)+(X^2/2)-[e^{-X}-(e^{-2X}/2^2)+(e^{-3X}/3^2)\ldots\ldots]\} \quad (12)$$

where $X = \dfrac{h\nu - h\nu_0}{kT} = \dfrac{E - \Phi}{kT}$.

For $X \gg 1$ or $\nu \gg \nu_0$, $Y = M(E-\Phi)^2/2k^2$, so that a plot of $Y^{\frac{1}{2}}$ versus E (the light energy) gives from the intercept Φ and from the slope $(M/2k^2)^{\frac{1}{2}}$. Four types of experiment were carried out: (a) Au on top of Pt, (b) Pt on top of Au, (c) simultaneous evaporation of the metals, and (d) CO chemisorption on the alloys. Only in the case of (c) did equilibration with the formation of an alloy occur in a reasonable time at 573 K and we will consider here only those surfaces formed by simultaneous deposition. Figure 10 shows Φ and M data as a function of Pt % composition for fresh and sintered films (573 K); it should be noted that Φ for the sintered surface was ~ 5.35 eV and independent of the composition between 10 % and 90 % Pt. It should be noted that the sintered alloys in this composition range have a work function lower than either pure metal, whereas the work functions of the fresh surfaces increase almost linearly with increasing Pt concentration. In other words, the surfaces in this case contain both Au and Pt in about the same proportions as indicated on the abscissa. Sintering at 573 K results in the formation of the stable alloy phase, either gold-rich or platinum-rich.

A most interesting phenomenon observed with CO is that the emission

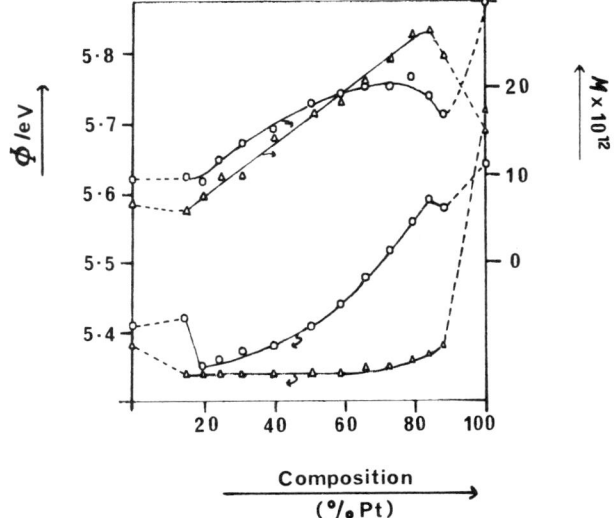

Figure 10 *Work functions* [27] *and emission constants (upper curves) of a film prepared by simultaneous deposition of gold and platinum* (○ *fresh film;* △ *sintering at* 573 K)

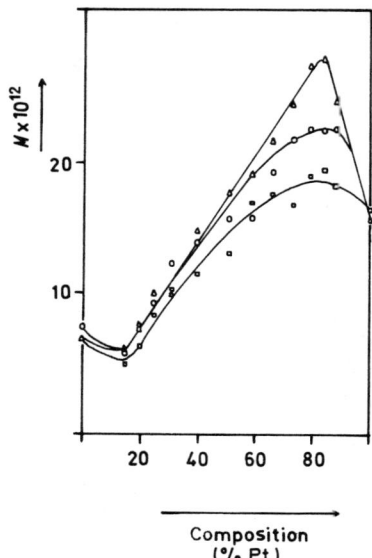

Figure 11 *Emission constants* M *of an equilibrated* Pt–Au *film after treatment* [27] *with* CO; △ *film sintered at* 573 K; ○ *after* 90 h *exposure to* 10^{-4} Torr CO *at* 293 K; □ *after* 80 h *exposure to* 10^{-4} Torr CO *at* 373 K

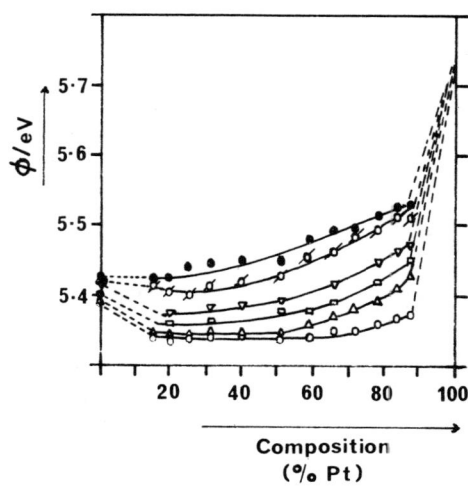

Figure 12 *The work functions* Φ *of an equilibrated* Pt–Au *film after treatment with* CO; ○ *film sintered at* 573 K; Φ *increases with increasing exposure and temperature of exposure;* ● *after an* 80 h *exposure at* 10^{-4} Torr CO *at* 373 K

constant M and the work function Φ were found to vary with prolonged exposures (Figures 11 and 12). Bouwman and Sachtler suggest that in the presence of CO(g) the alloy surfaces became slowly enriched with Pt, the diffusion of Pt being enhanced due to the creation of lattice defects by the corrosive chemisorption of CO, and as soon as a Pt atom arrives at the surface it is 'locked' at the surface by chemisorbing CO. The variation of M is compatible with some changes in the bulk during treatment with CO. The model proposed appears to be general for alloys possessing a miscibility gap and having considerably different diffusion rates. The rate of alloying depends on temperature, the diffusion path-length, and the gaseous ambient.

Seib and Spicer [28] have recently made photoemission ($h\nu \lesssim 11$ eV) and optical studies of bulk Cu-Ni alloys after ion bombardment and annealing at 925 K. It appears that for nickel-rich Cu-Ni alloys the density of states is unchanged below 20% Cu and identical to that of pure nickel. Also they suggest that for alloys containing up to 60% nickel the copper d-states can be identified at energies greater than 2 eV below E_F. Clift, Curry, and Thomson [29] have drawn qualitatively similar conclusions from soft-X-ray spectra; that in Cu-Ni alloys there is little sharing of electrons between the two kinds of atoms and there is no common d-band, although there are differences in detail between their data and those of Spicer. Eastman [30] has recently reviewed the photoemission spectroscopy of metals and alloys and we will not consider these further in this article.

Studies of Silicon.—Silicon has been the solid most extensively studied by photoemission, and many of the current ideas in the field stem from these investigations. Gobeli and Allen [4] combined work-function, photoelectric spectral, and energy-distribution data in their investigations of cleaved, heated, and sputtered silicon surfaces. A conclusion which was particularly pertinent to photoelectric studies of metal+oxygen systems was that the small photoelectric yield from an oxide (thickness t)-covered surface could not be accounted for merely in terms of the higher photoelectric threshold of the oxide. Gobeli and Allen concluded that the yield could be fitted to equation (13) if $\Phi_{ox} \approx 5.1$ eV and $e^{-t/l} \approx 0.06$, where l is the mean free path of the excited electron.

$$Y_{ox} = K e^{-t/l}(h\nu - \Phi_{ox})^3 \qquad (13)$$

It should be noted that $e^{-t/l}$ is the fraction that is transmitted of the electrons incident on the oxide from the underlying silicon. If $t \approx 20$ Å and $l \approx 10$ Å then it can be shown that 94% of the electrons incident from the underlying silicon are trapped by the oxide and do not emerge as photoelectrons. In a separate study of doped silicon, Gobeli and Allen [4] concluded that in going from extreme p-type to n-type semiconduction the work function (Fermi

---

[28] D. H. Seib and W. E. Spicer, *Phys. Rev.*, 1970, **2**, 1694.
[29] J. Clift, C. Curry, and B. J. Thomson, *Phil. Mag.*, 1963, **8**, 593.
[30] D. Eastman (in the press).

level) remained virtually invariant with doping, while the photothreshold first increased from ~ 4.8 to 5.15 and then decreased with the extreme n-type sample to ~ 4.74 eV. These experiments clearly have implications for doping studies in the general field of surface chemistry and catalysis.

As already discussed, photoelectric emission from semiconductors would be expected to obey the power law shown in equation (14), the value of the

$$Y = C_n(h\nu - E)^n \tag{14}$$

exponent n depending on the excitation and escape mechanisms involved. In general, the yield spectra are as in Figure 13; for photo energies ~ 1 eV above the threshold $n \approx 3 \pm 0.5$ but at higher energies, $Y \propto (h\nu - E)$. The transition from the cubic to the linear law is usually smooth. The total emission can be written as:

$$Y = C_i(h\nu - h\nu_i)^3 + C_d(h\nu - h\nu_i)$$

where C_i and C_d are constants containing the light intensity and the absorption efficiencies for the production of excited electrons with final states lying above the vacuum level. In other words, the yield rises from the indirect threshold and then increases abruptly as the direct photothreshold is exceeded and the more efficient process predominates. There is at present little ambiguity regarding the interpretation of the linear law ($n = 1$), Kane's theory [7] indicating (Table 1) direct optical transitions from the valence band and emission of electrons without scattering. On the other hand, a number of possible interpretations exist for $n \approx 3$. Firstly, excitation of electrons out of surface states (Scheer and van Laar [31]); secondly, excitation of electrons from

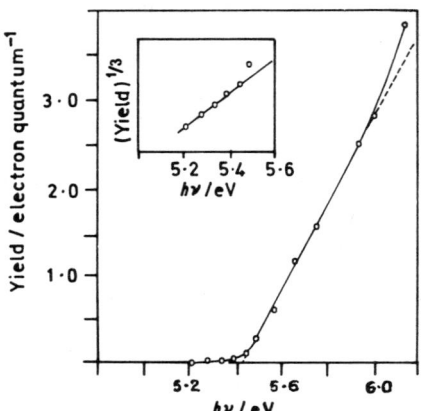

Figure 13 *Linear variation of yield with light energy for p-type silicon.[4] Inset shows cube-root plot for low-energy data (5.2—5.5 eV)*

[31] J. J. Scheer and J. Van Laar, *Phys. Letters*, 1963, **3**, 246; *Surface Sci.* 1965, **3**, 1189.

the top of the valence band with simultaneous exchange of normal momentum with the surface (Gobeli and Allen [32]); thirdly, and the most favoured, the whole yield spectrum (linear and cubic) may result from direct transitions of electrons from the valence band (Fischer, Allen, and Gobeli [33a]).

Further diagnostic information on emission mechanisms is available from energy distribution studies and in this an important new experimental approach has emerged. This entails the deposition of caesium (or indium) on the surface so that the work function is reduced, thus extending the range of electron energies observable for any given light energy. A number of groups have now used this for investigating energy structure of photoelectrons from silicon; these include Spicer and Simon,[33b] Scheer and van Laar (who used In), and Gobeli and Allen. Clearly a relevant question is whether in addition to lowering the work function the caesium influences the EDC by energy losses occurring within the layer. There is no evidence on this point as yet.

More recently, Broudy [34] has very carefully investigated various planes of silicon [(111), (100), and (110)]. The photoelectric thresholds for these planes were 4.60, 5.11, and 4.73 eV, respectively. Broudy points out that he, in contrast to van Laar and Scheer, and Gobeli and Allen, was unable to fit the yield data to a single- or double-law expression over the whole photon energy. Moreover the linear region ($n = 1$, $h\nu \gtrsim 5.6$ eV) is not observed. Broudy claims that Kane's theory has been used beyond its range of applicability and he suggests that only one mechanism is required to explain all the data, namely, direct transitions from valence to conduction band. Thus, there is only a single threshold, which he suggests should be determined by matching observed yields to a linear exponent ($n = 1$) for a small range in $(h\nu - h\nu_0)$. Another interesting feature of Broudy's work is a study of the crystallographic dependence of the yield; he argues that more accurate interpretations of Y versus $(h\nu - h\nu_0)$ must consider the position in the Brillouin zone of the optical transition.

Nickel + Oxygen and Copper + Oxygen Systems.—A fundamental problem in the understanding of metal–oxygen interaction is to diagnose the early stages of oxide formation, and in particular where chemisorption goes over to lattice penetration (incorporation) leading to oxide growth. In addition, we would like to know whether the oxide grows patch-wise or uniformly; is the oxide structurally, crystallographically, and electronically comparable with the bulk oxide, or is there a gradual transition from $M—O_{chemi} \longrightarrow M_xO_y \longrightarrow MO$, where MO is the bulk oxide? Anderson and Klemperer,[23b] in the early sixties, applied photoelectric measurements to study the $Ni + O_2$ system, using metal films as substrates. They plotted their spectral sensitivity data as $Y^{\frac{1}{2}}$ versus $h\nu$ (i.e. assuming applicability of the Fowler theory) and

---

[32] G. W. Gobeli and F. G. Allen, *Surface Sci.*, 1964, **2**, 402.
[33] (a) T. E. Fischer, F. G. Allen, and G. W. Gobeli, *Phys. Rev.*, 1967, **163**, 703; (b) W. E. Spicer and R. E. Simon, *J. Phys. and Chem. Solids*, 1962, **23**, 1817.
[34] R. M. Broudy, *Phys. Rev. (B)*, 1970, **1**, 3430.

interpreted composite curves as being due to two surface patches of different work function. When the surfaces (after reacting with oxygen) were stood *in vacuo* at 296 K the work function of the patch with the lower work function decreased to below the original value for the clean metal value, and this was interpreted as surface regeneration by oxygen incorporation. These conclusions were supported by capacitor studies of a number [35-37] of metal + O_2 systems including nickel, but there still remained some ambiguity regarding the interpretation, and in order to explore the observed phenomena further we obtained information on both the photoelectric yield and the energy distribution of the photoelectrons as a function of oxygen exposure and temperature.

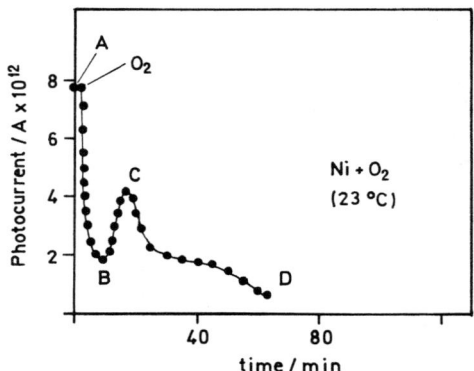

Figure 14 *Variation of the photocurrent ($\lambda = 2150$ Å) from a clean nickel ribbon* [26] *during exposure to oxygen at* 296 K

Yield. Variation of the photoelectric yield during oxygen interaction with 'clean' polycrystalline nickel and copper ribbons at 296 K is shown in Figure 14. With nickel there is clear evidence for two processes, one that increases the surface barrier for emission (decrease in photocurrent) and at a greater oxygen exposure a second surface process that leads to recovery of the photocurrent. The process responsible for the recovery of the yield is activated, since it is only observed at temperatures above *ca.* 153 K. In the case of copper [23c], there is no evidence from yield data alone (Figure 15) of two processes, but the energy distribution of the photoelectrons (see below) provides unambiguous information on oxide formation. Enhancement of the yield during oxygen interaction has also been recognized [38] in semiconductors, *e.g.* Cs_3Sb.

---

[35] C. M. Quinn and M. W. Roberts, *Trans. Faraday Soc.*, 1964, **60**, 899.
[36] *Nature*, 1963, **200**, 648.
[37] T. Delchar and F. C. Tompkins, *Proc. Roy. Soc.*, 1967, **A300**, 141.
[38] R. N. Bloomer and B. M. Cox, *Brit. J. Appl. Phys.*, 1965, **16**, 605.

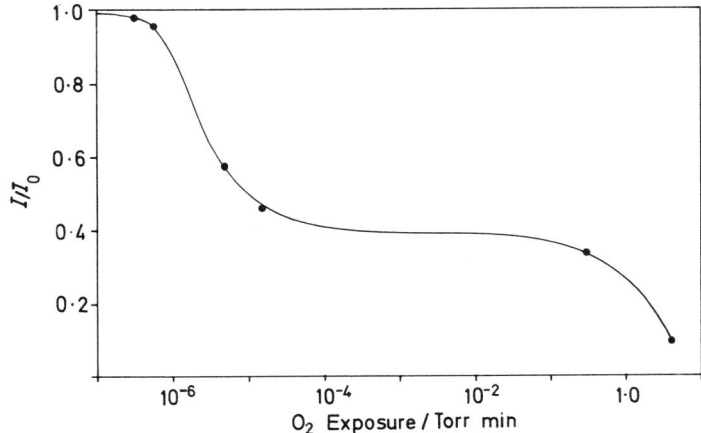

Figure 15 *Variation of the quotient* I/I_0 (I_0 = *photocurrent at zero oxygen exposure*) *during the interaction of* O_2 *with copper at* 296 K

Energy Distribution. The energy distribution of electrons emitted from a surface by photons of energy $h\nu$ can be ascertained by determining the variation of the photocurrent as a function of the voltage applied between the collector and the emitter surface. Equations (15) and (16) give the relationships between the stopping potential (V_0), saturation potential (V_s) (Figure 16), and the work function of the emitter (φ) and of the collector (φ_c).

$$eV_0 = \varphi_c - h\nu \qquad (15)$$

$$eV_s = \varphi_c - \varphi \qquad (16)$$

On the other hand, if the highest filled electron states are not at the Fermi level (which they are in the case of metal) then the stopping potential in this case (V_0') is given by equation (17), where δ (eV) is the distance of the highest

$$eV_0' = \varphi_c - h\nu \pm \delta \qquad (17)$$

filled electron states above or below the Fermi level. We therefore have in principle a method of distinguishing, during the course of metal–gas interaction, between the formation of a chemisorbed surface bond and the incipient growth of a new surface phase with its own characteristic electron band structure. In the case of chemisorption, the highest filled states would clearly remain unchanged and $V_0 = V_0'$; on the other hand, for a surface oxide we would *in general* anticipate a shift in V_0. A detailed interpretation of δ must consider possible energy losses which the emitted electrons undergo and this is not easy. Equation 17 in that case would take the form of equation (18),

$$eV_0' = \varphi_c - h\nu - E_{loss} \pm \delta \qquad (18)$$

where E_{loss} is the energy lost by the electron in escaping through the 'oxide'.

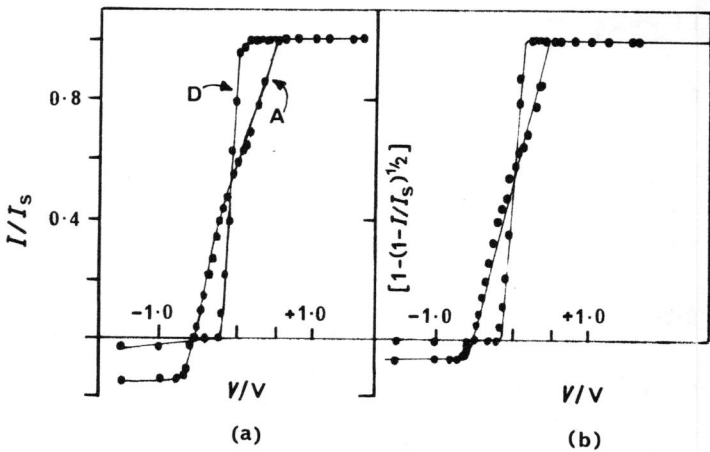

Figure 16 (a) *Relative photocurrent I/I_s (I_s is the saturation value at $+2$ Volts) as a function of the voltage V applied between the nickel emitter and the gold cylindrical collector;* (b) *parabolic plot of data in* (a). A *refers to the clean metal, and* D *the metal after oxygen exposure* (see Table 2)

But what is unambiguous is that if a shift in V_0 is observed then we have a means of distinguishing between true chemisorption and a surface whose band structure is perturbed in depth, although that depth may be no greater than 5—10 Å. It is this aspect of energy distribution data which we have found particularly useful in studies of metal–oxygen interaction.

When the energy distribution of the photoelectrons was studied after various oxygen exposures (points B, C, and D in Figure 14), changes in V_0' [equation (17)] were first detected at exposures corresponding to point C ($\sim 10^{-6}$ mmHg min). After an oxygen exposure of $\sim 10^{-4}$ mmHg min (D), a shift δ of ~ 0.35 eV was observed (Table 2). In a sense we can regard the shift in V_0, δ, as a 'chemical shift' reflecting the creation of new electron levels arising from oxygen incorporation, even though the total extent of oxygen uptake is not much in excess of a monolayer. It will be particularly interesting to consider in the same way oxygen interaction with individual crystal planes of nickel.

Table 2 *Values of V_0 and δ for Photoelectrons* [36] *from Ni at 5.76 eV as a function of O_2 exposure at 296 K*

Stage (see Figure 14)	V_0/V	δ/eV	Exposure/mmHg min
A	−0.50	—	0
B	−0.50	—	$\sim 10^{-7}$
C	−0.45	0.05	$\sim 10^{-6}$
D	−0.10	0.4	$\sim 10^{-4}$

The variation of the photoelectric yield during the interaction of oxygen with copper (Figure 15) did not show [23c] any maxima or minima (cf. Figure 14) but there was an extensive plateau where the yield did not vary very much although the exposure increased a hundred-fold. There is little doubt that during this region of *constant yield* extensive oxygen uptake occurs, as indicated by the microbalance studies of Jennings and Stone.[39] The reason for this is as yet not clear, but there are a number of possibilities and these will be discussed elsewhere.[23c]

One of the intriguing problems in the $Ni+O_2$ work [26] was that a shift in V_0 [equation (15)] was in fact observed at all when the total oxygen uptake was equivalent to no more than two atomic layers (O_2 exposure $\sim 10^{-4}$mm min). Volume effects in photoemission from metals and semiconductors had only just been accepted in the early sixties, but in all cases electron escape depths were quoted as several hundred ångströms except for one case [4] (silicon), where it was ~ 20 Å. We suggested that oxygen interaction with nickel at 296 K and 10^{-4}mmHg O^2 pressure led to the formation of Ni_xO and that this 'oxide' was active in the emission process. An alternative possibility was that we were seeing electrons emitted from the metal, but that these electrons lost energy, E_{loss}, in escaping through the 'surface oxide'. Whichever was the process involved, a surface oxide (not NiO) was essential to the argument. It is relevant to reconsider these conclusions in conjunction with those of a recent and interesting paper by Broudy,[40] who introduces the concept of surface-enhanced optical absorption (SEOA), in which for certain transitions excited by light polarized perpendicular to the surface, optical absorption in the region of the surface and immediate sub-surface is increased as much as a hundred-fold. It is suggested, and evidence is given in support, that the band structures near the solid–vacuum interface are perturbed by the presence of the surface in such a manner that enhanced absorption occurs for certain electron states and that SEOA is fundamental to the understanding of vectorial effects in photoemission. Although qualitative theoretical predictions, and confirming experimental data, have been presented for the influence of crystal symmetry on optical absorption and electron emission, it is likely that selection rules for absorption of light (which predict strong dependence of excitation probabilities on the direction of polarization) can probably be ruled out as the explanation for Broudy's vectorial observations. The main argument used was the influence of the surface and this is not encompassed in a volume selection rule.

The Aluminium–Oxygen and Aluminium–Water Interfaces.—The emission of electrons from aluminium has been investigated in some detail, much of the impetus for this study having come from a need to unravel the mechanism of what has been termed exo-electron emission. These exo-electrons were first

---

[39] T. J. Jennings and F. S. Stone, *Adv. Catalysis*, 1957, **9**, 441.
[40] R. M. Broudy, *Phys. Rev. (B)*, 1971, **3**, 3641.

observed during the deformation and abrasion of metals and were claimed [41] to arise from excitation by low-energy photons (visible region of the spectrum), their origin being suggested to be low-energy-level trapping sites in the oxide. Ramsey,[42] however, concludes that it is the ambient, particularly the water vapour present, which is chiefly responsible for the observed phenomena. The water molecules are suggested to orientate themselves with the positive end of the dipole outwards, thus lowering the work function to such an extent that radiation of wavelength 'in excess of 3450 Å' will be sufficient to excite photoelectrons. The photothreshold for 'clean' aluminium is at $\sim$2900 Å. This interpretation is consistent with the 'capacitor' work-function studies of Huber and Kirk [43] and Roberts and Wells [44] of the interaction of oxygen and water vapour with aluminium films. More recently, Batt and Mee,[45] using a novel variant of the usual photoelectric approach, have confirmed the capacitor data.

Studies by Delchar [46] have indicated quite clearly that, at least for nickel, copper, and tungsten, emission is clearly induced by oxygen chemisorption. The electron yield is $\sim 10^{-9}$ electrons per oxygen molecule adsorbed; the yield was temperature dependent in that it 'increased slightly with increasing temperature'.

Density of States.—The determination of the density of states from photo-emission data depends very much on the validity or otherwise of Koopmans' theorem, since the measuring process is essentially disruptive. If, however, the theorem holds, then it follows that the photon energy necessary to excite an electron from state ε_i to state ε_k is just the difference between the one-electron energies of the two states. However, if the eigenfunctions of other states are modified in the excitation, Koopmans' theorem will not be valid, and it has been suggested [47] that the density of states obtained from both photoemission and optical studies should be referred to as the optical density of states unless it is conclusively shown that there is no significant deviation from the ground state, *i.e.* the unperturbed density of states. Recently, for example, the density of states obtained for Ni by photoemission has been compared [48] to that determined by X-ray studies; it may in fact be dangerous to assume that the unperturbed valence-band density of states is given directly by X-ray emission spectra, since many-body effects (and therefore the breakdown of Koopmans' theorem) may occur. Fadley and Shirley [49a] and Spicer [49b] have recently critically reviewed the determination of density of

[41] L. Grunberg and K. H. R. Wright, *Proc. Roy. Soc.*, 1955, **A232**, 403.
[42] J. A. Ramsey, *Surface Sci.*, 1967, **8**, 313.
[43] E. E. Huber and C. T. Kirk, *Surface Sci.*, 1966, **5**, 447.
[44] M. W. Roberts and B. R. Wells, *Surface Sci.*, 1967, **8**, 453; 1969, **15**, 325.
[45] R. J. Batt and C. H. B. Mee, *Appl. Optics*, 1970, **9**, 79.
[46] T. A. Delchar, *J. Appl. Phys.*, 1967, **38**, 2403.
[47] W. E. Spicer, *Phys. Rev.*, 1967, **154**, 385.
[48] J. R. Cuthill, A. J. McAllister, and M. L. Williams, *Phys. Rev. Letters*, 1966, **16**, 993.
[49] (a) C. S. Fadley and D. A. Shirley, *J. Res. Nat. Bur. Standards*, 1970, **74A**, 543;
 (b) W. E. Spicer, *ibid.*, p. 397.

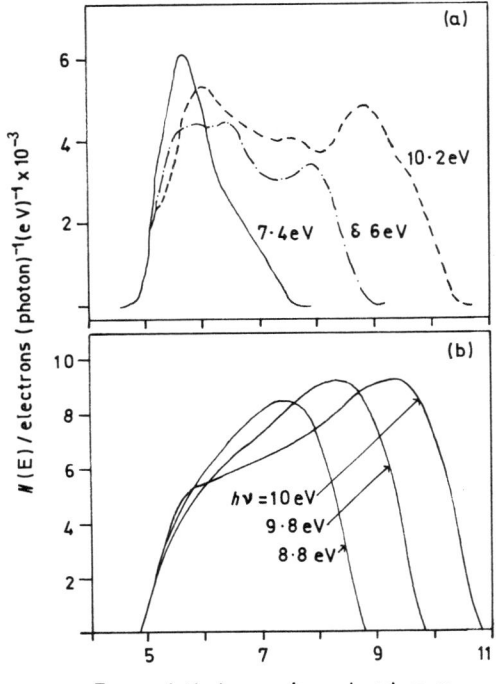

Figure 17 Photoelectron[49b] *EDC's for* (a) *cleaved germanium single crystal and* (b) *amorphous germanium film.* N(E) *is the number of electrons per absorbed photon per eV. The horizontal axis gives the electron energy relative to the maximum in the valence band. Note the sharp structure in* (a) *due to direct transitions*

states by photoelectron spectroscopy. Difficulties arise in considering electron scattering but Smith assumes, with some justification, that electrons escape from the noble metals without appreciable scattering. The most interesting conclusion from Smith's work [50] is that he observed clear evidence for direct transitions in caesiated copper. It appears that poor sample preparation prevented Berglund and Spicer [51] seeing direct transitions and contributed to Spicer advocating the non-direct model. It should be emphasized, however, that the non-direct-transition model provides [51,52] a simple way of analysing photoemission data to obtain an ODS.* Once this has been achieved, EDC's can be calculated and compared with experimental curves, and the relevance of the non-direct approach can be assessed. An interesting experiment for distinguishing between direct and indirect transitions is to examine [49b] the

---

* Optical density of states.
[50] N. V. Smith and M. M. Traum, *Phys. Rev. Letters*, 1970, **25**, 1017.
[51] C. N. Berglund and W. E. Spicer, *Phys. Rev.*, 1964, **136**, 1044.
[52] W. F. Krolikowski and W. E. Spicer, *Phys. Rev.*, 1969, **185**, 882.

effect of reducing or destroying the periodicity of the lattice. Since conservation of the wave vector k is imposed by the lattice periodicity, destroying it should remove k conservation as a selection rule, and in the case of complete disorder k will lose meaning as a quantum number. Germanium in fact provides a good example of a case where the u.v. EDC's change drastically when the long-range order is destroyed by forming amorphous Ge. There is here apparently clear evidence for direct transitions (crystalline) going over to non-direct transitions (amorphous), the loss of sharp structure reflecting the loss of long-range order (Figure 17).

Permission to reproduce various Figures in this chapter is acknowledged from Dr. G. W. Gobeli, Dr. C. H. B. Mee, Dr. F. J. Pipenbring, Professor Dr. W. M. H. Sachtler, and Professor W. E. Spicer.

6
The Application of Electron Spectroscopy to Surface Studies

BY C. R. BRUNDLE

1 Introduction

The potential of electron spectroscopy in the field of surface chemistry is reviewed. Each of the several branches of electron spectroscopy gives somewhat differing information and it has come about that separate experimental arrangements of widely differing capabilities have been used in each case. There seems little reason why all branches should not be carried out in a general purpose high-resolution electron spectrometer, based on one of the several research and commercial designs available.[1] A comparison of the surface information obtained by the various branches would then be possible; a situation which one hardly obtains at the present time because of the variation in conditions used (*e.g.* vacuum, surface cleanliness, specimen control and manipulation, sensitivity, resolution).

The ideal surface spectroscopic technique would be capable of answering the following questions:

(*i*) How clean is the 'clean' surface being studied? What are the impurities, how are they distributed between the surface and sub-surface, and in what concentrations?

(*ii*) Can the surface coverage be determined during adsorption?

(*iii*) For single crystal planes, how does the bonding site repeat itself over the crystallographic surface?

(*iv*) Is there strong chemisorption occurring?

(*v*) Is there surface incorporation occurring?

(*vi*) For (*i*), (*iii*), (*iv*), and (*v*) what is the nature of the bonding involved? Can changes in stoicheiometry be followed? Can the dissociation of surface species be studied?

With the exception of (*iii*), which in principle can be answered by LEED,[2] the other problems can be explored by the combined techniques of electron spectroscopy. In Section 3 the factors influencing the likelihood of our obtaining the required answers by electron spectroscopy are discussed, while in Section 2 the principles behind the techniques are outlined.

---

[1] Vacuum Generators Ltd., England; A.E.I. Ltd., England; Varian Associates Ltd., USA.
[2] See, for example, J. W. May, *Adv. Catalysis*, 1970, **21**, 151.

2 Principles of Electron Spectroscopy

Summary of the Techniques Involved.—When monochromatic radiation of sufficient energy, $h\nu$, strikes a free atom or molecule, electrons will be ejected from that atom or molecule with energies, $E_n = h\nu - I_n$ where I_n is the ionization potential to the nth ionized state of the system. Each ionization potential corresponds to the removal of an electron from a bound electron energy level (Molecular Orbital, MO). Determination of I_n by measurement of E_n has come to be known as photoelectron spectroscopy.[3—7] It is one branch of the more general field of electron spectroscopy, which includes any technique involving the analysis of electron energies. If a monochromatic electron beam replaced $h\nu$, the inelastic scattering of the beam by the sample

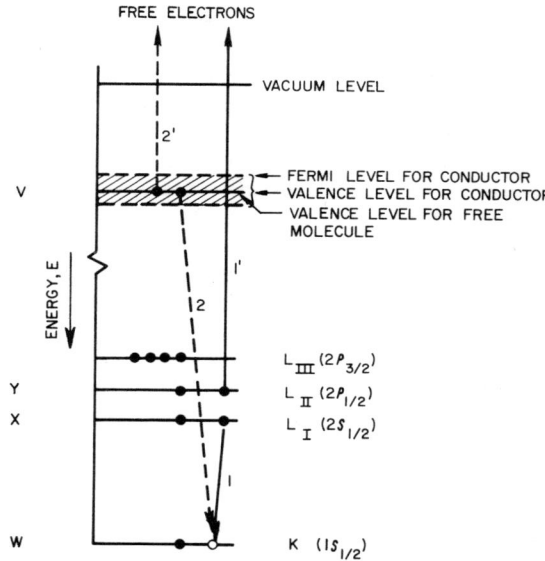

Figure 1 *Energy level diagram for Auger processes. The terminologies* W, X, Y, *and* V *are general ones.* W, X, *and* Y *represent any inner shell levels, with* W *the one carrying the original electron vacancy, and* V *is a valence level or band*

---

[3] K. Siegbahn *et al.*, 'Electron Spectroscopy for Chemical Analysis', Nova Acta Regiae Societatis Scientorium Upsaliensis, Ser. IV, 1967, Vol. 20.
[4] K. Siegbahn *et al.*, 'ESCA Applied to Free Molecules', North Holland, Amsterdam, 1969.
[5] D. W. Turner, in 'Physical Methods in Advanced Inorganic Chemistry', ed. H.A.O. Hill and P. Day, Interscience, London, 1968, p. 74.
[6] D. W. Turner, A. D. Baker, C. Baker, and C. R. Brundle, 'Molecular Photoelectron Spectroscopy', Wiley, New York, 1970.
[7] C. R. Brundle and M. B. Robin, in 'Determination of Organic Structures by Physical Methods', ed. F. Nachod and G. Zuckermann, Academic Press, New York, 1971, Vol. 3, p. 1.

could be studied, the technique generally being referred to as electron impact energy loss spectroscopy.[8] The processes occurring may be represented by the following equations:

$$e_1 + M \rightarrow M^* + e_2$$

and/or

$$e_1 + M \rightarrow M^+ + e_2 + e_3$$

where e_1 is the primary electron, e_2 the same electron inelastically scattered, and e_3 an ejected electron.

The Auger process [9, 10] can be studied by using monochromatic or polychromatic radiation or electron beams. It is a secondary electron process which follows the ejection of an electron (*i.e.* a 'photoelectron spectroscopy' electron) from an inner shell level, W (Figure 1). The hole is filled by an electron falling to the vacant level, which provides energy for another electron (the Auger electron) to be ejected. This is represented by processes 1,1′ or

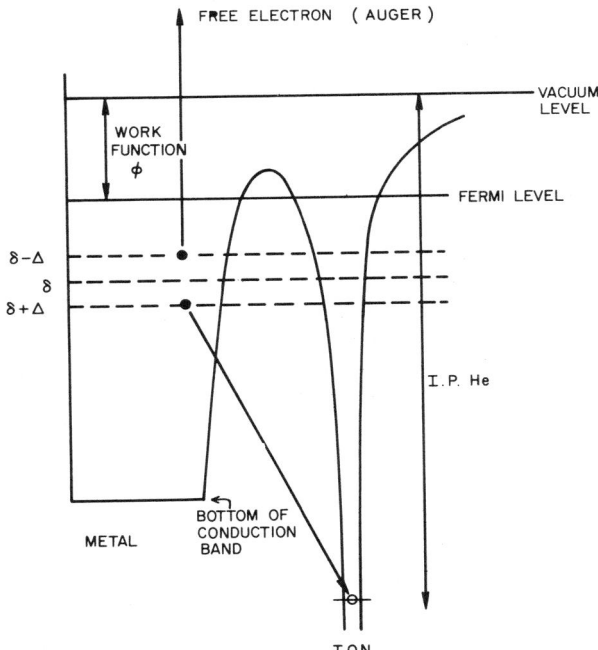

Figure 2 *Energy level diagram for the ion neutralization process*

[8] S. Trajmar, J. K. Rice, and A. Kupperman, *Adv. Chem. Phys.*, 1970, **18**, 15.
[9] N. J. Taylor, *Vacuum*, 1969, **19**, 575; *J. Vacuum Sci. Tech.*, 1969, **6**, 241.
[10] C. C. Chang, *Surface Science*, 1971, **25**, 53.

2,2′ in Figure 1. The energy of the Auger electron is independent of the energy of the impacting electron or photon, depending only on the electron energy levels involved in the secondary process. It should be appreciated that peaks corresponding to Auger processes are therefore also going to appear in photoelectron spectra, if $h\nu$ is of sufficiently high energy to cause a W hole, since this is a major decay process for M^+ in an excited state.

Ion neutralization spectroscopy [11] is a branch of Auger spectroscopy not involving initial ejection of an inner shell electron. The Auger process is induced by bombarding the sample with monoenergetic noble gas ions. An outer electron is transferred to the bombarding ion, and an Auger electron is ejected by the energy so released (Figure 2).

Chemical Information from Electron Spectroscopy.—*Photoelectron Spectroscopy*. Photoelectron (p.e.) spectroscopy can be divided into two classes, depending on the type of radiation, either u.v. and vacuum-u.v.[5-7] ($\leqslant 60$ eV), or soft X-rays [3, 4, 7] (*ca.* 1000 to 20 000 eV), used to produce ionization.

The vacuum-u.v. study of gases has been termed molecular p.e. spectroscopy. The most important aspect of such spectra is that, to a certain level of approximation, they are a direct display of the various molecular orbital energies, and their bonding characteristics, in a molecule. Figure 3 illustrates this for CO.[6, 12] The first three ionization potentials are shown, corresponding to the removal of an electron in turn from the three most weakly bound molecular orbitals. In addition to providing the orbital energies, considerations of the band shapes and vibrational structure indicate [6] that the highest σ-levels are nearly non-bonding in character, and that the π-level is bonding. This is in agreement with theoretical calculations.[7, 13] Over the past five years, great use has been made of the technique in elaborating the detailed electronic structure of organic,[14] inorganic,[15] and organometallic [16] molecules.

For solids there have been many studies [17] on metal surfaces using radiation of up to 11.6 eV energy (cut-off value for transmission through an LiF window) under ultra-high vacuum conditions. These are generally referred to as photoemission studies and the energy distribution curves (EDCs) of the ejected electrons give information on the band structures of the metals concerned. A typical EDC for clean Ni [17] is shown in Figure 4(a). Work

[11] H. D. Hagstrum, *Phys. Rev.*, 1966, **150**, 495.
[12] W. C. Price, in 'Molecular Spectroscopy', ed. P. Hepple, Inst. Petroleum, London, 1968.
[13] D. Neumann and L. C. Snyder, private communication.
[14] See, for example, A. D. Baker, C. Baker, C. R. Brundle, and D. W. Turner, *Internat. J. Mass Spec. and Ion Phys.*, 1968, **1**, 285.
[15] See, for example, C. R. Brundle, M. B. Robin, and G. R. Jones, *J. Chem. Phys.*, 1970, **52**, 3383.
[16] See, for example, S. Evans, J. C. Green, M. L. H. Green, A. F. Orchard, and D. W. Turner, *Discuss Faraday Soc.*, 1969, **47**, 112.
[17] D. E. Eastman, 'Photoemission Spectroscopy of Metals', to appear in 'Techniques in Metals Research VI', ed. E. Passaglia, Interscience, London and New York; W. E. Spicer in 'Survey of Phenomena in Ionized Gases', Internat. Atomic Energy Agency, Vienna, 1968, p. 271.

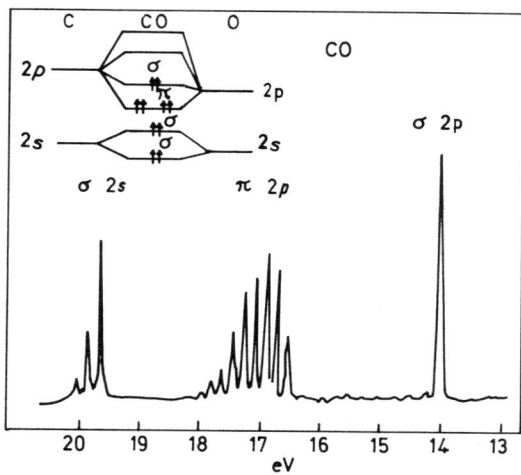

Figure 3 *Molecular photoelectron spectrum of* CO, *using the* HeI *resonance line* (21.2 eV) *as ionizing source. The three lowest* I.P.'s *are revealed, together with the vibrational structure for each state of the ion*
(Reproduced by permission from 'Molecular Photoelectron Spectroscopy', Wiley, New York, 1970)

of this nature is identical to molecular photoelectron spectroscopy, but because of the band nature of the valence levels of a solid, as opposed to the discrete molecular orbitals of gaseous molecules, the resulting spectra are very different (*cf*. Figure 3). Molecular photoelectron spectroscopy has been developed and used primarily by chemists and molecular physicists, whereas the photoemission work on solids has been the domain of the solid-state physicist and the metallurgist. There has been relatively little interaction between the two groups.

The use of soft X-rays to determine ionization potentials has come to be known as electron spectroscopy for chemical analysis (ESCA)[3,4] though other names have also been used [X-ray photoelectron (X—p.e.) spectroscopy, induced electron emission, IEE]. In addition to ejecting electrons from the valence shell orbitals, the X-rays have sufficient energy to eject electrons from some of the inner shells. For several reasons[3] the energy resolution in this work is much inferior to that of the vacuum-u.v. work, and for the valence shell region, intensities of ionization are very weak. The major usefulness of the ESCA approach is therefore in the inner shell region beyond the range of the vacuum-u.v. radiation. Inner shell orbitals are essentially atomic in nature, being tightly bound to individual atoms in the molecule, and as such the ionization potentials from these orbitals are characteristic of the atom concerned rather than the molecule of which it forms part. Thus, all C $1s$ ionizations fall at approximately 290 eV, all N $1s$ at 400 eV, and all O $1s$ ionizations at 530 eV. The atoms present in a molecule, or

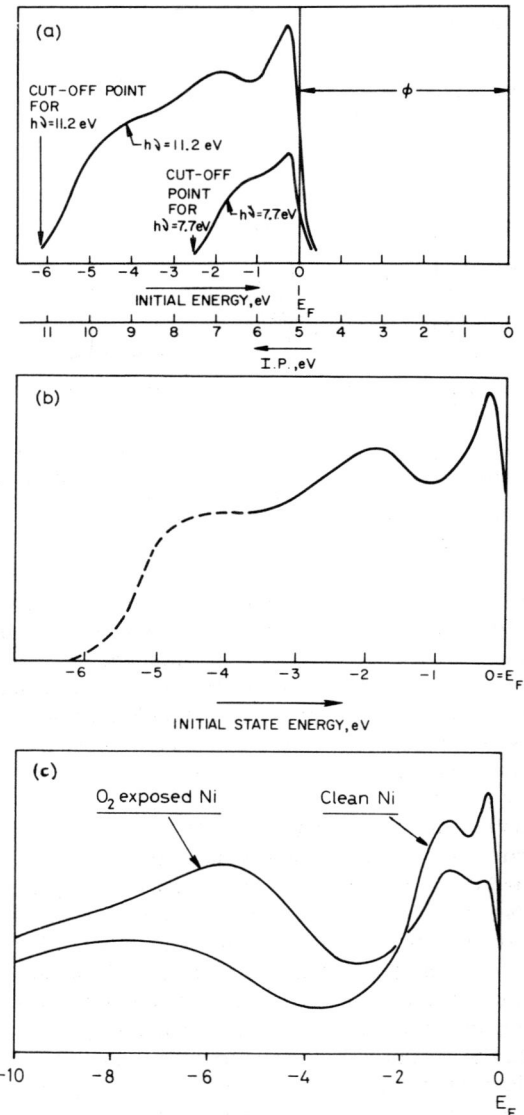

Figure 4 (a) *EDC of a clean nickel surface taken with* 11.2 eV *and* 7.7 eV *photon energies.* (b) *Optical Density of States (ODS) curve for nickel, derived from the* 11.2 eV *EDC of (a). The dotted portion indicates uncertainty in this region.* (c) *EDC of clean nickel, and the EDC of nickel plus adsorbed oxygen, recorded using* 21.2 eV *photon energy*
[Reproduced by permission from 'Techniques in Metals Research, VI', Interscience, London and New York, and *Phys. Rev. (B)*, 1971, **3**, 1769]

mixtures of molecules, can, therefore, be identified. The technique has the advantage that *all* elements except hydrogen (no atomic-like inner shell) may be detected in this fashion. The exact value of an atomic-like inner shell ionization does have some dependence on the atom's environment, however, giving rise to 'chemical shifts'. For example, the C 1s ionization of

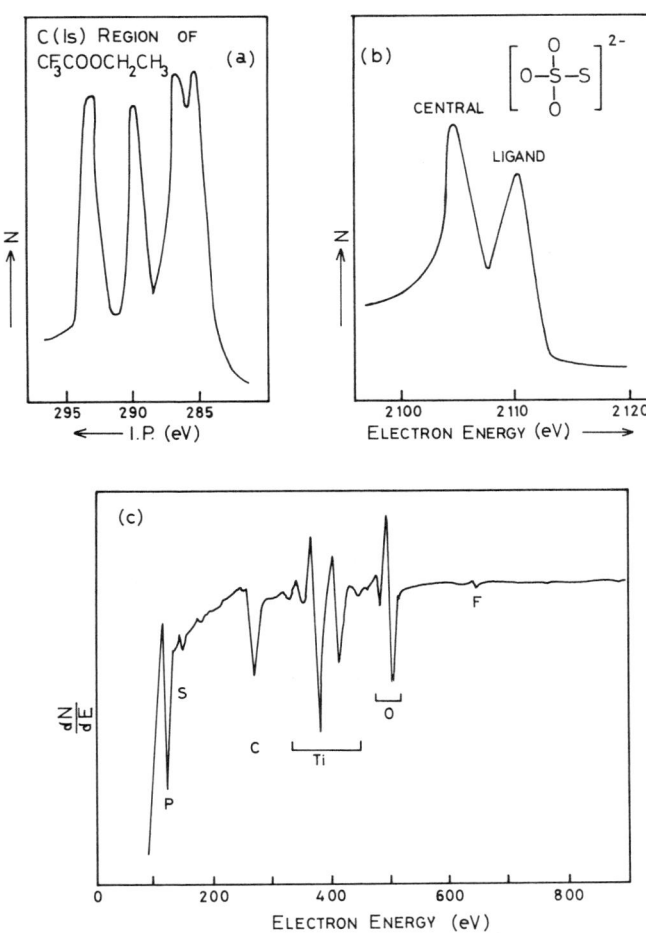

Figure 5 (a) *ESCA C 1s spectrum for* $CF_3COOCH_2CH_3$. (b) *KLL Sulphur Auger spectrum of sodium thiosulphate, X-ray induced.* (c) *Complete derivative Auger spectrum of etched titanium, electron impact induced*
[Reproduced by permission from 'Electron Spectroscopy for Chemical Analysis', Nova Acta Regiae Societatis Scientorium, Upsaliensis, Ser. IV, 1967, Vol. 20, and *Vacuum*, 1969, **19**, 575]

methane is 290.8 eV, but in CF_4 is 301.8 eV.[18] The values of the various fluoromethanes fall in between. For the molecule $CF_3CO_2CH_2CH_3$ four distinct C $1s$ ionizations are detected [7] (Figure 5a). The chemical shifts can be related, with varying degrees of success, to atomic oxidation states, formal oxidation numbers, or calculated atomic charges; *i.e.* one is gaining information on electron charge distribution within the sample.

Energy Loss Electron Spectroscopy. Here one is looking at inelastically scattered electrons. Much work has been done on gases [8] and the most useful information obtained relates to excitation processes not resulting in ionization. The technique spans the range of i.r. (vibrational exitation), visible, u.v., and vacuum-u.v. (electronic excitation) absorption spectroscopy, and provides similar information to these. The energy loss range for such transitions covers only about 0—40 eV. It will be appreciated that for incident electron energies above ionization limits, orbital electrons will be ejected, as in photoelectron spectroscopy, and at high enough electron energy, Auger and other secondary electron processes can also occur. These additional processes lead to much structure in the energy loss spectrum at energy loss values above the first ionization potential.

For solids,[19] discrete electron energy losses arise also from such processes as phonon and plasmon excitations as well as the interband electronic transitions.

Electron Impact Auger Spectroscopy. In Figure 1 an Auger process for a free molecule was depicted. In the KL_IL_{II} Auger transition depicted in process 1,1′ the energy of the ejected Auger electron is given approximately by $E_K - (E_{L_I} + E_{L_{II}})$. This energy will be characteristic of the atom involved, as in ESCA, since K and L_I are inner shell atomic-like orbital levels. Generally the 'chemical shift' behaviour will be more complex than that of ESCA because it represents the difference of the individual level shifts which show up in the ESCA spectrum. In cases where the upper levels involved in the Auger process are valence shell levels (process 2,2′ in Figure 1) there will be no simple correlation of the shifts observed in Auger peaks, on going from one molecule to another, to atomic charges *etc.*, because of the complex changes in the valence shell structures. Such Auger transitions are often named *WVV* transitions (see Figure 1). Examples of this type of transition are the *KLL* Auger spectra of CH_4, C_2H_6, and C_6H_6,[8] which show complicated band envelopes. If, however, all the levels involved in the Auger process are of the inner shell type, (process 1,1′ in Figure 1—named *WXY* Auger) then chemical shifts similar to ESCA shifts may be observed. In the simple theory of chemical shifts,[3] all inner shell level shifts for an atom in a molecule should be approximately the same, since they are all electrostatically affected by the charge of the valence shell to about the same extent (see Section 3).

[18] T. D. Thomas, *J. Amer. Chem. Soc.*, 1970, **92**, 4184.
[19] O. Klemperer and J. P. G. Shepherd, *Adv. Phys.*, 1963, **12**, 355.

Thus in a case where the atomic chemical shift in an inner shell electron binding energy is $+\Delta E$ in one molecule compared to another, the Auger $KL_\mathrm{I}L_\mathrm{II}$ electron energy is given approximately by:

$$E_K + \Delta E_K - (E_{L_\mathrm{I}} + \Delta E_{L_\mathrm{I}} + E_{L_\mathrm{II}} + \Delta E_{L_\mathrm{II}})$$

which, for $\Delta E_K = \Delta E_{L_\mathrm{I}} = \Delta E_{L_\mathrm{II}}$, is:

$$E_K - (E_{L_\mathrm{I}} + E_{L_\mathrm{II}}) - \Delta E_K$$

i.e. the same chemical shift, ΔE_K, as in the ESCA K level spectrum is observed. In practice, of course, ΔE_K will not be exactly equal to ΔE_L, but there are cases where similar shifts are observed in both ESCA and Auger spectra. The KLL Auger spectrum of sulphur in sodium thiosulphate [3] is an example. The chemical shift between the two types of sulphur ($+6$ and -2 oxidation states) in the Auger spectrum is 4.6 eV (Figure 5b), whereas the ESCA shift in the K shell is 7.0 eV, and in the L shell 6.0 eV. Thus the expected Auger shift is

$$\Delta E = \Delta E_K - 2\Delta E_L = 5.0 \text{ eV}$$

compared to the observed 4.7 eV value.

Many studies on both gases and solids have been carried out using Auger spectroscopy. Most of the gaseous studies [4, 20] have been made under high-resolution conditions and detailed studies have been largely confined to atoms and diatomic molecules. The reason for this is that owing to coupling schemes and the large number of transitions possible, an Auger spectrum is likely to be very complex if there is sufficient resolution available to separate all the lines present. Auger peaks as narrow as 0.15 eV (vibrational structure) have been observed in some cases for gases.[4] Factors which can broaden Auger peaks besides poor instrument resolution include lifetime limitations on the final-state doubly-ionized ion. In studies in the solid state, broadening occurs owing to solid-state effects, particularly when the valence level is involved. The best-known area in solid work is for the monitoring of surface impurities.[9, 10] Figure 5(c) shows a typical Auger spectrum of this type of work. It is of an etched 'clean' sample of titanium [9] and the Auger peaks are broad (5—20 eV half-width) and unresolved, but still highly characteristic of the elements present. (The spectrum shown is the derivative of the electron energy distribution, see later.) Clearly there is less chance of observing chemical shifts and being able to assign elements to specific molecules in such low resolution spectra. Figure 5(c) was taken using a polychromatic electron beam. If monochromatic X-rays had been used the Auger spectrum would appear superimposed on the normal ESCA spectrum.

Ion Neutralization (I.N.) Spectroscopy. It is generally accepted that i.n. spectroscopy is a genuine surface technique, and as it has been used solely

---

[20] T. A. Carlson, W. E. Moddeman, B. P. Pullen, and M. O. Krause, *Chem. Phys. Letters*, 1970, **5**, 390.

for this purpose major discussion will be postponed to Section 3. It has been applied to clean metal surfaces [21] to yield band structures (*cf.* photoemission studies), and it has been suggested that structure relating to the molecular orbitals of surface compounds can be seen in the electron energy distribution curves.[22]

3 Application to Surface Studies

In this section each branch of electron spectroscopy is taken in turn, the surface-related work already available reviewed, and suggestions made as to what further one might hope to achieve. It is necessary first to discuss a factor relevant to all the techniques when evaluating applications to surface studies.

Mean Free Path Lengths of Electrons.—Inelastic scattering can occur through a variety of interactions, the net result in an electron spectrum being a loss of intensity from the electron peak in which we are interested, and the production of peaks and continua at lower energies. These represent the various energy losses incurred and the secondary ionizations produced by the scattering processes. The scattering interactions can involve phonon excitations, single electron interactions, and plasmon excitations (collective electron vibrations of the Fermi electron gas), plus interband transitions. Phonon energies are small ($\leqslant 50$ meV) and so in many cases can be approximated to an elastic scattering process. Interband transitions are either of the valence band to higher band type (typical energy 1—8 eV for metals and semiconductors), or from deeper lying levels to unoccupied higher states (typically 15 eV). Plasmon energies are typically 5—30 eV, and multiple plasmon excitations may also occur. The total amount of electron scattering, and which process is dominant, is a function of the electron energy. The concept of electron mean free path, L, (the mean distance travelled by an electron before it is inelastically scattered; equal to the distance at which $1/e$ of the electrons have not suffered a collision) is useful here. Quinn [23] has derived an equation for the mean free path, L_{pl}, for plasmon scattering, as a function of electron energy, E, which is:

$$L_{pl}(E) = 2a_B \frac{E}{\hbar\omega_p} \left[\ln \frac{(1+y_p)^{\frac{1}{2}}-1}{x-(x^2-y_p)^{\frac{1}{2}}} \right] \quad (1)$$

where

$$x = (E/E_F)^{\frac{1}{2}} \qquad y_p = \frac{\hbar\omega_p}{E_F}$$

---

[21] H. D. Hagstrum and G. E. Becker, *Phys. Rev. Letters*, 1966, **16**, 230; 1967, **159**, 572.
[22] H. D. Hagstrum and G. E. Becker, *Phys. Rev. Letters*, 1969, **22**, 1054; *J. Chem. Phys.*, 1971, **54**, 1015; H. D. Hagstrum, *Phys. Rev.*, 1966, **150**, 495.
[23] J. J. Quinn, *Phys. Rev.*, 1962, **126**, 1453; N. V. Smith and W. E. Spicer, *Phys. Rev.*, 1969, **188**, 593.

and a_B = the Bohr radius. E is the energy measured from the bottom of the free-electron band, $\hbar\omega_p$ is the plasmon energy, and E_F is the Fermi energy. The function is infinite at, or less than, the plasmon formation energy $E_F + \hbar\omega_p$, and has an approximate E dependence at $E \gg E_F$.

A plot of L_{pl} versus E for aluminium from equation (1) is shown in Figure 6(a). Experimental data confirm its essential correctness.

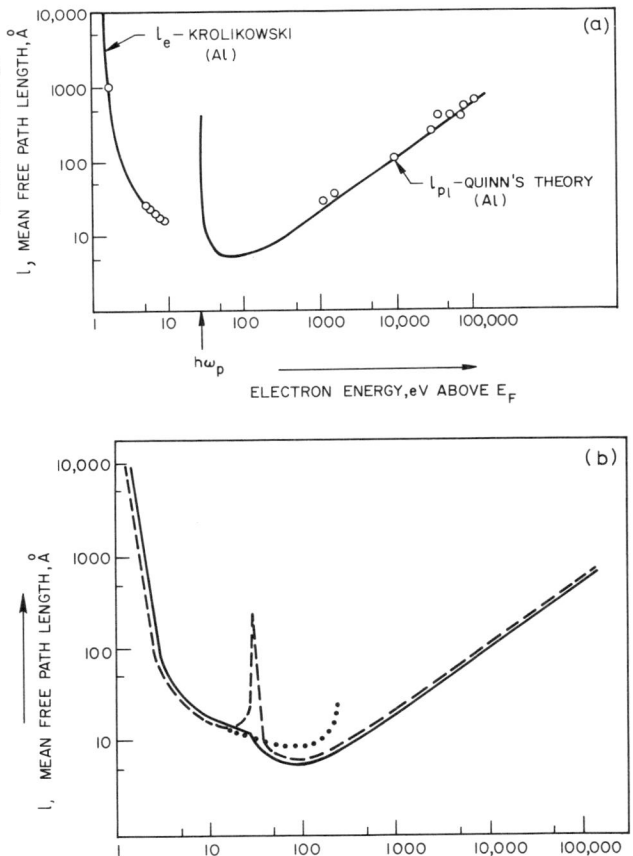

Figure 6 (a) *Plasmon and single electron scattering mean free path-length* vs. *electron energy curves.* (b) *Composite inelastic scattering (plasmon and single electron) curve for an 'average' metal.* See also ref. 23
[Reproduced by permission from 'Elementary Excitations in Solids', Benjamin, New York, 1964, and *Phys. Rev.* (B), 1970, 1, 522, 2357]

There is also an approximate formula for the single electron–single electron scattering path length L_e:

$$L_e(E) \simeq \frac{32 a_B}{Ky} \left[\frac{x^2}{(x^2-1)^2} \right] \qquad (2)$$

where $K = \arctan y + y/(1+y^2)$, $a = (4\pi/9)^{\frac{1}{3}}$, and $y = (\pi/a r_s)^{\frac{1}{2}}$, and where $a_B r_s$ is the radius of a sphere whose volume equals the volume per electron. The equation has a *ca.* $1/(E-E_F)^2$ dependence. The treatment seems to be valid at least over the energy range for which there are available data (Figure 6) though equation (2) is really not appropriate for energies well above the Fermi level, for which L_e will be grossly underestimated. It has been indicated [24] that, for energies above $E_F + \hbar\omega_p$, plasmon production is the dominant scattering process, and it is known that at $E \gg E_F + \hbar\omega_p$ this is true.[19] If the curves in Figure 6(a) are combined to cover the whole energy range above the Fermi level, and if it were assumed that as soon as plasmon interaction started, single electron interaction ceases, a curve something like the dashed curve in Figure 6(b) is obtained. Such enormous discontinuities do not occur in scattering measurements or in photoemission data in this region,[23] so it must be assumed that L_e remains dominant above $E_F + \hbar\omega_p$ for some distance and then takes an upward turn [the dotted line in Figure 6(b)]. A reasonable representation of the total mean free path, L_T, is then given by the solid curve in Figure 6(b). In fact it is known [25] that, in the intermediate energy range, single-electron and plasmon processes are not completely separable, and that no discontinuity at all occurs in the scattering curve at the plasmon energy.

Some experimental figures for mean free path lengths are given in Table 1. We would like some estimate of contribution of the other scattering processes involved. Kanter [26] obtained estimates for phonon scattering of Au $\sim$250 Å, Ag $\sim$400 Å, Al $\sim$250 Å in the energy region 5.5—7.5 eV. Phonon scattering is expected to be independent of energy to a first approximation and therefore will only become a dominant form of scattering below about 2 eV above the Fermi level. In the intermediate region $\sim$10—200 eV it matters little whether other scattering in addition to plasmon and electron–electron scattering occurs (data are scarce for this region) since L_T is so small as to be totally dominant. In the higher energy region, Kanter gives values for L_{elastic} of 26 Å and 37 Å for Al at 1.5 and 2.0 keV respectively. The 'elastic' term includes all other forms of inelastic scattering apart from the plasmon scattering. These are known not to be negligible,[26] but would be expected to have a similar E dependence to plasmon scattering.

Summing up this section, we can say that (for metals) at electron energies up to a few eV above the Fermi level, the mean free path, L, is a rapidly varying function of energy dropping from the 10^3 Å range to the 10^1 Å range when $E \sim$5—10 eV. Over the range $E \sim$200—$\sim$100 000 eV, L increases approximately linearly over a similar Å range. In the intermediate energy

---

[24] H. Thomas, *Z. Physik*, 1957, **147**, 395.
[25] D. Pines 'Elementary Excitations in Solids,' Benjamin, New York, 1964, p. 177.
[26] H. Kanter, *Phys. Rev. (B)*, 1970, **1**, 522, 2357; *Phys. Rev.*, 1961, **121**, 461.

Table 1 Some experimental inelastic electron mean free path lengths[a]

Energy above Fermi level	L_e(Å) 6 eV	L_e(Å) 7 eV	L_e(Å) 9 eV	L_{pl}(Å) 1.5 keV	L_{pl}(Å) 2.0 keV
Au	40	34	21	—	—
Ag	38	30	20	—	—
Al	50 at 5 eV[b]			70	88

[a] Ref. 26.
[b] An independent measurement using a different technique gives an average 4 Å $< L_e <$ 12.8 Å for all electrons of energy between vacuum level and 21.2 eV above Fermi level. T. F. Gesell and E. J. Arakawa, *Phys. Rev. Letters*, 1971, **26**, 377.

range, experimental data are largely lacking but it seems likely that L becomes fairly constant with energy and could be less than 5 Å, (see also Chapter 5, p. 155).

Electron Impact Auger Spectroscopy —Auger spectroscopy is taken first because it is already a well-known surface technique and there is greater understanding of its capabilities than of those of the other branches of electron spectroscopy. It will be assumed that assignments of peaks in Auger spectra are straightforward and unambiguous, although this is not always true. Assignment of a peak to an element is fairly easy because of the documented data accumulated from elemental materials,[3, 10, 27] but assignment of the peak to a specific transition is often made empirically, with help from X-ray tables.[27, 28] Chung and Jenkins have discussed the factors involved in calculating Auger energies.[29]

Sensitivity is governed by the rate of escape of Auger electrons from the solid, and the efficiency of the detecting system. The rate of escape depends on the rate of Auger electron production and the scattering rates. For E_W (Figure 1) $<$ 500 eV the competing decay process of X-ray emission is negligible and so the rate of production of Auger electrons is determined by the W shell ionization rate. At higher values of E_W, X-ray emission becomes important and Auger emission is weakened (the two are equivalent at about 2000 eV). The rate of W shell ionization is dependent on the energy of the incident primary electron beam, E. Experimentally, maximum ionization is found at $E_0 \sim 3E_W$.[30] Scattering effects in the solid will affect the final intensity of the Auger spectrum, but the effect is a very complex one because of the numerous processes possible and quantitative estimates are difficult to make. More important is that scattering controls the depth, d, of material probed (hereafter referred to as the 'escape depth') by making only the surface layers accessible. From the earlier discussions on mean free path lengths it is clear that in the energy range used for E_0 ($\sim 3E_W$, and E_W usually less than 700 eV, therefore E_0 about 400—2100 eV) the mean free path

[27] T. W. Haas, J. T. Grant, and G. J. Dooley, *Phys. Rev.* (*B*), 1970, **1**, 1449; R. D. Hill, E. L. Church, and J. W. Minelich, *Rev. Sci. Instr.*, 1952, **23**, 523.
[28] P. W. Palmberg and T. N. Rhodin, *J. Appl. Phys.*, 1968, **39**, 2425.
[29] M. F. Chung and L. H. Jenkins, *Surface Science*, 1970, **22**, 479.
[30] H. E. Bishop and J. C. Riviere, *J. Appl. Phys.*, 1969, **40**, 1740.

length of the primary beam should be energy dependent (*ca.* $\propto E_0$). Only Auger electrons which have not been inelastically scattered contribute to the measured Auger peak. The Auger peaks of most interest lie between 20 and 700 eV, thus covering the region (Figure 6b) where scattering mean free paths are smallest. In practice, the primary beam energy used is usually $\geqslant$ 1000 eV which means the energy of the escaping Auger electron, (E_{WXY} or E_{WYV}), is generally going to be the dominant factor in defining the escape depth of the technique, with E_0, for which L is longer, only of secondary importance. Experimentally, estimates can be made for d by comparing the signal from an atomic monolayer with that from the bulk material. Gallon [31] has given the theory for this treatment. Measurements of the mean free path lengths, L, in Ag, Au, and Cu are given in Table 2. As a rough guide, d can be taken as three times L. This gives $d \sim$12 Å for a 72 eV electron in Ag, and $d \sim$24 Å for a 362 eV electron. Chang [10] suggests a range for d of 5—50 Å for $E_{WXY} \leqslant$ 500 eV. A Z dependence [30] is expected for $Z = $ 3—30, such that d is smaller for high Z materials. Above $Z = $ 30 little decrease with Z is expected.

Table 2 *Mean free path lengths found in Auger spectroscopy*[28]

Element	Electron energy (eV)	Mean free path length (Å)	No. of monolayers
Ag	72	4	2
	362	8	4
Au	72	4	2
Cu	950	6	4

Summing up the practical aspects of this section it can be said that Auger spectroscopy is limited to the first 50 Å and often much less, and that low-energy Auger electrons from high Z materials are likely to have the smallest escape depth.

The multiple scattering of the Auger (and other) electrons before being emitted (often referred to as secondary electron diffusion) results in a broad background spectrum on which sit the small unscattered Auger peaks [see Figure 5(c)]. Thus, though scattering gives the technique its surface nature, it also causes signal-to-background intensity ratio problems. Most data have been collected using retarding grid electron energy analyser systems which give integrated electron energy distribution curves as primary data. The first derivative of these curves gives the electron energy distribution, but in practice the second derivative is usually taken,[9—10] to help locate the small Auger peaks [*e.g.* Figure 5(c)]. The main sensitivity limitation to such a detection system is shot noise. By using a high-luminosity cylindrical electrostatic electron energy analyser instead of a grid system, a signal-to-noise ratio improvement of up to three orders of magnitude can be obtained.[32] Also the electron energy distribution is obtained as primary data, not the integrated

[31] T. E. Gallon, *Surface Science*, 1969, **17**, 486.
[32] P. W. Palmberg, G. K. Bohn, and J. C. Tracy, *Appl. Phys. Letters*, 1969, **15**, 254.

spectrum, and the instrumental resolution can be made better than for a retarding gird system.

Calculations show that a 'surface' concentration of $n \sim 10^{12}$ atoms cm^{-2} should be the limit of detection for grid systems.[10] (About 10^8 atoms are required for the signal. The true meaning of n depends on the escape depth and the depth distribution of the sample concerned.) This corresponds to about 1% 'surface' concentration. The cylindrical electrostatic analyser should be able to improve this by a factor of about 100, *i.e.* a 0.01% 'surface' concentration; this is probably comparable with the limit of surface cleanliness obtainable in most systems.

The main use of Auger spectroscopy has been to identify impurities at or near the surface, in terms of the elements involved,[9—19] while subjecting the surface to a variety of treatments, *e.g.* cooling, etching, and electron or ion bombardment. For single crystals, Auger spectroscopy provides analytical 'eyes' for the LEED technique.[28] In a study of a copper (100) crystal[33] it was demonstrated that with 'normal' cleaning procedures carbon and sulphur were still present even though the LEED pattern obtained had the features expected for a clean surface. Special cleaning procedures had to be adopted to prevent this contamination (segregation from the bulk), which is a quite common occurrence. On adsorbing CO on to the clean surface, only the oxygen Auger peak (525 eV) appeared. A tentative conclusion is that adsorption with the oxygen atom outermost occurs, in contrast to CO adsorption on to (100) tungsten.[34]

Harris[35] has shown that, by measuring the Auger spectrum at large angles from normal emittance, the surface features are emphasised even more with respect to the bulk. By varying the angle a rough measure of the depth distribution of the elements can be obtained. By studying Auger lines of different kinetic energies similar information should be obtainable.

The two types of 'chemical shift' expected in Auger work were discussed earlier. Nordling *et al.*,[36] observed chemical shifts in Auger peaks from heavily oxidized metal surfaces using *X*-ray excitation. Haas and Grant[37] have demonstrated by LEED that three different surface oxide structures of less than one monolayer coverage can be formed on the tantalum (110) crystal face. In the Auger spectrum, the $N_V N_{VI} N_{VI}$ tantalum peak (intermediate between WVV and WXX type transitions) shifts by 0.2, 0.5, and 1.5 eV respectively, compared to clean tantalum for each of the three oxide structures. The final shift on heavy oxidation is 6 eV. A heavily oxidized layer of tungsten was found to have undergone a shift of 4 eV.[37] These are important results

---

[33] R. W. Joyner, C. S. McKee, and M. W. Roberts, *Surface Science*, 1971, **26**, 303; 1971, in the press; C. S. McKee, Inst. of Petroleum Conference, London, 1970.
[34] J. H. Pollard, *Surface Science*, 1970, **20**, 26.9
[35] L. A. Harris, *Surface Science*, 1969, **15**, 77.
[36] C. Nordling, E. Sokolowski, and K. Siegbahn, *Arkiv Fysik*, 1958, **13**, 483.
[37] T. W. Haas and J. T. Grant, *Phys. Rev. Letters*, 1969, **30A**, 272; *J. Vacuum Sci. Tech* 1970, **7**, 43.

because they demonstrate that shifts which are related to the bonding characteristics of the surface phases can be detected for less than one monolayer coverage.

Amelio and Scheibner [38] studied the KLL Auger transition (WVV type) in graphite single crystals, which appears at about 270 eV, using a grid system. Because of the two-electron nature of the Auger process, a WVV transition in a solid (Figure 1) produces an Auger peak approximately twice as wide as the valence band width. In this case the Auger peak width is about 35 eV. By a mathematical 'unfolding' process, one can transmute the shape of this Auger peak into a band about 17 eV wide representing a transition density which is the product of the occupied density of states (valence band, $\sim$17 eV wide in graphite) and the Auger transition probability. A simple way of approximating the unfolding process is to take the differential of the Auger electron energy curve, which in any case is the way the spectrum is normally recorded. The approximate transition density curve is thus given twice, once as an 'up' peak and once as a 'down' peak. The important point here is that the transition density found is a *surface* (or near-surface, depending on d) density and as such should be highly sensitive to any surface changes affecting the valence band electronic structure. Chang [10, 39] has looked at the WVV transition for silicon (65–95 eV) and was able to relate changes in the differentiated energy spectrum to the surface oxidation of Si to SiO_2. In the same study, a 'stray' peak near 250 eV was observed whose position and intensity fluctuated from run to run. It was suggested that it represents chemisorbed carbon, as CO or CO_2, (*cf.* 270 eV transition for graphite) since the peak is sharp (*i.e.* the carbon has a narrow or discrete valence level characteristic more of a free molecule than a solid), and disappears on heating to 1073 K, with a corresponding increase in the silicon Auger peak.

Haas and Grant [40] have shown that the shape of the KLL Auger carbon spectrum of CO adsorbed on to molybdenum is different from that of carbon segregated at the surface by heating (carbide), and that the KLL Auger spectrum of graphite is quite different from either of these. On the other hand, the spectrum of carbon on platinum (100) [40] is quite similar to that of graphite, supporting earlier conclusions [41] that the carbon is present as graphite overlayers. Coad and Rivière [42] have similarly shown that carbon segregated to the surface of nickel foil is present as Ni_3C below 673 K, but a transition to graphite occurs above 673—873 K.

The above discussion demonstrates that Auger spectroscopy is a surface-sensitive technique, and by alteration of some of the parameters can be made even more so. It is a high-sensitivity technique, and beside elemental identi-

[38] G. F. Amelio and E. J. Scheibner, *Surface Science*, 1968, **11**, 242.
[39] C. C. Chang, *Surface Science*, 1971, **23**, 283.
[40] T. W. Haas and J. T. Grant, *Appl. Phys. Letters*, 1970, **16**, 172; J. T. Grant and T. W Haas *Surface Science*, 1971, **24**, 332.
[41] P. W. Palmberg in 'The Structure and Chemistry of Solid Surfaces,' ed. G. A. Somorjai, Wiley, New York, 1969.
[42] J. P. Coad and J. C. Riviere, *Surface Science*, 1971, **25**, 609.

fication (which should reach the quantitative stage soon) both true chemical shift data and valence band shape changes can be used to study bonding effects on the surface. The main emphasis to date has been on sensitivity rather than resolution. For much solid-state Auger work the lack of resolution is unimportant since bands are usually broad (>3eV). However, it may be a problem when looking for small chemical shifts and small changes in WVV band shapes, and here the high-resolution analyser has the advantage. Since it is known that gaseous WVV spectra often exhibit vibrational fine structure [4, 20] it is not impossible that some surface structures may retain some of this, *i.e.* the molecular orbital levels are more or less discrete and characteristic of a 'surface compound' rather than a solid.

One final point is that there is evidence [43] that the line profiles of WVV Auger transitions in ionic crystals are more a reflection of the coupling of the final doubly-ionized state of the molecule with lattice vibrations than of the density of states of the valence band. Coupling in metals will be much smaller and the density of states treatment valid. Covalent crystals may provide an intermediate case.

Ion Neutralization Spectroscopy.—I.n. spectroscopy is less well known than the Auger technique; detailed papers by Hagstrum and Becker [21, 22] present all the salient features of this work.

Referring to Figure 2, a clean metal is represented with the inert gas ion potential well outside the surface. He^+, Ar^+, and Ne^+ have been used in these experiments. The Auger process involves two electrons at energies $\delta - \Delta$ and $\delta + \Delta$ in the valence band, one of which drops into the He^+ well, neutralizes it, and releases energy (I.P.$_{\text{effective}}$ $He - \varphi - \delta - \Delta$) eV. The ejection of the Auger electron requires $(\varphi + \delta - \Delta)$ eV energy, leaving [I.P.$_{\text{effective}}$ $He - 2(\varphi + \delta)$] eV to be carried off by the Auger electron. Except for the lack of an initial W shell vacancy, the process is analogous to a WVV Auger transition. Consequently, the resulting information is of the same type, requiring a similar unfolding procedure. There are two important differences from the electron impact Auger work, however. The first is that i.n. spectroscopy is restricted in the depth of the valence band that can be studied by the energy (effective I.P.) of the incident ion used, E_0. The cut-off limit is at $E_0 - 2\varphi$. The second difference is that in i.n. spectroscopy there is apparently no ambiguity as to how many layers of surface are being sampled. Hagstrum and Becker have listed [22] the features which lead one to conclude that the electron interactions take place *at, and just above, the surface*. Physically it can simply be expressed by saying that the ion is outside the surface when the neutralizing Auger process occurs, so the density of states sampled is that in the tails of the wavefunctions projecting into vacuum from the surface. There are insufficient data on WVV transitions in electron impact Auger spectroscopy to see whether there are differences in the transition density compared to i.n. spectroscopy which reflect the finite thickness involved for electron impact.

[43] J. A. D. Mathew, *Phys. Rev. Letters*, 1970, **32A**, 261.

Figure 7(a) shows the i.n. folded function for the (100) of nickel, and the unfolded transition density curve (Hangstrum and Becker suggest that the 'transition density' is essentially the valence band density of states at the surface). It is known from LEED work that two different monolayer crystallographic structures of sulphur can be formed on nickel (100), a $c(2\times2)$ and a $p(2\times2)$ phase. $c(2\times2)$ and $p(2\times2)$ phases can be formed also with selenium and oxygen. The i.n. folded and unfolded spectra [22] for the two selenium phases are shown in Figure 7(b) and 7(c). The d-band structure of the underlying nickel has been almost entirely suppressed, and structure related to the surface phases has appeared. The resemblance between the features in the $c(2\times2)$ selenium spectrum and the orbital energies of free H_2Se [as found by photoelectron spectroscopy and marked with arrows on Figure 7(c)] suggest that, in the surface phase, a selenium atom is bonded to two nickel atoms in C_{2v} symmetry (there are additional reasons for arriving at this conclusion) [22] and that this bonding is stronger than the corporate bonding throughout the surface, so that the concept of a 'surface molecule', Ni_2Se, is valid. Arguments [22] were also advanced for suggesting that the $p(2\times2)$ selenium structure, which from the i.n. spectrum is obviously different from the $c(2\times2)$, is a Ni_4Se pyramidal type. In both these cases the selenium atoms are sited above the nickel surface. A similar structure for the Ni–S system was suggested, but for Ni–O the i.n. spectrum is entirely different (much broader peaks) and a surface reconstructed phase was proposed. It must be stressed that at this stage the assignments of the observed structure to specific orbitals are tentative and that the case shown, Ni–Se, is the most convincing example. The important features of the study are

(*i*) A genuine surface layer is observed,

(*ii*) The surface 'electronic structure' as revealed by i.n. spectroscopy can be different for different crystallographic phases, affording a route to study the bonding differences,

(*iii*) The multipeak i.n. structure observed *in some cases* suggests that a molecular orbital approach (*i.e.* discrete energy levels) rather than a solid-state band structure approach may be valid, *i.e.* the bonding is such that the concept of surface compounds is valid.

The instrumentation for i.n. spectroscopy has been described fully in the literature.[22] The electron energy analysis is done by applying a retarding potential between the target crystal and a collection sphere surrounding it, and the resolution of such a system is unlikely to be better than 0.2 eV. Single crystal work only has been performed. There are insufficient data available to make a direct comparison of the sensitivity of i.n. spectroscopy for revealing valence band structure of solids with that of the WVV electron impact Auger work, but several points can be made. The i.n. spectroscopy procedure is a high-efficiency technique with one electron being detected for about every 10 incident ions. There appear to be very few scattered electrons present (because surface penetration is less than the electron scattering mean free path length), and so i.n. spectroscopy does not consist of weak peaks

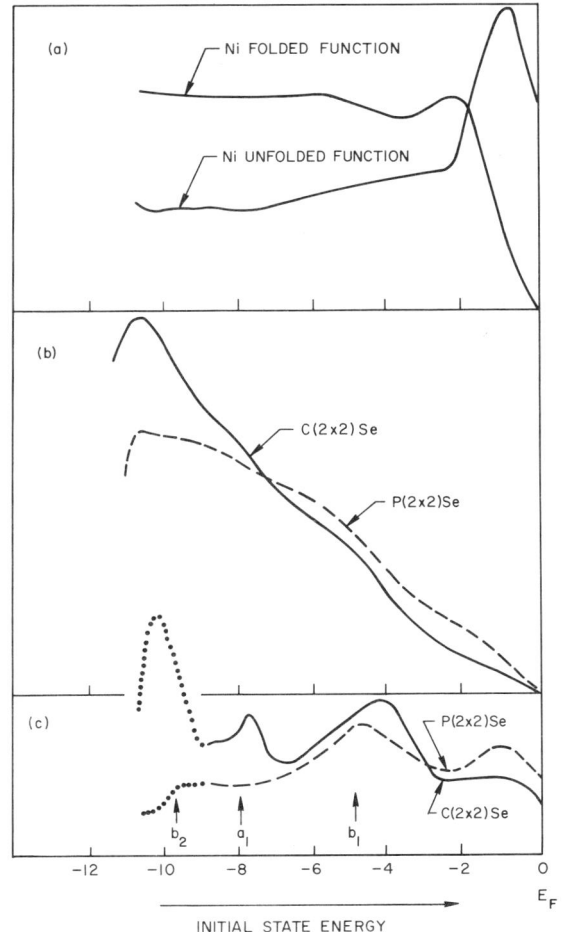

Figure 7 (a) *Fold and unfold ion neutralization spectra of a clean* Ni (100) *face.* (b) *Fold spectra of* c(2×2)Se *and* p(2×2)Se *monolayer structures on* Ni (100) *face.* (c) *Unfold spectra of* (b)—'*Transition density*' *at the surface. The arrows mark the free molecule,* H_2Se, *orbital energies*
(Reproduced by permission from *Phys. Rev. Letters*, 1969, **22**, 1054)

sitting on a high background. Bombarding intensities can be made much higher in the electron impact Auger work, but this sometimes induces changes in the surface structure being observed.[44] For these reasons it would seem that

[44] J. C. Tracy and P. W. Palmberg, *J. Chem. Phys.*, 1969, **51**, 4852.

perhaps i.n. spectroscopy is more suitable for the valence band studies than *WVV* Auger studies.

The original lack of interest in i.n. spectroscopy by chemists can probably be traced to the apparent complexity of the apparatus and data-transforming procedures. The mathematical unfolding can be approximated, however, as for *WVV* Auger transitions, by taking the first derivative of the EDC. After all, the features in Figure 7(c) are evident in the folded curves of Figure 7(b).

At present the results are tentative and most weight should be placed on the observation of change in the i.n. spectrum as adsorption occurs rather than in the specific form of a spectrum. A complete range of interactions are to be expected, from chemisorption where an individual orbital treatment of the 'surface compound' is valid, to complete incorporation into the substrate surface layers (*e.g.* $Ni + O_2$) where such a treatment would not be valid. For the former, one expects narrow resonances in the i.n. spectra and, for the latter, much broader genuine solid-state band structure.

Photoelectron Spectroscopy.—*Molecular Photoelectron Spectroscopy: Photoemission Studies.* As mentioned earlier, the term molecular p.e. spectroscopy has generally been used for free-molecule studies, and photoemission studies for the solid-state work. The primary aim of most of the photoemission studies has been to examine the valence band structures of pure metals [17, 23, 45] (and semiconductors [46]) in detail by obtaining an experimental EDC and transforming this into the 'optical density of states' (ODS) curve, assuming that the non-direct transition mechanism [47] describes the photoexcitation process. The word 'optical' is introduced to imply that the density of states is that found by photoemission.

Most of the experiments have been performed using an incident light energy of $hv \leqslant 11.6$ eV. The usual procedure is to shine light from a vacuum-u.v. monochromator through the LiF window on to a sample in an ultra-high vacuum chamber, and analyse the energies of the electrons by a simple retardation process, as for i.n. spectroscopy. In Figure 4(a) the *d*-band structure from a clean nickel polycrystalline sample is clearly evident in the EDC. To relate these curves to an ODS curve it is necessary to return to the scattering curve of Figure 6. The energies of the electrons represented in Figure

---

[45] C. N. Berglund and W. E. Spicer, *Phys. Rev.*, 1964, **136**, 1030, 1044.
[46] T. E. Fischer, *Surface Science*, 1969, **13**, 30.
[47] The non-direct transition mechanism assumes that *k*-conservation (where *k* is the Bloch wave-vector) is unimportant, and therefore peaks in the 'optical density of states' relate directly to peaks in the density of states of the filled bands of the metal. There is by now much evidence (see ref. 6, 48) that this hypothesis is incorrect in some cases. When *k*-conservation becomes important the relationship between the 'optical density of states' and the 'true density of states' becomes much more complex.
[48] J. F. Janak, D. E. Eastman, and A. R. Williams, *Solid State Comm.*, 1970, **8**, 271; W. E. Spicer, *Optics Comm.*, 1969, **1**, 157; D. E. Eastman and J. K. Cashion, *Phys. Rev. Letters*‘ 1970, **24**, 310.

4(a) are in the region where L_e is dropping rapidly with increasing electron energy, and this must be allowed for in a correction of the experimental EDC. Also, since we are in a region of comparatively large L_e, it is usually assumed that it is the bulk property being measured. A second allowance has to be made for the 'threshold function' for escaping electrons. This relates to the elastic reflection of electrons at the surface (only one electron is detected for every 100 photons incident), which is a function of energy acting in the opposite direction to the L_e correction. It turns out that the combined correction [49] for electrons escaping at energies significantly above the vacuum level is approximately zero. The 2 eV region immediately above the vacuum level is not very meaningful anyway since it will include multiply-scattered electrons from a variety of processes which cannot be adequately corrected for. In the region where data can be obtained and are meaningful, the ODS curve is very similar to the observed EDC. To illustrate this, Figure 4(b) shows the ODS curve derived from the EDC of Figure 4(a). In general, peak locations in the ODS curve are fixed to within ± 0.1 eV in the EDC curve [49] (provided the non-direct mechanism holds).[47]

We are now in a position to compare directly the photoemission results with the unfolded 'transition density' of i.n. spectroscopy. The unfolding process of WVV Auger and i.n. spectroscopy is not required of the p.e. curves because of the one-electron nature of the process involved. We see that the distributions in Figure 4(a) [or (b)] and in Figure 7(a) are in fact different, and this is interpreted as meaning that the d-band density of states at and just outside the surface (i.n.) is different from that in the bulk (p.e.).[22] The question now is can p.e. spectroscopy produce anything like the structure observed in i.n. spectra for surface adsorbed species? If so, the technique is likely to have an advantage over i.n. spectroscopy because of the simplicity of the experiment and the lack of unfolding required. At first sight it would appear that the chances are not good, because the 'escape depth' will be fairly large for the electron energies of the ejected electrons ($L_e = 40$ Å for Ag, Au at 6 eV, see Table 1). However, a set of measurements made by Eastman [50] yielded $L_e \sim 10$ Å for Y, Gd, and Ni, at 7 eV above the Fermi level, and Smith and Fisher [51] have proposed values of little over 1 Å for the alkali metals. If one used a vacuum-u.v. resonance line of high energy as the ionizing source the ejected electrons would have higher energies and shorter mean free paths. The most commonly used resonance line in molecular p.e. spectroscopy [6] is the HeI resonance line at 21.2 eV, which would place the energies of electrons ejected from the valence band in the region 15—21 eV above the Fermi level (arbitrary 6 eV valence band width) where L_e would be expected to be very small. The very recent measurement of attenuation length for photoelectrons excited in aluminum by 21.2 eV photons, based on the measurement of photoelectric yield as a function of angle,, suggest that this

---

[49] W. F. Krolikowski, PhD. Thesis, Stanford University, 1967.
[50] D. E. Eastman, *Solid State Comm.*, 1970, **8**, 41.
[51] N. V. Smith and G. B. Fisher, *Phys. Rev. (B)*, 1971, **3**, 3662.

is true. A value of 4 Å $< L_e <$ 12.8 Å was found for such electrons.[52] Use of the HeII line at 40.8 eV might reduce the escape depth even further.

There have been only two published studies on the p.e. spectroscopy of adsorbed species. Bordass and Linnett [53] studied methanol adsorbed on to tungsten using HeI radiation and a high-resolution electrostatic deflection analyser. The poor vacuum conditions of the experiment were rather unsuited to surface work. A methanol spectrum differing somewhat from the gas-phase spectrum was observed but no real conclusions can be drawn as to the nature of the metal–methanol bonding, owing to the rather undefined nature of the tungsten surface.

Eastman [54] has looked at the EDC of evaporated nickel films, using HeI radiation and a high-resolution analyser, in a vacuum system in the 10^{-7} N m^{-2} range. The EDC of clean nickel slowly changes with time as contamination of the surface occurs, but there is ample time to adsorb a known quantity of a gas and record the resultant EDC. On adsorption of less than a monolayer of oxygen, the nickel d-band intensity decreases and an additional band appears at about 5.4 eV below the Fermi level [Figure 4(c)]. On adsorbing further oxygen this band increases its intensity and the nickel d-band intensity falls further to a very low level. On heating the surface, the oxygen band intensity decreases and the p.e. spectrum reaches a terminal stage which indicates that oxygen is still present on the surface. The position and width of the band (3 eV) corresponding to oxygen adsorption are almost identical with those from the oxygen adsorption on to both the $p(2 \times 2)$ and $c(2 \times 2)$ (100) nickel surfaces as recorded by i.n. spectroscopy. Hagstrum has interpreted [22] the i.n. spectra as indicating oxygen atoms incorporated into the nickel surface planes, which rationalizes the fact that the structure observed for polycrystalline surfaces by p.e. spectroscopy should be so similar. Clearly, this is a case at the strong end of the interaction range and so a discrete MO approach to the Ni–O bonding will be invalid. It would very interesting to see what results are obtained for selenium and sulphur adsorption by p.e. spectroscopy.

Another interesting point here is that early EDCs of 'clean' nickel showed the structure described above, and an attempt was made to explain this theoretically [55] in terms of discrete surface states present for pure nickel. It seems clear now that the anomaly relates to adsorbed oxygen as suggested earlier by Hagstrum,[56] and not to intrinsic nickel band structure.

Preliminary studies [57] on the adsorption of CO on to nickel have been made and it was found that adsorption of CO on to the clean surface yielded two bands at 7.5 and 10.7 eV below the Fermi level attributable to the CO. Preliminary studies [57] of the Pd–H$_2$ system have also been made and structure

[52] T. F. Gesell and E. T. Arakawa, *Phys. Rev. Letters*, 1971, **26**, 377.
[53] W. T. Bordass and J. W. Linnett, *Nature*, 1969, **222**, 660.
[54] D. E. Eastman, *Phys. Rev. (B)*, 1971, **3**, 1769.
[55] F. Forstmann and V. Heine, *Phys. Rev. Letters*, 1970, **24**, 1419.
[56] H. D. Hagstrum, *J. Appl. Phys.*, 1969, **40**, 1398.
[57] D. E. Eastman, personal communication.

relating to the hydrogen is clearly observable in the EDC with a width of about 1.5 eV and an energy about 3 eV lower than the binding energy of H_2. Studies of nickel laid down on the gold substrates have shown that 6 Å of nickel can be detected by observing the nickel d bands superimposed on the gold EDC.

The results of Eastman remove the doubt that vacuum-u.v.—p.e. spectroscopy can be applied to the determination of surface structures. The ease of the technique has much to recommend it but it is unlikely to be as strictly a surface technique as i.n. spectroscopy. Bordass and Linnett [53] were unable to see any 'adsorbed gas signal' in their instrument unless grazing incidence was used. The sensitivity of this instrument is a factor of about 100 less than that of the one used by Eastman, who was able to record a complete Ni–O spectrum in about three minutes with adequate counting statistics, so signal intensity will probably not be a major problem. Differences in the exact shapes of the peaks observed by i.n. and p.e. spectroscopy will probably occur because i.n. spectroscopy samples the density of states at and just outside the surface layer. The possibility of observing vibrational structure under high resolution for the surface compound type of interaction must be considered for p.e. spectroscopy.

X-Ray Photoelectron Spectroscopy, ESCA.—Though the majority of ESCA studies have been carried out on solids,[3] there have been a few attempts to evaluate its usefulness in surface chemistry and physics. The escape depth, d, is of primary concern again, and since the X-ray penetration depth is deep compared to the mean free path length of the escaping electrons, it seems that the problem can be treated in a manner analogous to electron impact Auger spectroscopy, where the ejected electron mean free path length was also the dominant factor. The lowest energy, and the most commonly used, X-rays in ESCA work are Mg$K\alpha$ and Al$K\alpha$ at 1254.6 eV and 1486.6 eV respectively. Ionization of the valence band electrons will therefore produce ejected electrons with approximately these energies above the Fermi level, *i.e.* in the region in Figure 6 where mean free path lengths are in the > 50 Å range. Electrons ejected from inner shell levels will have energies ranging from near zero to within about 50 eV of the X-ray energy used, depending on the binding energy of the particular inner shell. Thus there will sometimes be ESCA peaks corresponding to electrons with kinetic energies in the critical region where mean free paths are shortest. For these peaks there is nothing to indicate that ESCA should be any less of a surface technique than is Auger spectroscopy, unless the depth of incident electron penetration in the Auger work is more important than is generally thought. Accepting this then, the question is how many of the elements of interest in surface studies give ESCA peaks in the critical energy range? The tables of electron binding energies [3] indicate that we are often limited to ejected energies which are rather high. In a comparison with information obtainable from WXY Auger transitions we are likely therefore to have the advantage that chemical shift behaviour is less complicated for ESCA, (since only one level is involved) and a disadvantage in that often the escape depth will be greater than we would like. Presumably, both grazing angle of incidence and of emittance can be utilized to increase total ionization near the surface and reduce the escape depth respectively, as has been done for p.e. and Auger spectroscopy.

There has been only one published estimate of escape depth in ESCA. This was a measurement [58] made on iodostearic acid for which an 'average depth' of emission of 'less than 100 Å' for an 860 eV electron was obtained. It is unfair to compare results for an organic material with data existing for Auger spectroscopy, since the latter (Table 2) refer to metals having shorter scattering mean free path lengths. If major differences in d are found between ESCA and Auger spectroscopy in a similar electron kinetic energy range for similar materials, this will be an indication that X-ray and incident electron beam penetration depths are more important than is thought.

The interpretation of chemical shifts,[3] which was avoided in the treatment of WXY Auger spectroscopy, is briefly discussed here because shifts in ESCA are generally more clearly observed experimentally, and are more simply treated theoretically. Also, a wealth of experimental data exists for ESCA shifts in solids,[3] but there is much less for Auger work.

Simple classical models can be used for a qualitative understanding of the relationship between binding energy shifts and the charge on atoms. Considering a free atom, A, the inner shell electron under consideration sits inside a charged shell made up of the valence electrons. If n electrons from the valence shell were removed to infinity, the energy of the inner shell electrons would be lowered by $E = ne^2/r$, where r is the radius of the valence shell electron. Thus the binding energy of the inner shell electron would be ne^2/r greater in A^+ than in A. This 'chemical shift' would be the same for *all* inner shell electrons, and the magnitude of the calculated shift is *ca.* 10 eV per unit charge. In making a chemical bond between two atoms, A and B, electrons are not moved to infinity, but in a fully ionic approximation they may be considered as having been transferred from the valence shell of one atom to that of the other. The chemical shift in inner shell binding energy in the molecule A^+B^- compared to the free atom, A, can therefore be represented by the modified equation

$$E = e^2 \left(\frac{1}{r} - \frac{1}{R}\right)$$

where R is the internuclear distance. For solids (ionic crystals) a further modification is required to include the Madelung constant effect. Few crystals can be considered as approaching full ionicity and the chemical shifts calculated by such procedures are far larger than the experimental values. Estimation of the percentage ionic character by quantum mechanical methods, by using Pauling electronegativities, or by using full-scale SCF procedures brings calculated results more into line with expected values, and correlations

---

[58] Ref. 7, p. 139

show for many cases a roughly linear relationship between calculated charge on an atom and the measured binding energy of inner shell electrons, for a series of compounds. ESCA natural linewidths in solids can range from probably a few tenths of an eV to two or three eV, but present instrumental resolution is restricted to a best half-width of about 0.9 eV, mainly owing to the width of the X-ray line.[3] (This is still narrower than Auger peaks where two energy levels are involved). This means that, for chemical shifts of the order of 1 eV per unit change in oxidation state, there are problems of resolution when states of an atom differing by one unit of oxidation are involved.

The major drawback to the few published results on the application of ESCA to surface work has been that the equipment was neither designed for maintaining clean surfaces nor had the flexibility of sample handling required. Fadley and Shirley [59] found that an iron sample in their spectrometer showed a doublet in the Fe 3p peak [Figure 8(a)] and that a strong O 1s peak was also detectable. On heating to 1103 K in 1 Nm^{-2} H$_2$ the oxygen peak largely disappeared and the Fe 3p level reduced to one line [Figure

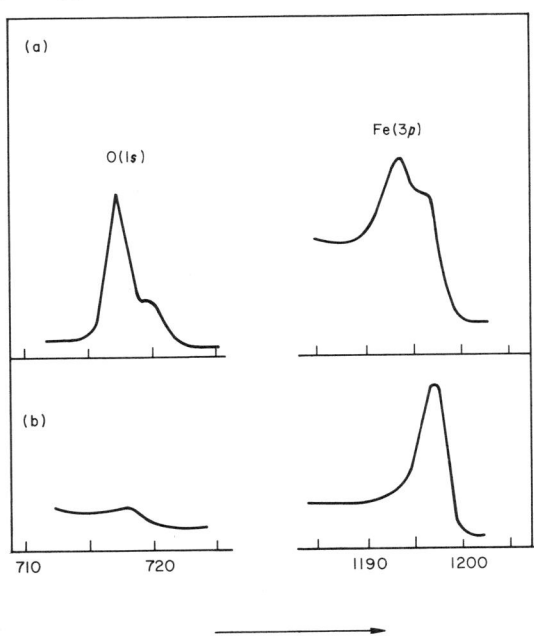

Figure 8 *ESCA Fe 3p and O 1s lines from* (a) *a sample of iron at* 298 K (b) *the same sample heated to* 1103 K *in* 1 Nm^{-2} *of* H$_2$
(Reproduced by permission from *Phys. Rev. Letters*, 1968, **21**, 980)

[59] C. S. Fadley and D. A. Shirley, *Phys. Rev. Letters*, 1968, **21**, 980.

8(b)]. The extra Fe line is clearly due, therefore, to a chemical shift effect (about 2.5 eV to higher binding energy) caused by oxidation of the surface. The O $1s$ line of Figure 8(a) also shows two peaks. This may be due to adsorbed H_2O, O_2, or CO_2 as well as the oxygen incorporated into the iron surface. No attempts were made at estimating the thickness of the oxide layer present.

Delgass et al.,[60] in an extended series of experiments, attempted to study surface effects by ESCA in three areas: adsorption; the behaviour of supported metals compared to unsupported metals; and crystalline oxides. Some of their results are reviewed briefly here.

An attempt to observe the adsorption of CO on platinum failed because, with the inadequate vacuum, strong concentrations of C and O were already present on the Pt surface before CO adsorption. Nitrogen adsorption experiments were more successful. An 85% exchanged NH_4-Y zeolite wafer showed a N $1s$ line at 492 eV binding energy, but the signal was so weak that 7 hours were required for recording. Heating reduced the signal intensity, indicating desorption, and a slight increase in linewidth was also recorded, which may indicate the presence of more than one state of nitrogen. On exposing the surface to pyridine (5 Torr) at 393 K after heating in H_2 at 723 K, a peak due to adsorbed pyridine appeared with a chemical shift of 4.5 eV to lower binding energy. The N $1s$ peak from the zeolite also moved 1.1 eV lower. It was thought that this may be due to the effect of the removal of H_2O from the co-ordinating sphere of NH_4^+ ions. For detailed studies higher sensitivities are required.

Dispersion of active metals on supports is a method widely used to improve the efficiency of a catalyst. Figure (9a) reproduces the spectra recorded for the $4f_{5/2}$ and $4f_{7/2}$ lines of Pt.[60] A chemical shift of 3.8 eV compared to PtO_2 is observed [Figure 9(b)]. The Pt foil had been handled in air and probably therefore had a surface oxide layer. A small amount of PtO could account for the difference in intensity ratio of $4f_{7/2}$ to $4f_{5/2}$ lines in the Pt spectrum compared to the PtO_2 spectrum. Figure 9(c) shows the spectrum of 5% Pt dispersed on SiO_2 (56% dispersion). No charging corrections were made (see later) to the absolute energies of the sample, so chemical shifts cannot be compared with Figure 9(a) or 9(b), but the high intensity of the left peak of the doublet clearly indicates a considerable amount of oxidation of the dispersed Pt, even though the sample was being heated in a low pressure of H_2. The spectrum can be taken as qualitative indication of the effect of dispersion on the oxidizability of dispersed Pt. Studies of 5% Cu dispersed on MgO, and 6% Ni on silica–alumina were also carried out. Wolberg et al.[61] have also studied the application of ESCA to supported metals. They looked at the Cu $2p_{3/2}$ peak in $CuAl_2O_4$ and in CuO (chemical shift ca. 1 eV), and then studied samples of ca. 10 wt % copper on alumina of 301 m^2g^{-1} and 72 m^2g^{-1} surface area calcined at 783 K. In this fashion they were able to show that the high surface area alumina assumed the $CuAl_2O_4$

---

[60] W. N. Delgass, T. R. Hughes, and C. S. Fadley, *Catalysis Rev.*, 1971, **4**, 179.
[61] A. Wolberg, J. L. Ogilvie, and J. F. Roth, *J. Catalysis*, 1970, **19**, 86.

structure, whereas the low surface area alumina formed a copper oxide surface phase. Upon calcination of the latter at 1173 K the Cu peak energy moved to that of $CuAl_2O_4$ showing that an aluminate phase forms at this higher temperature.

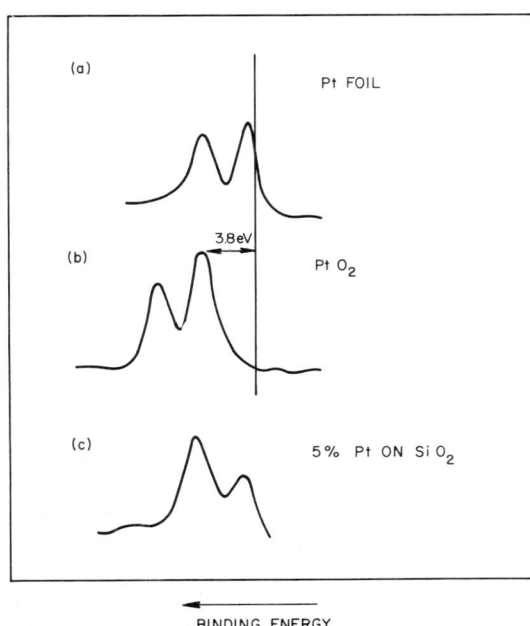

Figure 9 *ESCA* Pt 4f *lines from* (a) *a sample of* Pt *foil* (*handled in air*) (b) *a sample of* PtO_2 (c) 5% Pt *on* SiO_2 (56% *dispersion*)
(Reproduced by permission from *Catalysis Rev.*, 1971, **4**, 179)

Among the crystalline oxides studied [60] was the spinel catalyst, FeV_2O_4. The spectra of the unused catalyst and a sample used for the dehydration of cyclohexane at 698 K were taken. The vanadium lines (3s and 3p) in the used sample were observed to shift to higher binding energies ($\sim$1 eV) and the oxygen lines (1s and 2s) split into doublets and moved to lower energies ($\sim$1 eV). A major change in the bulk sample had not occurred since the *X*-ray diffraction pattern remained unchanged. As the catalyst activity is known to decrease with use, the changes in the ESCA spectrum probably reflect the chemical changes in the surface layers causing the loss of activity.

Several statements have already been made comparing *WXY* Auger results with ESCA results. They may be summarized as follows:

(*i*) for equivalent photoelectron energies the two techniques should be of comparable surface effectiveness;

(ii) ESCA peaks are often restricted to electron energies where (i) does not hold. Also there may be some question as to the importance of the penetration depths of X-rays and electron beams for the two techniques. The general conclusion is that for many cases ESCA is only a semi-surface technique;

(iii) ESCA peaks are usually narrower than Auger peaks;

(iv) chemical shifts are less straightforward in Auger spectroscopy;

(v) the combination of (iii) and (iv) makes the study of chemical shifts less attractive in Auger work than in ESCA.

Further comments are made below relevant to the merits of the two techniques for surface chemistry studies. Since the rate of production of Auger electrons depends partly on the rate of W shell ionization, the intensity of the electron impact Auger spectrum compared to the ESCA spectrum depends partly on the relative intensities of the electron and X-ray beams used, and on their relative cross-sections for W shell ionization (in practice ionization of the W shell by secondary electrons can enhance the Auger spectrum, but since each W shell ionization can lead to several subsequent Auger processes the rate of an individual Auger process will represent only a fraction of the W shell ionization rate). Ionization cross-sections for X-ray processes are much higher than for electron impact, particularly when the X-ray energy approaches the threshold energy for ionizations,[62] but the electron beams used in the Auger work are more intense. The background of scattered electrons in the ESCA work for gases is also very much lower than in the electron impact Auger work. The consequence of these factors for gases is that the signal-to-background ratio is little different in the two techniques. Similar results might be expected for solids, in which case the ESCA technique should be of comparable sensitivity to the Auger for general bulk studies. From (ii) above this does not imply equal surface sensitivity in all cases. An advantage in the electron impact Auger work is that the electron beam can be focused on to much smaller surface areas than the X-ray beam. A disadvantage is that high-intensity electron beams can often cause damage at the surface being studied. Radiation damage is also possible when high-power X-ray sources are used.

A further point is that one can also study the Auger peaks accompanying the normal ESCA peaks in an X-ray induced spectrum.[3] There is evidence that they are more prominent features in the EDC than for an electron impact induced spectrum, under comparable experimental conditions,[63] but one is still left with doubt concerning escape depths.

As stated in Section 2, ESCA is most useful for studying the inner shell levels not accessible to the higher resolution, higher sensitivity, vacuum-u.v.—p.e. spectroscopy. ESCA can record the valence levels, however, though the signals are often very weak. Figure 10 shows the ESCA valence band spec-

---

[62] T. A. Carlson, personal communication.
[63] N. G. Nakodkin and P. V. Mel'nik, *Soviet Physics—Solid State*, 1964, **5**, 1779; 1965, **6**, 1462.

trum for nickel, which may be compared to the i.n. spectrum of Figure 7 and the p.e. spectra of Figure 4. Apart from a smearing out of the spectrum, owing to the loss of resolution in the X-ray experiment, the structure is quite similar to that in the p.e. curve. (The higher intensity in the latter below the d-band, compared to the ESCA spectrum, may be due to the presence of an excessive number of scattered and secondary emitted electrons in this region which have not been properly allowed for). There seems to be some evidence for the presence of oxygen on the nickel surface in the ESCA spectrum as there is a shoulder at about 5 eV below the Fermi level [$cf.$ Figure 4(c)], though the spectrum has recently been restudied and the shoulder interpreted by Fadley and Shirley [64] as a very weak characteristic energy loss peak. Since we know that the p.e. spectrum is sensitive to small quantities of adsorbed oxygen (using 21 eV photons), what is required is that the ESCA spectrum be studied under the same good vacuum conditions and the O 1s line and the band 5 eV below the Fermi level looked for. This would give some idea as to the surface sensitivity of the technique compared to p.e. spectroscopy.

The ESCA valence band spectra of many other metals have been studied and detailed comparisons with p.e. spectroscopic results made.[64] There is good agreement in some cases, and only poor agreement in others, which reinforces the argument for studies under well-defined surface conditions.

In summary, it seems likely that ESCA will be only a semi-surface technique in many cases, particularly for valence level studies, though comparative data are scarce because of the lack of ultra-high vacuum ESCA equipment. Because of this it will probably find its major applications in the study of thick surface phases. The valence shell studies are a useful addition to those from WVV Auger, i.n., and p.e. spectroscopy, though we are limited presently for ESCA to an instrument resolution of about 0.9 eV. The most useful information, however, will undoubtedly come from the study of the chemical shift behaviour of the inner shell ionizations, and the work of Delgass et al.[60] is a useful start here. When there is sufficient sensitivity and a small enough escape depth, coupled with apparatus designed for clean vacuum work and sample manipulation, results in this area are likely to be more fruitful than those from electron impact Auger spectroscopy. The reduced sensitivity for published ESCA work (for surface studies) compared to published Auger electron impact results is a combination of instrumental set-up and the difference in surface effectiveness (escape depths). Most of the Auger work has been done with high-luminosity instruments under low-resolution conditions, whereas the ESCA machines have usually been of very low luminosity but higher resolution. Also the differentiating method of recording Auger spectra is designed to pick up small peaks. More escape depth studies must be made. This can be done by putting down controlled thicknesses and monitoring attenuations, and also by comparing X-ray Auger spectroscopy

[64] C. S. Fadley and D. A. Shirley, *J. Res. Nat. Bur. Stand.*, 1970, **74A**, 543.

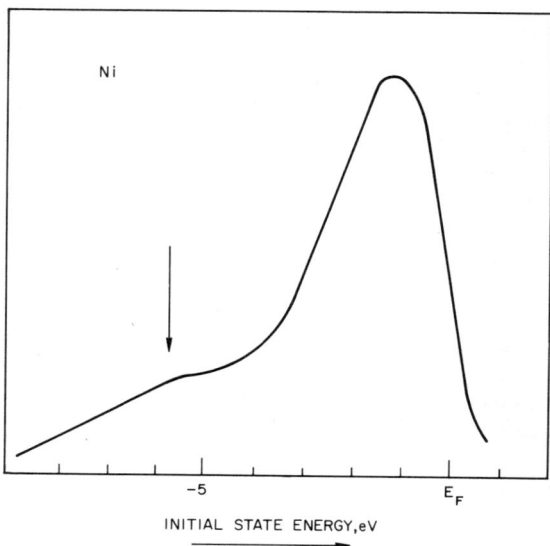

Figure 10 *ESCA spectrum of the valence region of* Ni (d-*bands*) (Reproduced by permission from *Phys. Rev. Letters*, 1968, **21**, 980)

sensitivity to monolayer coverages with that from electron impact induced Auger; and ESCA valence band sensitivity with that of p.e. spectroscopy.

Nothing has been said at all so far about charging effects. Insulating materials may become charged by several volts inside the spectrometer, and this will hinder the determination of chemical shifts. The possibility of band bending through the surface layers of the sample must also be considered. Delgass *et al.* mentioned these problems briefly, making the point that further work is required to overcome calibration problems in such cases. The same problem will apply to insulators in all the electron spectroscopy techniques, the reasons it has not caused many problems so far being that most of the work has been performed on conducting samples, and that chemical shift measurements have not been the fashion outside ESCA. A start has been made on this problem in a few cases. Wei [65] showed that stable LEED patterns and reproducible energy loss spectra could be obtained (prevention of charging) for the insulators LiF and NaF, provided the impinging primary beam was above a critical energy, which was temperature dependent. Other LEED–Auger studies of LiF, NaF, and KCl have also been made.[66]

Energy Loss Spectroscopy.—The interesting energy loss region for electrons

[65] P. S. P. Wei, *Surface Science*, 1971, **24**, 1.
[66] P. W. Palmberg and J. N. Rhodin, *J. Phys. Chem. Solids*, 1968, **29**, 1917; T. E. Gallon, *Surface Science*, 1970, **21**, 224.

once-scattered inelastically from surfaces (characteristic energy loss region) is that from 0 to about 40 eV from the incident electron energy. Included in this region are energy loss peaks due to collisions producing

(i) phonon excitations, i.e. interactions with the crystal lattice. Since phonon energies are small ($\leqslant 50$ meV), such collisions will tend only to broaden the inelastically scattered electron peaks, rather than show discrete energy loss peaks;

(ii) plasmon excitations, i.e. interactions with the electron plasma of the solid;

(iii) surface plasmon, or 'lowered' plasmon excitations, which theoretically [19] are $\sqrt{2}$ of the bulk plasmon value for an ideal surface and $\sqrt{3}$ for a surface consisting of spherical grains, for free-electron metals;

(iv) electronic excitations (interband transitions) from filled to unfilled levels, equivalent to the spectroscopic transitions observed in the gas-phase work.[3]

It will be appreciated that these features will appear in an electron impact Auger spectrum EDC, along with other scattered and multiply scattered electrons which make up the background. In experiments in which energy loss peaks are investigated, the incident electron beam must have a narrow energy spread since, unlike the secondary Auger process, this spread contributes to the resolution of the energy loss peaks. The peaks of the characteristic energy loss region do not generally interfere with the Auger peaks because they fall in a different energy region of the spectrum (Figure 11). The two can be identified fairly easily anyway by changing the incident electron energy and seeing which peaks move with this energy (energy loss peaks), and which maintain a fixed kinetic energy (Auger peaks). This separation can be achieved automatically by means of oscillating and lock-in amplifier circuitry. The experimental set-up has been described in a preliminary report [67] as a means of separating Auger peaks from energy loss peaks resulting from the inelastic processes which caused the original W shell ionizations, i.e. not any of the processes (i)—(iv) but the energy losses equivalent to the binding energies of the inner shell levels. This branch of energy loss spectroscopy has been given the name ionization spectroscopy,[67] but it is too early to say whether it will prove fruitful for surface work. It has been claimed that the peaks are of comparable intensity to the smaller Auger peaks, and that they are often narrower than the Auger peaks. Chemical shifts would be easier to interpret since only one level is involved (electron impact equivalent of ESCA). However, there is some ambiguity concerning the actual process involved, since it is not clear whether inner shell ionization actually occurs, with the ejection of an electron, or whether the inner shell electron is promoted to a vacant level between Fermi and vacuum levels. If the former is correct, then the role of the ejected electron in the EDC cannot be ignored.[68]

[67] R. L. Gerlach, J. E. Houston, and R. L. Park, *Appl. Phys. Letters*, 1970, **16**, 179; J. E. Houston and R. L. Park, *Appl. Phys. Letters*, 1969, **14**, 358.
[68] C. R. Brundle, *Surface Science*, 1971, **27**, 681.

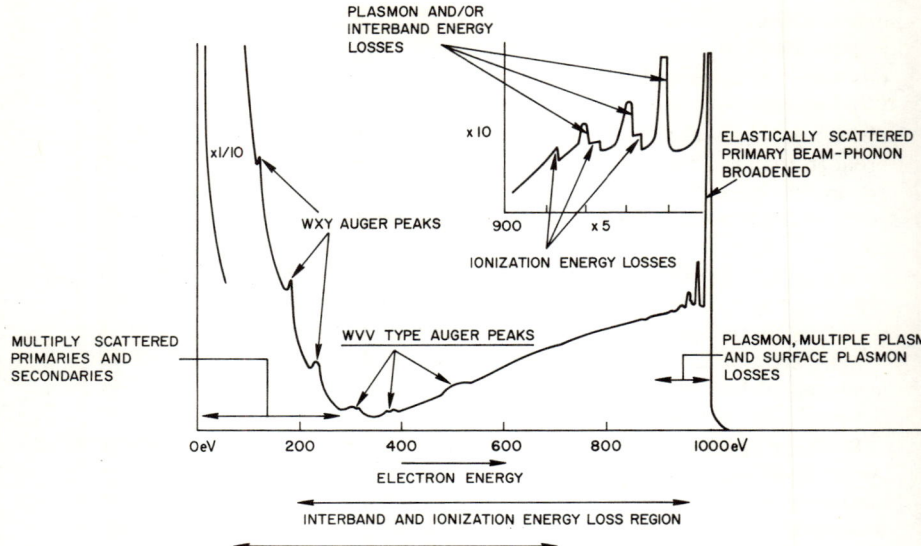

Figure 11 *'Typical' EDC obtained from* 1000 eV *primary beam impact on a solid sample*

Returning to energy losses caused by processes (*i*)—(*iii*), these have been studied primarily to understand the nature of phonon and plasmon excitations, rather than using the characteristic values in any fashion for chemical analysis. Studies have been made mostly on metals, and for a long time experimental results were in serious disagreement with each other, probably owing to severe surface contamination. The experimental data have been reviewed by Klemperer and Shepherd.[19] More recent studies have attempted to use plasmon spectra for chemical and surface chemical studies. Lander and Morrison[69] examined the tungsten (100) face characteristic energy loss region using 10 eV primary electrons and a resolution of about 0.5 eV, while increasing a caesium coverage from zero to a saturated 'duolayer'[70] coverage. A strong surface plasmon loss of 1.5 eV was observed at the duolayer coverage, which shifted gradually to a bulk 2.4 eV value on increasing to a multilayer coverage. The strength of the transition suggests that one should also expect such effects in photoemission data, when it would be easy to misinterpret such loss features as genuine band structure. (Plasmon peaks have, in fact, been observed[71] in photoemission data for potassium, rubi-

[69] J. J. Lander and J. Morrison, *Surface Science*, 1969, **14**, 465.
[70] A. K. MacRae, K. Miller, J. J. Lander, and J. Morrison, *Surface Science*, 1969, **15**, 483.
[71] N. V. Smith and W. E. Spicer, *Phys. Rev. Letters*, 1969, **23**, 769.

dium, and caesium.) Simmons [72] has shown that, for titanium and chromium, surface plasmon loss peaks can be distinguished from the bulk losses for clean samples, and that they are sensitive (reduction in intensity) to as little as 0.04 monolayer coverage of oxygen. Hartley and Swan [73] attempted to use the characteristic loss region to study the structure of different percentage composition alloys. While changes with changing composition were observed, the differences were insufficient to relate these to the properties of the alloys. It was suggested that a higher resolution study (1—2 eV used here) might be more successful.

Energy loss spectroscopy has been used successfully for the study of adsorbed species on surfaces.[74] Low incident energy (4.5 eV) was used and the electron beam was monochromatized by passage through an electrostatic electron energy analyser. The combination of low incident energy and careful monochromatization allowed quite high resolution (0.05 eV) studies to be made, whereas most previous work on solids has been of rather low resolution ($\geqslant 0.5$ eV). Propst and Piper [74] were able to observe energy losses in the 0—800 meV region for the tungsten (100) face with H_2, N_2, CO, and H_2O adsorbed on to the surfaces, and were able to relate the peaks to the vibrations of the species on the surface. From the values of the vibrational frequencies it was possible to deduce something about the adsorption process. For instance, for H_2 and N_2 adsorption, no frequencies were observed which could correspond to the molecular vibration, indicating that the molecule was dissociated on the surface, and the small vibrational frequencies observed corresponded to vibrations of the adsorbed atom in the surface bond. From the frequencies and the desorption energies, estimates of the number of substrate atoms involved in a bond with a single ad-atom were made. In the case of H_2O adsorption, O–H and O–W frequencies were observed, O–H appearing only after saturation of the O–W peak with time, suggesting that a surface oxide phase was formed and then H_2O or O–H bonded to the new surface.

In summary, the study of the vibrational structure of adsorbed gases by energy loss spectroscopy seems to have great potential in determining in which molecular form the species are present on the surface. In principle, the same information can be obtained from i.r. techniques,[75] but there are great experimental limitations. The instrumentation used by Propst and Piper was fairly elementary and both resolution and signal intensity are capable of improvement.[76] Also no attention was given to escape depths. An increased surface effectiveness might be possible if the energy of the impacting electron beam was raised to, say, 10 eV.

It would seem likely that study of plasmon features under high resolution

[72] G. W. Simmons, *J. Colloid Interface Sci.*, 1970, **34**, 343.
[73] B. M. Hartley and J. B. Swan, *Austral. J. Phys.*, 1970, **23**, 655.
[74] F. M. Propst and T. C. Piper, *J. Vacuum Sci. Tech.*, 1966, **4**, 53.
[75] R. P. Eischens and W. A. Pliskin, *Adv. Catalysis*, 1958, **10**, 1; N. J. Harrick, *Ann. New York Acad. Sci.*, 1963, **101**, 928.
[76] P. S. P. Wei, Bell Laboratories Tech. Memo 69–1133–3.

might be more generally useful than the low resolution work done at present. More data on ionization spectroscopy features are required before we can tell whether it can effectively compliment Auger spectroscopy as a surface technique.

The author is pleased to acknowledge helpful discussion with T. A. Carlson, J. K. Cashion, W. N. Delgass, D. E. Eastman, H. D. Hagstrum, C. S. McKee, N. V. Smith, and J. C. Tracy, and is grateful to these authors and also to C. C. Chang for access to data prior to publication.

7
Exchange and Equilibration Reactions on Metal Surfaces

BY R. P. H. GASSER

The use of isotopic substitution as an aid to the elucidation of reaction mechanisms is well established, both for homogeneous and for heterogeneous reactions. Recently, the particular interest in these latter reactions, when they involve the interchange of isotopic partners on a metal surface, has arisen through the light which the study of these processes may shed on the nature of the interactions between the adsorbed gas and the surface. In addition, the mechanisms of reactions taking place on a surface are of intrinsic interest, and isotopic substitution can make a valuable contribution to the determination of these mechanisms. It is, perhaps, necessary to make a semantic distinction between isotopic exchange reactions, in which one isotope replaces another in a compound, and isotope equilibration reactions, in which the composition of a mixture of separated isotopes approaches its equilibrium proportions. In experimental practice both types of experiment are performed and both will be discussed.

The relationship between catalysis and chemisorption is a close yet elusive one. Some interesting experiments have been reported in the last decade, the period with which we shall principally be concerned, which have had as their aim the investigation of this relationship. The use of isotopes may yield a more detailed insight into the atomic nature of the events occurring during the process of adsorption on the metal surface. It will be necessary, therefore, from time to time to discuss some aspects of chemisorption. In this Report, the gas/metal-surface systems which are to be discussed are grouped by the gaseous component, and the reactions of the various metals with this component are then described.

1 The Hydrogen–Deuterium Reaction

Chemical processes in which the isotopes of hydrogen are involved have been extensively studied using a great variety of surfaces, both metallic and non-metallic. A considerable diversity of mechanisms seems to be possible or at least has been postulated for the equilibration reaction. The most important mechanisms have been summarized [1,2] and are as follows:

[1] D. D. Eley and P. R. Norton, *Discuss. Faraday Soc.*, 1966, **41**, 135.
[2] J. J. F. Scholten and J. A. Konvalinka, *J. Catalysis*, 1966, **5**, 1.

(I) Bonhoeffer–Farkas, chemical mechanism:

$H_2(g) + D_2(g) + 4M(surf) \rightarrow 2M-H(surf) + 2M-D(surf) \rightarrow 2HD(g) + 4M(surf)$

(IIa) Rideal mechanism, which requires empty sites adjacent to chemisorbed atoms:

$$D_2(g) + \begin{array}{c} M-H(surf) \\ | \\ M \end{array} \rightarrow \begin{array}{c} M\text{---}H \\ | \quad \diagdown \\ | \quad \quad D \\ | \quad \diagup \\ M\text{---}D(surf) \end{array} \rightarrow \begin{array}{c} M \\ | \\ M-D(surf) \end{array} + HD(g)$$

(IIb) The Eley mechanism, which postulates a triatomic complex without the need for an empty adjacent site

$$D_2(g) + M-H(surf) \rightarrow M \overset{H}{\underset{D(surf)}{\diagup\diagdown}} D \rightarrow M-D(surf) + HD(g)$$

(III) The Boreskov version of (IIa, b), in which hydrogen atom recombination is followed by desorption from a low energy site on the surface

(IV) The Schwab mechanism, in which a bimolecular surface reaction takes place between molecularly adsorbed species.

Distinguishing between these mechanisms is by no means easy and the problems involved have been fully discussed.[3] A useful aid to deciding between alternatives may be the measurement of the rate at which the chosen catalyst will cause essentially pure *para*-hydrogen or *ortho*-deuterium, either of which can be obtained by low-temperature equilibration of the normal mixtures, to approach their equilibrium proportions. All of the mechanisms (I)—(IV) are available and, in addition, a purely intramolecular nuclear spin re-orientation process can be catalysed by a paramagnetic surface site.

The reactions of a mixture of hydrogen and deuterium have been measured on palladium in the form of a wire or sponge.[2] A difficulty arose from the readiness with which palladium absorbs hydrogen to form a 'hydride'. This process greatly complicated the situation and was the cause of the difficulty in making an unambiguous assignment of the mechanism. The preparation of a truly clean nickel surface is much more difficult than is the preparation of the clean surfaces of the much-studied refractory transition metals of the second and third rows. These latter elements have unstable oxides which can be removed from the surface by heating *in vacuo*. However, the use of sufficiently rigorous oxidation/reduction cleaning procedures can probably produce truly clean nickel surfaces.[1] The properties of a carefully cleaned nickel filament have been compared with the properties of nickel films laid down under u.h.v. conditions,[1] from which it appeared that the wire was the more reactive. When the mixture of hydrogen and deuterium was at a pressure of *ca.* 1 Torr (1 Torr = 133.3 N m^{-2}) the results for the equilibration reaction could be grouped into three temperature ranges. On a wire at 330—400 K the apparent activation energy, as deduced from Arrhenius plot of the absolute reaction rate *vs.* reciprocal temperature, was 7.57 kcal mol^{-1}.*

* 1 cal = 4.1868 J

[3] G. C. Bond, 'Catalysis by Metals,' Academic Press, London and New York, 1962.

The order of the reaction depended on the pressure, falling from 0.76 (at 1—10 Torr) to 0.35 (at 0.3—1.0 Torr) and finally to zero (0.03–0.3 Torr). These last results were, however, based on scanty data and too much reliance cannot be placed on them. There was satisfactory numerical agreement between these results and some earlier experiments in which argon-ion bombardment was used to clean the nickel,[4] *e.g.* at 293 K and a pressure of 1 Torr the two rates were 2×10^{-17} and 8.8×10^{-16} molecules cm^{-2} s^{-1}. In the temperature ranges 200—300 K and 77—150 K the activation energy fell to *ca.* 2.4 kcal mol^{-1}, whilst the order became independent of pressure and was 0.4 at 2.73 K or 0.3 at 77 K. At temperatures above 200 K the rate of equilibration of hydrogen and deuterium was similar to the rate of the spin reversal reaction, but at 77 K this latter reaction was a thousand times faster than the chemical process. These results were interpreted using a potential energy diagram for the bonding of hydrogen to nickel in which three types of binding were considered: Type A, strongly bonded atoms; Type C, weak chemisorption either as atoms or molecules; van der Waals, molecular adsorption. A change of mechanism from the Bonhoeffer–Farkas process at the highest temperatures to the Eley–Rideal process below 300 K was postulated. At the lowest temperatures, where exchange was much slower than spin reversal, the latter process was taking place by a molecular mechanism on paramagnetic centres.

The catalytic efficiency of a nickel wire, cleaned by a process similar to that used in ref. 1, for hydrogen/deuterium equilibration has been measured at low hydrogen pressures ($p < 10^{-5}$ Torr) and compared with the rate of adsorption of hydrogen by the clean filament.[5,6] The technique used was typical of the way in which very low pressure studies of isotope equilibration reactions are made. A mixture of approximately equal pressures of hydrogen and deuterium was allowed to stream over the filament and out of a calibrated orifice to the u.h.v. pumps. An *in situ* mass spectrometric analysis of the isotopic composition of the gas was made under steady-state conditions at each chosen filament temperature. The gas pressure was between 1.5×10^{-7} and 2.5×10^{-6} Torr. In contrast with the higher pressure results, the rate of production of HD molecules was directly proportional to the rate of impingement of the unequilibrated gas. The rate of equilibration at a filament temperature of 150 K was below the limit of sensitivity of the apparatus, but the rate rose rapidly above 280 K to a peak at *ca.* 500 K. The catalytic efficiency of the surface was defined as the ratio of the measured rate of production of HD to the rate which would be observed if every collision were effective in equilibration. At the peak, this efficiency was 11% and it then declined linearly to 7% at 1100 K. A mechanism based on a model similar to that used in ref. 1 was proposed.

The equilibration of a mixture of hydrogen and deuterium, either by an

---

[4] D. Shooter and H. E. Farnsworth, *J. Phys. and Chem. Solids*, 1961, **21**, 219.
[5] R. P. H. Gasser, K. Roberts, and A. J. Stevens, *Surface Sci.*, 1970, **20**, 123.
[6] R. P. H. Gasser, K. Roberts, and A. J. Stevens, *Trans. Faraday Soc.*, 1969, **65**, 3105.

initially polycrystalline wire or by a ribbon of molybdenum, has been studied, using u.h.v. techniques, by two authors.[7, 8] A flow technique such as that described earlier was used to identify the rate-determining step in the equilibration process; this step was adsorption. The pressure range investigated was 10^{-8}—10^{-5} Torr, whilst the filament temperature was between 298 and 1073 K. The main result of these experiments was a convincing demonstration that the molybdenum surface made no distinction between the isotopes. Thus, sticking probabilities, measured both by isotope equilibration and by direct observation, and saturation uptakes were the same for both species. The characteristic, essentially temperature-independent initial sticking probability, s_0, of hydrogen on a metallic surface was also illustrated by these experiments. Thus, at room temperature, a conventional gas adsorption experiment gave an s_0 value of *ca.* 0.4—0.5, which is very similar to the value of 0.4 above 513 K derived from the isotope reaction. A remarkable feature of these experiments is the still-rapid isotope reaction at room temperature, where the filament was fully covered with chemisorbed gas; the net rate corresponded to a sticking probability of 0.15. The mechanistic implications of this result are interesting. Thus, whereas the high-temperature, low-coverage exchange could be attributed to the atomic mechanism (I), at lower temperatures an alternative mechanism involving low-energy states was invoked. Although these latter states could not be identified closely, this suggestion is clearly in conformity with a mechanism of type (IIa, b).

An element of complexity was introduced by later work [8] into the satisfactory picture of the reaction of hydrogen with molybdenum so far observed. A different experimental procedure was used, in which a mixture of hydrogen and deuterium was adsorbed at room temperature and then flash desorbed from the ribbon. The composition of the desorbed gas was determined mass spectrometrically. Over a wide range of initial conditions the desorbed gas had the theoretical equilibrium proportions. However, surprisingly, it appeared that the relative amounts of the various isotopes in the puff of desorbed gas were essentially unrelated to the composition of the mixture which had been adsorbed. This result could, erroneously, be interpreted as the consequence of a difference between the sticking probabilities of the isotopes, but since the apparent sticking probability ratio could be varied at will by suitable pre-treatment of the ribbon, the conclusion drawn was that the desorbed gas emanated largely from the bulk. It was pointed out that if the participation of processes within the bulk of a catalyst was common, this would have considerable implications for theories of catalysis. However, it is worth noting that a later paper, dealing with the non-steady-state behaviour of hydrogen and deuterium on a molybdenum surface, reported no evidence for the participation of the bulk of the metal in the observed phenomena.[9]

[7] B. Bergsnov-Hansen and R. A. Pasternak, *J. Chem. Phys.*, 1966, **45**, 1199.
[8] G. E. Moore and F. C. Unterwald, *J. Chem. Phys.*, 1968, **48**, 5378.
[9] B. Bergsnov-Hansen and R. A. Pasternak, *Surface Sci.*, 1969, **17**, 402.

The desorption of hydrogen from tungsten, following saturation at 77 K, occurs in three separate temperature regions.[10] Consecutive adsorptions of hydrogen and deuterium showed that at the lowest temperature the γ state (desorbing at 77—190 K) was essentially molecular in nature, since no isotopic mixing took place.[11] By contrast, the two upper states, α (desorbing at 190—300 K and β (desorbing at 300—600 K), were largely mixed. Mixing also took place between α and β forms. Both of these latter states were thought to result from dissociative adsorption. In striking contrast to the complete absence of any isotope effect in the adsorption of hydrogen on molybdenum, the adsorption of hydrogen on a (100) oriented tungsten surface showed a considerable kinetic isotope effect.[12] The recorded ratio of the sticking probabilities, $s_{D_2} : s_{H_2}$ was 1.4. This value is, significantly, close to the ratio of the molecular velocities, ($\sqrt{2} = 1.41$). The exchange of previously-adsorbed hydrogen by deuterium subsequently admitted was also studied. No exchange took place at 77 K or 195 K. At 300 K exchange occurred slowly whilst at 350 K it was more rapid. The activation energy for the replacement process was ca. 5 kcal mol^{-1}. These measurements relate only to hydrogen or deuterium because the pressure of the mixed molecule HD was not monitored during the desorption process. The mechanism postulated for the replacement of one adsorbed species by the other isotope was by way of a weakly-bound intermediate state, rather than *via* direct gas phase replacement.

Although the participation of the bulk metal in the reactions of hydrogen isotopes with molybdenum is doubtful, no such doubt exists in the cases of the reactions of hydrogen (and other gases) with tantalum; absorption undoubtedly occurs. In two experiments designed to investigate the rate processes in these reactions, isotope exchange between tantalum loaded with interstitial hydrogen and gas phase deuterium was measured at temperatures between 803 and 913 K and at a pressure of ca. 100 Torr.[13, 14] The activation energy for the underlying rate process was 14.9 kcal mol^{-1} and this value probably sets the upper limit to the activation energy for the diffusion of hydrogen atoms through the lattice. The first-order kinetics of the exchange process were taken as evidence that desorption of molecules from the surface was the rate-determining step. This in turn suggests the Bonhoeffer–Farkas mechanism for the reaction. The very low reactivity of germanium towards molecular hydrogen has been confirmed in two studies of hydrogen isotope equilibration.[15, 16] Neither a Ge(100) crystal nor an evaporated film had detectable activity at room temperature; the rate constant k_m was, therefore, $< 5 \times 10^{11}$ mol cm^{-2} s^{-1}. Doping germanium to produce *p*-type semi-conductivity increased the activity, but even the maximum rate was scarcely more than a

[10] F. Ricca, R. Medana, and G. Saini, *Trans. Faraday Soc.*, 1965, **61**, 1492.
[11] F. Ricca, R. Medana, and G. Saini, *Trans. Faraday Soc.*, 1966, **62**, 2273.
[12] P. W. Tamm and L. D. Schmidt, *J. Chem. Phys.*, 1970, **52**, 1150.
[13] F. J. Cheselke, W. E. Wallace, and W. K. Hall, *J. Phys. Chem.*, 1959, **63**, 505.
[14] W. K. Hall, W. E. Wallace, and F. J. Cheselke, *J. Phys. Chem.*, 1961, **65**, 128.
[15] G. E. Moore, H. A. Smith, and E. H. Taylor, *J. Phys. Chem.*, 1962, **66**, 1241.
[16] D. Shooter and H. E. Farnsworth, *J. Phys. Chem.*, 1962, **66**, 222.

hundred times the rate of reaction at the glass walls of the reaction chamber. By contrast, n-type doped germanium was completely inert. A u.h.v.-evaporated platinum film was reactive towards hydrogen, and the kinetics of the replacement of adsorbed hydrogen by deuterium have been reported.[17] Not unexpectedly, gold is an ineffectual catalyst for hydrogen/deuterium exchange [18] and even such activity as exists might be due to active impurities. In a molecular beam study of the scattering from an Ag(111) surface it was not possible to measure the distribution of desorbed molecules directly by use of an isotope exchange technique, though this was possible for Ni(111).[19]

The use of molecular beam techniques holds considerable promise for the elucidation of the details of the atomic processes which occur at the gas–solid interface and the utilization of isotopic exchange reactions will undoubtedly be helpful in interpreting the results. For example, in a recent study of the interaction of hydrogen with tantalum, a modulated beam of either hydrogen or deuterium molecules impinged on a heated tantalum target.[20] The signal, which was detected mass spectrometrically, from diatomic molecules, declined smoothly above a filament temperature of 400 K until a constant, lower plateau region was reached at 2300 K. The diminution in the reflected signal was attributed in part to surface diffusion away from surface regions accessible to the detector and in part to the desorption of atoms, a process which rose rapidly in importance above a target temperature of 1600 K. The separate isotopes behaved similarly and a surprising result for this work was the absence of any HD when the beam consisted of equal pressures of hydrogen and deuterium. This absence of equilibration on an otherwise reactive surface was attributed to completely independent surface interactions by the two components. The surface diffusion away from the area on which the beam impinged was thought to lower the surface concentration of adatoms so much that the second-order process necessary for recombination of an H and a D atom became of negligible probability. In marked contrast to the observations on tantalum, a molecular beam experiment in which deuterium molecules bombarded a heated (111) nickel surface, maintained in a low pressure (6×10^{-6} Torr) of hydrogen, showed that extensive HD formation occurred on the surface.[19] Indeed, the results of the measurements of the angular dependence of desorbed HD molecules made an important contribution to the discussion of the interactions at the surface, because these molecules can only be formed on the surface. Their distribution is, therefore, unambiguously identified with the true desorption process and there can be no contribution from reflection processes.

In view of the extensive study of the hydrogen/deuterium reaction on metals, it is of interest to compare, briefly, the behaviour of other types of surface. The most frequently studied non-metallic surfaces are metallic

[17] H. Geustch, *Z. phys. Chem. (Frankfurt)*, 1962, **35**, 69.
[18] H. Wise and K. M. Sancier, *J. Catalysis*, 1963, **2**, 149.
[19] R. L. Palmer, J. N. Smith jun., H. Saltsburg, and D. R. O'Keefe, *J. Chem. Phys.*, 1970, **53**, 1666.
[20] K. A. Krakowski and D. R. Olander, *J. Chem. Phys.*, 1968, **49**, 5027.

oxides, on which the reaction can, in principle, proceed by any of the mechanisms (I)—(IV). Catalytically active oxides include those of the transition metals [21] and the rare earth metals.[22] These surfaces tend to be inert at low temperatures, though a few transition metal oxides retain some activity down to 77 K. However, all the oxides are able to flip a nuclear spin and convert *para* hydrogen to *ortho* at this temperature. When the oxide is reactive at 77 K the equilibration reaction probably proceeds by the Eley–Rideal mechanism (IIa, b), whilst a dissociative, chemical mechanism may operate at room temperature. It is noteworthy that only a small proportion of the surface is catalytically active. An investigation of the surface properties of the interstitial-type carbide of tantalum, TaC, has been made.[23] The formation of the carbide eliminated the absorption processes characteristic of the metal, and the catalytic activity of TaC for hydrogen-isotope equilibration started only above room temperature. Indeed, a temperature similar to that required to desorb the chemisorbed gas was needed, and the Bonhoeffer–Farkas atomic mechanism was, therefore, postulated.

2 Nitrogen Isotope Reactions

The thermodynamic properties of $^{15}N_2$ and $^{14}N_2$ are not significantly different, so that the most convenient, and exclusively used, method of analysing a mixture at a low pressure is mass spectrometrically. One continuing thread in experimental observations of the nitrogen isotope equilibration reaction is the attempt to determine the part played by adsorbed nitrogen in the ammonia-synthesis reaction. The first experimental observation of nitrogen isotope equilibration used a doubly-promoted iron catalyst which became catalytically active at *ca.* 723 K.[24] The subsequent history of investigations of the reaction up to about 1962 has been thoroughly reviewed.[3] The difficulties and discussions which have been associated with the ammonia synthesis reaction are well illustrated by the fact that different conclusions have been drawn from the same experimental data. Thus, whilst the original authors [25] concluded from an analysis of the kinetic data and the inverse hydrogen/deuterium kinetic isotope effect that the predominant surface species on a doubly-promoted iron catalyst during ammonia synthesis were —NH radicals, a later re-analysis of the same data supported the view that the major surface species were adsorbed nitrogen atoms.[26] The replacement of $^{15}N_2$ pre-adsorbed on an unpromoted iron catalyst by gaseous $^{14}N_2$ led to the appearance of both $^{14}N^{15}N$ and $^{15}N_2$ in the gas phase.[27] It was thought that there may be two processes at work on the surface, isotopic displacement and isotopic mixing. An analysis of the time-dependence of the gas phase

[21] D. R. Pearce, P. C. Richardson, and R. Rudham, *Proc. Roy. Soc. A*, 1969, **310**, 121.
[22] D. R. Ashmeed, D. D. Eley, and R. Rudham, *J. Catalysis*, 1964, **3**, 280.
[23] A. J. Clark, R. P. H. Gasser, and P. Whitehead, *Trans. Faraday Soc.*, 1970, **66**, 2085.
[24] G. G. Joris and H. S. Taylor, *J. Chem. Phys.*, 1939, **7**, 893.
[25] A. Ozaki, H. S. Taylor, and M. Boudart, *Proc. Roy. Soc. A*, 1960, **258**, 47.
[26] S. R. Logan and J. Philip, *J. Catalysis*, 1968, **11**, 1.
[27] Y. Morikawa and A. Azaki, *J. Catalysis*, 1968, **121**, 145.

composition suggested also that under the experimental conditions, 653 K and 357 Torr, *ca*. 80% of the adsorbed nitrogen was in an undissociated form. The effect of substituting deuterium for hydrogen during the synthesis of ammonia over an unpromoted catalyst was to accelerate the reaction,[28] an effect which had also been found in the earlier work on the doubly-promoted catalyst.[25]

The reaction of nitrogen with a tungsten surface has been the subject of intensive investigation since the advent of u.h.v. techniques. A wide range of effects has been studied though none has been more quantitatively reproducible than the equilibration reaction which is catalysed by a hot tungsten filament. The failure of the first attempt to observe this reaction, by making use of an electric light bulb filament in the temperature range 973—1173 K [24] can probably be attributed to a contaminated surface. Later work has shown that tungsten wires are liable to be superficially contaminated with carbon and that adsorbed oxygen is an effective catalytic inhibitor. The use of u.h.v. techniques and proper cleaning procedures has meant that truly clean tungsten surfaces can now be prepared, maintained, and regenerated at will.

The two main aims of recent nitrogen isotope exchange and equilibration experiments have been (*i*) the determination of the dependence of the rate of reaction on temperature and pressure and (*ii*) the characterization of the atomic nature of the binding states on a polycrystalline tungsten surface. These will be considered in turn. In experiments in which a mixture of $^{15}N_2$ and $^{14}N_2$ streamed over the tungsten filament, the isotope scrambling process became measurable only when the filament temperature exceeded *ca*. 1000 K.[29, 30] The activity then rose rapidly to a maximum at 1400—1500 K followed by a gentle decline. Under steady-state conditions at high filament temperatures the equilibrium coverage of nitrogen atoms is very low. Provided that there is full equilibration on the surface, the sticking coefficient, which is equal to the ratio of the flux of adsorbed molecules to the incident flux, can be equated with s_0, since the coverage is low. In two widely separated laboratories, quite striking numerical agreement has been reached on the values of the sticking coefficient over a range of temperatures.[29, 30] Thus, at 1500 K the values were 0.0144 and 0.0152; at 1600 K, 0.0136 and 0.0144; at 1700 K, 0.0125 and 0.0135. These results were obtained at different pressures, 2×10^{-7} Torr and 6×10^{-8} Torr. A slight pressure dependence of the sticking probability makes the values extrapolated to equal pressure even closer. This excellent agreement at high temperatures is in marked contrast to the range of initial sticking probabilities which has been recorded at room temperature. The onset of the isotope scrambling process occurred at a similar temperature to the onset of the desorption of the strongly

---

[28] K. Aika and A. Ozaki, *J. Catalysis*, 1969, **13**, 232.
[29] R. P. H. Gasser, C. P. Lawrence, and D. G. Newman, *Trans. Faraday Soc.*, 1965, **61**, 1771.
[30] T. E. Madey and J. T. Yates jun., *J. Chem. Phys.*, 1966, **44**, 1675.

bonded β form of adsorbed nitrogen. The high temperature required, and the near independence of the sticking probability from pressure, were taken as evidence that the reaction proceeded by the Bonhoeffer–Farkas atomic mechanism.

Flash desorption experiments have shown that there are three identifiable binding states for nitrogen on a polycrystalline tungsten surface; in order of increasing strength they are conventionally labelled γ, α, and β. Of these states, α and β are populated at room temperature but γ is found only after adsorption below 150 K. Furthermore, the β-state may split into two, β_1 and β_2. A full discussion of these results has been given.[31] The outstanding problem of interest about these states is the identification of those in which the nitrogen is dissociated. Although kinetic analysis of the desorption data seemed to point to the β-states containing nitrogen atoms, this conclusion has been contested. The adsorption of a mixture of isotopes followed by flash desorption led to complete equilibration of the β-state, which is consistent with atomic adsorption, but not of the α and γ. These latter are, therefore, probably molecular.

The dissociative adsorption of a diatomic molecule requires that the surface-to-adatom bond be strong enough to make good the energy lost in dissociating the molecule. Normal dissociative adsorption will thus be limited to the relatively strongly-bonded sites on the surface. The sensitivity of adsorption to surface structure is particularly marked in the case of nitrogen, with its large dissociation energy, 225 kcal mol^{-1}. A more complete picture of the surface states, which includes those of lower stability, may be obtained if the molecules are activated in some way (*e.g.* thermally or by electron bombardment) before adsorption. Isotope exchange experiments have played a significant role in elucidating the nature of these excited states. In one experiment,[32] a polycrystalline tungsten surface was first saturated with $^{15}N_2$ and then flashed to 1200 K to desorb any weakly-held species and leave only β-N_2. The residual gas was pumped away and $^{14}N_2$ was let in to give a pressure of 5×10^{-6} Torr. This gas was activated by electron bombardment, thus causing further adsorption to take place, into a state designated x-N_2. On flashing the filament, the nitrogen was desorbed in two temperature regions: at 900—1200 K the newly formed x-N_2 was desorbed, whereas at high temperatures the normal β-N_2 was desorbed. Kinetic analysis of the pressure record during desorption showed that desorption of x-N_2 was second order, whilst mass spectrometric analysis of the desorbed gas revealed complete isotopic equilibration of the successively adsorbed $^{15}N_2$ and $^{14}N_2$ (activated) in the x-state. The β-state, however, was richer in $^{15}N_2$, perhaps because there are some sites on a polycrystalline surface which do not participate in the $(\beta+x)N_2$ structure. These results are consistent with the suggestion that x-nitrogen is atomic and they are a close parallel to the

---

[31] G. Ehrlich, *Ann. Rev. Phys. Chem.*, 1966, **17**, 295.
[32] R. Matsushita and R. S. Hansen, *J. Chem. Phys.*, 1970, **52**, 3619.

observations of adsorption of ammonia and nitrogen on tungsten.[33] In these latter experiments, a filament was pre-dosed with $^{14}N_2$ and subsequently treated with $^{15}NH_3$. On flashing the filament a nearly random mixture of isotopes of mass 28, 29, and 30 was desorbed, again with $(x+\beta)$ states apparent, though the x-N_2 fraction was consistently richer in the pre-adsorbed ^{14}N. The structural consequences of these results are, however, in disagreement with evidence from LEED (low energy electron diffraction) on the reaction of ammonia with tungsten(100), which suggested that the adsorbed moiety was NH_2.[34]

Nitrogen isotope exchange and equilibration reactions have been used as a probe to investigate the relationship between the catalytically active sites and the binding sites on a tungsten surface.[35, 36] It was found that the reaction of a mixture of $^{14}N_2O$ and $^{15}N_2$ to produce $^{14}N^{15}N$ became measurable only at a tungsten filament temperature 700 K higher than the temperature at which $^{14}N_2$ and $^{15}N_2$ react to produce $^{14}N^{15}N$. The reason for the difference was attributed to the blocking of reactive sites by oxygen adsorbed from the N_2O molecule. At room temperature the $^{15}N_2$-saturated tungsten filament would adsorb the oxygen atom of the $^{14}N_2O$ molecule, but the nitrogen atoms appeared in the gas phase as $^{14}N_2$. No isotopic mixing took place, from which result it was suggested that adsorption of the oxygen atom took place by a simple cleavage of the N—O bond, without participation of the ^{14}N atoms in the adsorption process. The inhibiting effect of pre-adsorbed oxygen on nitrogen isotope equilibration was used to investigate the nature of the catalytically active sites on the tungsten surface. A near-linear decline in efficiency, which closely matched the decline in uptake of β-N_2, was attributed to an essentially catalytically uniform surface on which 'active sites' did not play a major role.

The general chemical and physical similarities between tungsten and molybdenum and, to a lesser extent, between tungsten and rhenium make comparative studies of the surface reactivity of these elements valuable. The nitrogen isotope equilibration reaction catalysed by molybdenum was closely similar in its dependence on filament temperature and on gas pressure to the reaction on tungsten.[37] However, participation of the bulk metal in the reaction process has also been postulated.[38] Multiple binding sites have been detected for nitrogen on a polycrystalline rhenium filament by the flash filament technique.[39] A low-temperature γ-state developed at 77 K and desorbed without any isotopic mixing taking place. The high temperature β-state desorbed in a completely equilibrated mixture following simultaneous adsorption of $^{14}N_2$ and $^{15}N_2$. However, a 'memory' effect was observed when

[33] K. Matsushita and R. S. Hansen, *J. Chem. Phys.*, 1970, **52**, 4871.
[34] P. J. Estrup and J. Anderson, *J. Chem. Phys.*, 1968, **49**, 523.
[35] R. P. H. Gasser, M. F. King, and P. R. Vaight, *Trans. Faraday Soc.*, 1968, **64**, 2852.
[36] R. P. H. Gasser and P. R. Vaight, *Nature*, 1969, **221**, 166.
[37] R. P. H. Gasser, A. Hale, and C. J. Marsay, *Trans. Faraday Soc.*, 1967, **63**, 1789.
[38] G. E. Moore and F. C. Unterwald, *J. Chem. Phys.*, 1968, **48**, 5393.
[39] J. T. Yates jun. and T. E. Madey, *J. Chem. Phys.*, 1969, **51**, 334.

adsorption of $^{14}N_2$ and $^{15}N_2$ was allowed to take place successively and it was attributed to the existence of a number of β-sub-states which were not resolved in the desorption peak. The rate of isotopic equilibration of a gas phase mixture of $^{14}N_2$ and $^{15}N_2$ streaming over the rhenium filament rose rapidly at filament temperatures above *ca.* 800 K. A plateau was reached at 1100 K, where the sticking probability s, calculated by equating s with the probability of equilibration, was 3.5×10^{-4}. The shape of this curve of sticking probability *vs.* temperature was very similar to the curve obtained from a series of quite separate measurements of isotope equilibration, and the high-temperature plateau value of s, *ca.* 3×10^{-4}, was also in excellent agreement.[40] These high-temperature sticking probabilities measured by isotope equilibration relate to surfaces on which the coverage of nitrogen is very low, $\theta < 0.1$. It is thus appropriate to compare the results with initial sticking probabilities measured at room temperature. When this is done, a marked decline in the initial sticking probability of nitrogen on tungsten, rhenium, and molybdenum is observed.

3 Carbon Monoxide Isotope Reactions

It is generally believed that carbon monoxide is adsorbed on transition metals without becoming dissociated. The principal evidence comes from kinetic analysis of desorption pressure bursts and from i.r. spectroscopic studies of the adsorbed species. All experiments under u.h.v. conditions involving carbon monoxide are hampered by the uncertainties which arise as the result of the reversible adsorption of carbon monoxide by the surfaces of the u.h.v. chamber. A detailed analysis of the consequence of the formation of this 'surface phase' has been given.[41] The relevant consequence of this experimental difficulty is that the quantitative agreement recorded between different workers for nitrogen isotope experiments should not be anticipated for carbon monoxide.

A kinetic study of the reaction

$$^{12}C^{18}O + {}^{13}C^{16}O \rightarrow {}^{12}C^{16}O + {}^{13}C^{18}O$$

catalysed by a tungsten filament was aimed at identifying the mechanism of isotopic mixing.[42] The technique used was the standard one of streaming a mixture of the isotopes through the u.h.v. chamber and recording mass spectrometrically the steady-state composition, as a function of filament temperature. When a comparatively low total pumping speed was used, 120 cm^3 s^{-1}, the percentage attainment of equilibrium due to the tungsten filament rose rapidly above 900 K to essentially 100% at 1300 K. Even when the experimental filament was inactive, the incandescent filament of the mass spectrometer caused 66% equilibration. The mechanism proposed

---

[40] C. Boon, R. P. H. Gasser, and H. Tovey, *Trans. Faraday Soc.*, 1969, **65**, 3111.
[41] J. P. Hobson and J. W. Earnshaw, *J. Vacuum Sci. Technol.*, 1967, **4**, 257.
[42] T. E. Madey, J. T. Yates jun., and R. C. Stern, *J. Chem. Phys.*, 1965, **42**, 1372.

for the surface reaction required the formation of a four-centre transition state complex of the form C---O. Exchange of partners is then possible

$$\begin{array}{c} \text{C---O} \\ |\quad | \\ |\quad | \\ \text{O---C} \end{array}$$

without dissociation of the molecules. A similar, subsequent experiment using a higher rate of flow of gas was in close agreement with the earlier measurement and gave a value for the plateau of the exchange efficiency of ca. 50%.[43] This latter experiment also recorded the effect of pre-adsorbed oxygen on both the efficiency of catalysis and the adsorption of carbon monoxide. As with nitrogen [36] the results were in accord with the properties expected of a uniformly reactive surface, though in this case a concavity in the efficiency curve was attributed to the greater number of surface sites required for the formation of the four-centre transition state.

The relationship between the type of bonding to the surface and isotope equilibration was explored both in the experiments already described and in the reaction of carbon monoxide with molybdenum.[44] On both tungsten and molybdenum two chemisorbed states have been identified. The lower temperature α-state desorbs at ca. 500 K whereas the more strongly bonded β-state, which may be sub-divided,[45] desorbs above 900 K. For neither metal did the α-state participate in the equilibration process. The plateau efficiency of molybdenum was identical with that of tungsten. A similar result was obtained when rhenium catalysed carbon monoxide isotope exchange, though in this system there was a significant decline in efficiency at the highest filament temperatures.[46] For all of the metals tungsten, molybdenum, and rhenium, the elevated temperatures needed to reduce catalytic activity result in very low coverages at the maximum efficiencies. As with nitrogen, therefore, if the sticking probability is equated with the efficiency the resulting number relates to an essentially clean surface. In marked contrast with nitrogen, the initial sticking probability of carbon monoxide on these three metals is independent of filament temperature.

The carbon monoxide isotope scrambling reaction appears to take place more readily at low temperatures on nickel than on the high-melting transition metals. Thus, a nickel catalyst prepared by reduction of the powdered oxide with hydrogen was able to catalyse the reaction at 298—358 K, though long contact times (up to 17 h) were necessary.[47a] Even so, the full statistical equilibrium was far from fully achieved. The reaction was probably associated exclusively with the reversible adsorbed gas taken up at the highest pressures, since exchange stopped when this gas was pumped away. However, a nickel film prepared by evaporation under u.h.v. conditions was not catalytically active at 273 K.[47b] The difference between the two surfaces

[43] R. P. H. Gasser and P. R. Vaight, *Nature*, 1970, **225**, 933.
[44] R. P. H. Gasser and C. J. Marsay, *Trans. Faraday Soc.*, 1968, **64**, 516.
[45] P. A. Redhead, *Trans. Faraday Soc.*, 1961, **57**, 641.
[46] J. T. Yates jun. and T. E. Madey, *J. Chem. Phys.*, 1969, **21**, 334.
[47] (a) J. T. Yates jun., *J. Phys. Chem.*, 1964, **68**, 1245; (b) R. Suhrman, Hj. Heyne, and G. Wedler, *J. Catalysis*, 1962, **1**, 208.

may stem from the participation of residual adsorbed moieties on the reduced oxide, since this method of preparation cannot be relied upon to produce the truly clean surface obtained by u.h.v. evaporation.[48] An alternative explanation is that, because the experimental conditions in the oxide experiment were not an optimum for observing exchange, this process was present but remained undetected.

The existence of a radioactive isotope of carbon, ^{14}C, with an experimentally convenient half life, provides an alternative and direct method of monitoring the adsorbed gas. In a study of the adsorption of ^{14}CO on polycrystalline molybdenum, no residual activity remained after the filament had been flashed above 1200 K.[49] This result rules out the occurrence of the surface disproportionation reaction

$$2^{14}CO(ads) \rightarrow {}^{14}C(ads) + {}^{14}CO_2(g)$$

and is entirely consistent with the view that the adsorption of carbon monoxide on refractory metals is molecular and non-dissociative. Complete and rapid exchange between adsorbed and gas-phase carbon monoxide has been found for the (110) and (100) planes of nickel.[50] The signal from ^{14}CO adsorbed on either plane was stable in time, but on admitting ^{12}CO to the crystal, with gas and metal at 298 K, rapid replacement took place. Similarly, when ^{12}CO was adsorbed and ^{14}CO was admitted, the surface signal increased rapidly, $\tau_i \sim 30$ s for $p_{14_{CO}} = 1.5 \times 10^{-3}$ Torr. The replacement reaction had first-order kinetics and it was thought to take place by the replacement of the whole carbon monoxide molecule in one exchange event. The rate corresponded to one exchange for every 250 collisions at 1×10^{-6} Torr or to one exchange for every 4000 collisions at 1.5×10^{-3} Torr. The decrease in efficiency at higher pressures was attributed to a decreasing free area for the adsorbing molecules as the pressure increased. The results of these isotope exchange experiments were in satisfactory agreement with the results of earlier measurements of the dependence of the rate of adsorption on surface coverage. The general chemical inertness of gold metal in bulk gives an opportunity of investigating the effect of dispersion on chemical reactivity. The reaction chosen to characterize this effect was [51]

$$^{14}CO_2(g) + {}^{12}CO(g) \rightarrow {}^{12}CO_2(g) + {}^{14}CO(g)$$

When sufficiently finely divided (diameter < 100 Å), gold supported on magnesium oxide became catalytically active at temperatures in excess of 623 K. However, the reactivity was sensitively dependent on the method of preparation of the gold catalyst and upon the support material employed.

4 Oxygen Isotope Reactions

The strong and irreversible adsorption of oxygen by most metals limits the

[48] M. W. Roberts and K. W. Sykes, *Trans. Faraday Soc.*, 1958, **54**, 1958.
[49] A. D. Crowell and L. G. Matthews, *Surface Sci.*, 1967, **7**, 79.
[50] K. Klier, A. C. Zettlemoyer, and H. Leidheiser jun., *J. Chem. Phys.*, 1970, **52**, 589.
[51] D. Y. Cha and G. Parravano, *J. Catalysis*, 1970, **18**, 200.

usefulness of isotope exchange and equibration reactions and, as a result, rather few have been reported. The technological importance of silver as an oxidation catalyst makes an understanding of the chemisorption of oxygen on this metal of interest. The results of earlier work have been conflicting, probably as the result of surface contamination. A more recent comparison of the rate of the isotope reaction

$$^{16}O_2 + {}^{18}O_2 \rightarrow 2\,{}^{16}O^{18}O$$

with the rate of desorption of chemisorbed gas has been made.[52] Above a temperature of *ca.* 433 K the chemisorption of oxygen becomes reversible, but there is also a high-temperature form of much higher binding energy. The rate of the exchange reaction, though slow, was recorded in the temperature range 433—473 K and compared with the rate of desorption. The two rates were the same, and it was concluded that at least part of the more strongly bonded oxygen participated readily in the equilibration reaction. The rate of equilibration was thought to be determined by the rate of desorption from the surface.

Under u.h.v. conditions oxygen is an intractable gas, since not only does it react with the surfaces of the apparatus but it also undergoes chemical reactions at the hot filaments normally present in a u.h.v. system. Moreover, it adversely affects the performance of ionization gauges.[53] The reactions of oxygen with some refractory metals have been reported, sometimes with conflicting results.[54] Perhaps the most striking feature of the desorption of oxygen in a flash filament experiment is the absence of molecular oxygen in the primary desorbed species. For example, mass spectrometric analysis of the gas desorbed from a rhenium filament covered with $^{16}O_2$ into an apparatus whose walls were covered largely by $^{18}O_2$ gave a mixture of $^{16}O_2$, $^{18}O_2$, and $^{16}O^{18}O$. This result suggests that the oxygen was desorbed from the filament in a highly reactive form, probably as atoms, which subsequently underwent reactions at the walls. The results of an electron impact desorption study of ^{16}O or ^{18}O adsorbed on tungsten provide a rare example of a measurable difference between isotopes other than those of hydrogen.[55] The ratio of the cross-sections for the desorption of the ions $^{16}O^+$ and $^{18}O^+$ was about 1.5, a result which was in good agreement with the theoretical treatment of the desorption process.[56]

5 Exchange of Hydrogen Isotopes in Organic Molecules

The range of potentially interesting molecules in which one or more hydrogen atoms can be replaced by deuterium is enormous. Coupled with the numerous metals which are capable of catalysing the exchange process there is a plethora

[52] Y. L. Sander and D. D. Durigon, *J. Phys. Chem.*, 1965, **69**, 4201.
[53] P. A. Redhead, *Vacuum*, 1962, **12**, 267.
[54] R. P. H. Gasser, *Ann. Reports (A)*, 1970, **67**, 213.
[55] T. E. Madey, J. T. Yates jun., D. A. King, and C. J. Uhlaner, *J. Chem. Phys.*, 1970, **52**, 5215.
[56] P. A. Redhead, *Canad. J. Phys.*, 1964, **42**, 886.

of systems available for study. The really rapid progress in this field followed the development of the mass spectrometer as a quantitative analytical instrument. An authoritative survey of the situation as far as the exchange reaction of hydrocarbons, to which much attention has been given, are concerned was published in 1959.[57] The topic is also dealt with in some detail in ref. 3. In brief, it appears that the conversion of hydrocarbons to the deuteriomolecules follows closely the expected product distribution. Thus, for example, the measured successive equilibrium constants for the substitution of propane with deuterium, a reaction which is catalysed by a tungsten film, were in agreement with the calculated values. Determinations of the rates of isotopic substitution and of the initial product distributions have allowed reactions to be classified as either (*i*) simple, stepwise processes in which one hydrogen atom at a time is replaced in the molecule or (*ii*) multiple exchange reactions in which more than one deuterium atom is introduced into the molecule at each surface event.

Some general features of the reactions were also described. These included (*i*) the observation that metal films are usually more reactive than the bulk metal, (*ii*) the order of reactivity of transition metals for the much-studied reaction of ethane with deuterium is W(Mo, Ta) > Rh > Pt(V, Zr) > Pd and this order is similar for other exchange reactions. The reactivities of the hydrocarbons increase in the sequence : methane < neopentane < ethane < propane < cyclic hydrocarbons. Subsequent work has extended the range of catalytic materials and the organic molecules studied.

Two interesting problems in catalysis relate to the roles played by surface topology (the 'geometric effect') and the electronic structure ('electronic effect') in the catalytic reactivity. One method of tackling this problem has been to use a range of alloys to catalyse a particular reaction. One reaction chosen was the exchange of hydrogen in methane with deuterium, which is catalysed by alloys of platinum with palladium and of palladium with rhodium.[58] A complete series of solid solutions is formed by both pairs of elements. In the Pd–Rh series the lattice constant changes from 3.890 Å in Pd to 3.804 Å in Rh whereas for Pt–Pd the constant changes only from 3.916 Å in Pt to 3.890 Å in Pd. For all of the alloys, platinum metal, and palladium metal, step-wise exchange was the most important mechanism, though multiple exchange increased for the alloys at high temperatures. On rhodium, stepwise exchange was predominant at high temperatures. The reactivity at 423 K for both pairs of alloys varied with composition, but in neither case was there a peak in the plot of reactivity *vs.* alloy composition. No single physical parameter was adequate to explain the observations, though there was a rough correlation with the (slight) changes in atomic spacings. By contrast with these results, the reactivity of Pt–Rh alloys went through a maximum at 60—70 wt. % rhodium, at which composition there is one

---

[57] C. Kemball, *Adv. Catalysis*, 1959, **11**, 223.
[58] D. W. McKee and F. J. Norton, *J. Catalysis*, 1964, **3**, 252.

unpaired d electron per metal atom.[59] It was thought likely that the variation of activity in this system could be attributed to the electronic factor, because the lattice constant varies only very slightly between platinum and rhodium and does so nearly linearly with composition. The relative reactivity of the transition metals for methane/deuterium exchange has been reviewed.[60]

Copper–nickel films, prepared by successive deposition of the two metals followed by sintering in u.h.v., are active in catalysing the exchange of deuterium with benzene at 314 K.[61] The two metals form only a limited series of solid solutions and the exchange experiments, taken in conjunction with adsorption and hydrogenation experiments, show that the copper-rich phase envelops the nickel-rich phase. The reactivity thus stays constant within the limited miscibility region over a wide range of overall alloy composition. Although the catalytic activity of transition metals for the hydrogenation of acetylene is well established, experiments using a mixture of acetylene and deuterium have shown that, for example, ruthenium and osmium catalyse isotope exchange as well as hydrogenation.[62]

Radioactive ^{14}C, ^{3}H, and ^{125}I have all been used to investigate catalytic processes, generally involving organic molecules. The transfer of hydrogen between cyclohexane and benzene is catalysed by several transition metals.[63] The stepwise reaction was written

$$C_6H_{12}(g) \rightarrow C_6H_6(g) + 6H(s)$$
$$^{14}C_6H_6(g) + 6H(s) \rightarrow {}^{14}C_6H_{12}(g)$$

Chromatographic separation of the reacting mixture of [$^{14}C_6$]benzene and cyclohexane, followed by radioactive analysis using liquid scintillation counting techniques, was used to follow the course of the reaction. The metal catalysts were supported on alumina or silica and the reactivities were in the sequence Pt > Pd > Ir > Ru > Rh. The efficiencies for the reaction were low, in the range 2.8×10^{-6} to 9.1×10^{-6}. Perhaps surprisingly, the reactivity of gold at 508 K seemed to be very roughly comparable with the reactivity of the other metals at 390 K.[64] The formation of an adsorbed layer of hydrogen atoms, needed for the proposed mechanism, at such a high temperature on gold is unexpected. A series of radiochemical experiments of chemisorption and catalysis has revealed that the support material used for the metal catalyst may itself participate in the reaction. For example, in an investigation of the exchange of [$^{14}C_2$]ethylene pre-adsorbed on a supported platinum catalyst with inactive ethylene, it was found that there was a large fraction of the initially adsorbed material which was non-exchangeable.[65] Use of tritium showed that the greater retention by alumina-supported platinum

---

[59] D. W. McKee and F. J. Norton, *J. Catalysis*, 1965, **4**, 510.
[60] A. Frennet and G. Lienard, *Surface Sci.*, 1969, **18**, 80.
[61] P. van der Plank and W. M. H. Sachter, *J. Catalysis*, 1968, **12**, 35.
[62] G. C. Bond, G. Webb, and P. B. Wells, *J. Catalysis*, 1968, **12**, 157.
[63] G. Parravano, *J. Catalysis*, 1970, **16**, 1.
[64] G. Parravano, *J. Catalysis*, 1970, **18**, 320.
[65] J. A. Altham and G. Webb, *J. Catalysis*, 1970, **18**, 133.

than by silica-supported platinum was due to the participation of the support. A similar conclusion had been reached previously in a radiochemical study of [$^{14}C_2$]ethylene adsorbed on a nickel film.[66] Only a fraction of the pre-absorbed $^{14}C_2H_4$ was removed during the subsequent hydrogenation of inactive ethylene. The heterogeneity of an evaporated nickel film towards hydrogen has been established by an ingenious replacement technique.[67] The film was first allowed to adsorb tritiated hydrogen which was then progressively replaced by mercury. During the first 10% of desorption the probability of replacement of hydrogen was 30% higher than the probability of replacement of tritium. For the remaining 70% this probability was 20% higher. It was also established that the principle of 'last-on first-off' applies, from which it must be concluded that not all of the hydrogen on the film is mobile. The exchange of radioactive ^{125}I between ethyl iodide and propyl iodide has been investigated over copper, iron, nickel, palladium, rhodium, and iridium powder catalysts.[68] On nickel and iron, iodide formation occurred and quantitative data were not collected. For the other metal catalysts the activation energy was approximately constant at *ca.* 12 kcal mol^{-1}. It was suggested that the reason for the constancy was that the activation energy refers to the same process in every case. This process could be the breaking of one C—I bond and one C—*I bond and the formation of two energetically equal bonds. Naturally for the overall process the enthalpy change is zero.

[66] S. J. Thomson and J. L. Wishlade, *Trans. Faraday Soc.*, 1962, **58**, 1170.
[67] G. K. L. Cranstoun and S. J. Thomson, *Trans. Faraday Soc.*, 1963, **59**, 2403.
[68] R. A. Morrison and K. A. Krieger, *J. Catalysis*, 1968, **12**, 25.

8
Infrared Spectra of Adsorbed Species on Metals

BY J. PRITCHARD

Until recently, i.r. spectroscopy has been the only spectroscopic method giving information about the nature and bonding of adsorbed species on metal surfaces. It remains the most widely used technique. Applications since a recent review [1] include co-adsorption of nitrogen and hydrogen onto supported iron catalysts as well as the adsorption of carbon monoxide onto a single crystal copper surface.

Transmission spectroscopy of supported samples provides the basis of most investigations. Several short-path-length cells suitable for pressed oxide pellets have been described.[2-4] A very short optical path has been achieved [3] in a cell which can be heated to 573 K. The sample can be moved for baking at 1073 K before being accurately replaced. Another short-path-length cell [4] designed for combined i.r. and catalytic studies of carbon monoxide oxidation on supported palladium utilizes radiant heating to control the sample temperature. Avery [5] has described a cell which allows rapid cooling of samples housed in a nickel-plated copper block attached with a kovar seal to a glass re-entrant. The nickel plating eliminated vapour transport of copper to the pellet. In another design [6] the sample is held in a metal carriage attached to a tubular silica frame which contains a heating element. *In situ* temperatures of 723 K are possible with this arrangement, and low temperatures can be obtained by passing a refrigerant through the silica tubes. Two more cells have been described [7] for use with powdered adsorbents.

Short path-lengths reduce the intensity of undesirable gaseous contributions to the measured spectra. Alternatively, a compensation cell may be placed in the reference beam of a double-beam spectrometer.[8] This gives the added advantage of largely cancelling the background absorption of the sample. A serious problem arises when very weak absorption bands are to be detected in the frequency ranges where water vapour absorbs, as it is difficult to completely

---

[1] J. Pritchard, *Ann. Reports* (A), 1969, **66**, 65.
[2] G. D. Chukin and L. A. Ignat'eva, *Zhur. priklad. Spektroskopii*, 1970, **13**, 89.
[3] H. Knözinger, H. Stolz, H. Bühl, G. Clement, and W. Meye, *Chem.-Ing.-Tech.*, 1970, **42**, 548.
[4] R. F. Baddour, M. Modell, and R. L. Goldsmith, *J. Phys. Chem.*, 1970, **74**, 1787.
[5] N. R. Avery, *J. Catalysis*, 1970, **19**, 15.
[6] P. Schürer, L. Kubelkova and M. Smrkovska, *Czech. J. Phys.* (*B*)., 1969, **19**, 1421.
[7] A. Buckland, J. Ramsbotham, C. H. Rochester, and M. S. Scurrell, *J. Phys.* (*E*)., 1971, **4**, 146.
[8] R. Lambert and N. Singer, *J. Phys.* (*E*)., 1970, **3**, 655.

annul water vapour effects even in a double-beam spectrometer. Hockey [9] has shown how water vapour peaks at 3751, 3747, and 3743 cm^{-1} can appear under high resolution conditions leading to an apparent but spurious resolution of the 3750 cm^{-1} band of the surface hydroxy-groups on silica. Accidental contamination of samples by volatile matter from waxes or resins used for attaching i.r. windows can also lead to spurious spectral features according to Bozon-Verduraz,[10] who suggests the use of Viton gaskets.

1 Carbon Monoxide

The spectra of carbon monoxide were the first to be obtained from metal surfaces and still receive more detailed study than those of any other adsorbate. A comprehensive review of carbon monoxide adsorption on the transition (and Group 1B) metals has appeared.[11] Although the broad features of the spectra on any one metal are often reproduced in different investigations, there may be considerable variation in the frequencies of band maxima and in relative band intensities, reflecting different distributions of adsorption sites. Site distributions may be affected by the particle size of the metal in supported samples or evaporated films, the nature of the support material, and the methods of preparation and pretreatment. Kavtaradze and Sokolova [12] compared the spectra of silica- and alumina-supported samples of copper and rhodium. The single band with copper differed in intensity but not in frequency (2110 cm^{-1}) on the two supports. With rhodium, which gives two bands, there was a slight frequency shift from 2060 and 1870 cm^{-1} on alumina to 2050 and 1860 cm^{-1} on silica, but a greater change in the relative intensities of the two bands. In reviewing earlier spectra with various support materials, the conclusion is reached [12] that variations in frequency are small, and often largely within experimental error. Large variations in site distributions, and hence in relative band intensities, may result from particle size and shape effects and from the chemical pretreatment of the metal. In a study of the 'break-in' of a silica-supported palladium catalyst for carbon monoxide oxidation [4] it was observed that the spectra of adsorbed carbon monoxide changed as the catalytic activity changed and that these changes could account for the apparent discrepancies between values of i.r. band maxima reported in the literature. The spectrum of carbon monoxide on palladium consists of a high-frequency band (above 2000 cm^{-1}) and a much stronger low-frequency band which is usually attributed to bridge CO groups. A freshly prepared and reduced catalyst exposed only to carbon monoxide gave bands at 1920, 1885, and 1865 cm^{-1} in the low-frequency region and a high-frequency band at 2035 cm^{-1}. With increasing coverage, bands appeared at 1942 and 1967 cm^{-1} and at high coverage the main bands were at 1965 and 1985 cm^{-1} in the low-frequency range and at 2078 cm^{-1} in the high-frequency range.

[9] J. A. Hockey, *J. Phys. Chem.*, 1970, **74**, 2570.
[10] F. Bozon-Verduraz, *J. Catalysis*, 1970, **18**, 12.
[11] R. R. Ford, *Adv. Catalysis*, 1970, **21**, 51.
[12] N. N. Kavtaradze and N. P. Sokolova, *Russ. J. Phys. Chem.*, 1970, **44**, 603.

After several alternating exposures to oxygen and carbon monoxide at room temperature followed by reduction, the spectra obtained with carbon monoxide had changed such that the 1965 cm^{-1} band was now rather less intense than the 1985 cm^{-1} band and the high-frequency band maximum was at 2087 cm^{-1}. After stabilization as a catalyst at 473 K for carbon monoxide oxidation, followed by reduction *etc.*, this change had become more emphasised with the 1965 cm^{-1} band reduced to a shoulder on a much stronger 1985 cm^{-1} band. The pre-'break-in' and post-'break-in' spectra correlate well with earlier results of different groups of workers. It is believed that structural rearrangement during catalysis resulting in differing distributions of exposed crystal planes is responsible for these effects. It is obviously desirable in any account of the spectra of adsorbed species on supported metals that the methods of preparation and pretreatment be clearly stated.

Ferreira and Leisegang [13] have reinvestigated the adsorption of carbon monoxide on silica-supported nickel and cobalt. Six bands (*A* to *F*) were observed on each metal. The low-frequency *A*-bands (1700 to 2000 cm^{-1} on nickel, 1800 to 2040 cm^{-1} on cobalt) are considered to be due to bridged carbonyl species, but there is no new supporting evidence, such as the observation of a band for the asymmetric stretching vibration, for convincingly rejecting the alternative linear possibility.[14] LEED studies [15] support the model of bridged carbonyl groups on palladium, however, and the observation of Ferreira and Leisegang that, at high coverage, the 'linear' bands grow at the expense of 'bridged' bands is similar to that of Baddour *et al.*[4] for supported palladium after break-in. Higher frequency bands on nickel at 2035 to 2050 cm^{-1} (*B*) and 2060 to 2090 cm^{-1} (*D*) are assigned to linear carbonyl groups, band *B* possibly being associated with sites of low co-ordination number. They were observed to shift to higher frequency with increasing coverage in accord with the molecular orbital model for carbon monoxide chemisorption proposed by Blyholder.[14] At high coverage, an additional sharp band (*C*) at 2058 cm^{-1} appeared. Although this coincides with the frequency of gaseous nickel tetracarbonyl, experimental evidence suggested that it was due mainly to a surface species. An assignment to structures with two or more ligands chemisorbed at a single site is suggested. Such structures are likely to be intermediates in the formation of nickel tetracarbonyl and it is interesting to note that the tricarbonyl, formed in rare-gas matrices at 15 K by photolysis of the tetracarbonyl,[16] absorbs at very nearly the same frequency. The additional bands *E* and *F* at 2100—2130 cm^{-1} and 2165—2195 cm^{-1} were observed at high coverages on incompletely reduced samples and, in agreement with earlier results, are assigned to molecules attached to nickel and oxygen ions of nickel oxide. The general features and interpretation of the spectra on cobalt are similar, with bands *B* at

---

[13] L. C. Ferreira and E. C. Leisegang, *J. S. African Chem. Inst.*, 1970, **23**, 136.
[14] G. Blyholder, *J. Phys. Chem.*, 1964, **68**, 2772.
[15] G. Ertl and J. Koch, *Z. Naturforsch.*, 1970, **25a**, 1906.
[16] A. J. Rest and J. J. Turner, *Chem. Comm.*, 1969, 1026.

2060—2073 cm^{-1}, C at 2070 cm^{-1}, D at 2090—2100 cm^{-1}, E at 2130 cm^{-1}, and F at 2180 cm^{-1}. The addition of potassium as a catalyst promoter lowered the frequencies of bands A, B, and D and reduced the intensities of B and D.

Hobert and Thieme [17] have combined i.r. spectroscopy with thermal desorption analysis to investigate the influence of mercury on the adsorption of carbon monoxide on silica-supported nickel. In agreement with earlier work, it is found that the low-frequency band at 1950 cm^{-1} in the spectrum of the mercury-free system is greatly reduced in intensity leaving an intense high-frequency band at 2060 cm^{-1}. An unexplained sharp band also appeared at 1600 cm^{-1}. In the thermodesorption spectra, peaks at 423 and 653 K were removed leaving the low-temperature peak at 383 K which, on continued treatment with mercury, was finally lowered to about 353 K. The weakening of the chemisorption indicated by the thermodesorption analysis is contrary to the interpretation [14] of the earlier i.r. data.

Co-adsorption of hydrogen and carbon monoxide is of considerable interest in relation to Fischer–Tropsch catalysis and has been investigated on an iron–magnesium oxide mixed catalyst.[18] The resulting i.r. spectrum is complex, with evidence for carbonate and carboxylate species as well as chemisorbed carbon monoxide. After exposure to the mixed gases at 453 K, bands appeared at 2840 and 2720 cm^{-1} which are attributed to aldehyde groups. Much of the complexity of the spectrum arises from interactions with the support. Thus carbon monoxide on pure magnesium oxide gave bands in the 1700—800 cm^{-1} region due to a variety of carbonate groups [19] and the carbonyl stretching vibration of the aldehyde groups was obscured. Moreover, the interaction of carbon monoxide alone with the mixed catalyst gave a complex spectrum.[20] Bands at 2010, 1925, and 1885 cm^{-1} are assigned to carbonyl groups on the iron. They were not influenced by recrystallization or by hydrogen pretreatment of the iron, whereas a band at 1825 cm^{-1} was shifted to 1855 cm^{-1} by hydrogen. It is suggested that this band may result from the dissociative adsorption onto the iron of carbon dioxide produced by the Boudouard reaction on the oxide support, the carbonyl frequency being lowered to 1825 cm^{-1} through the influence of an adjacent oxygen anion. Treatment with hydrogen could reduce the effect of the oxygen and thereby cause the frequency shift to 1855 cm^{-1}. The boundary between the iron particles and the magnesium oxide support may well be the active region. Another carbonyl band at 1950 cm^{-1} appeared after adsorption onto the freshly prepared catalyst. It could not be observed after recrystallization and is believed to be connected with sites of low co-ordination.

Magnesium oxide is an unusually transparent support material, usable [21] over the range 4000—300 cm^{-1}. Thus a band at 475 cm^{-1} was observed [18]

---

[17] H. Hobert and J. Thieme, *Z. Chem.*, 1970, **10**, 3467.
[18] H. Kölbel, M. Ralek, and P. Jiru, *Erdöl u. Kohle, Erdgas, Petrochem.*, 1970, **23**, 580.
[19] H. Kölbel, M. Ralek, and P. Jiru, *Z. Naturforsch.*, 1970, **25a**, 670.
[20] H. Kölbel, M. Ralek, and P. Jiru, *Coll. Czech. Chem. Comm.*, 1971, **36**, 512.
[21] P. Jiru, *Chimica e Industria*, 1970, **52**, 128.

with carbon monoxide on the iron catalyst and assigned to an Fe—C vibration. Silica absorbs much more strongly at low frequencies. Cant and Hall [22] have shown that the apparent increase in transmission through a pressed disc of silica-supported platinum at elevated temperatures [23] was spurious, and was a result of radiation from the heated sample. At even lower frequencies pressed silica discs recover their transparency, and spectra between 200 and 25 cm^{-1} before and after removal of hydrogen-bonded hydroxy-groups at 1073 K have been reported.[24] The difference spectrum appears to indicate an absorption band due to hydrogen-bonding species with a maximum at 150 cm^{-1}. However, the thickness of the discs must be comparable to the wavelength of the radiation and the apparent spectrum may have been strongly influenced by shifting interference fringes.

In spite of the opacity of silica, the band reported [23] at 476 cm^{-1} for adsorbed carbon monoxide on silica-supported platinum is confirmed in work by Blyholder and Sheets.[25] They employed the technique originated by Blyholder of depositing evaporated metal in a hydrocarbon oil film. After exposure to carbon monoxide, platinum deposited in this way gave bands at 2045, 1815, and 480 cm^{-1}. The interpretation of the last band as the Pt—C stretching vibration is supported by the frequency shifts of both the 2045 and 480 cm^{-1} bands when $C^{18}O$ was used. These shifts were consistent with the linear Pt—C—O structure for the adsorption complex. The band at 1815 cm^{-1} on a freshly formed platinum surface was unusually intense compared with bands reported on silica-supported platinum, but this is convincingly explained as due to a larger proportion of highly dispersed particles. Addition of oxygen led to an additional band at 560 cm^{-1} assigned to a surface oxide.

The same technique has also been used to investigate the interaction of carbon monoxide and ethylene on iron.[26] After the adsorption of hydrogen and ethylene, large amounts of carbon monoxide were adsorbed at room temperature without displacing ethylene. The carbonyl band appeared at 1900 cm^{-1} compared to 1950 cm^{-1} with carbon monoxide alone. Broad bands appeared at 1100 and 950 cm^{-1} after heating at 373 K. Deuteriated ethylene caused the bands to appear at 1035 and 885 cm^{-1}. These results are believed to indicate the formation of propoxide species.

A major advantage of the oil film technique is the wide spectral range available for transmission spectroscopy. The possible influence of the oil on the behaviour of the metal surface should not be ignored, but good agreement has generally been obtained with results from oxide-supported metals. As indicated above, it is also possible to work over a wide spectral range with magnesium oxide as a support, allowing a much wider range of

[22] N. W. Cant and W. K. Hall, *J. Phys. Chem.*, 1970, **74**, 1403.
[23] J. K. A. Clarke, G. M. Farren, and H. E. Rubalcava, *J. Phys. Chem.*, 1968, **72**, 327.
[24] I. A. Brodskii, A. E. Stanevich, and N. G. Yaroslavskii, *Russ. J. Phys. Chem.*, 1970, **44**, 998.
[25] G. Blyholder and R. Sheets, *J. Phys. Chem.*, 1970, **74**, 4335.
[26] A. J. Goodsel and G. Blyholder, *Chem. Comm.*, 1970, 1122.

temperatures to be used but admitting the possibility of concurrent chemical processes on the support. A genuinely inert matrix for metal particles has recently been described, namely condensed argon layers.[27] By first evaporating nickel into an argon matrix at 20 K and then condensing argon with 5% carbon monoxide on top of this, a spectrum of the condensed carbon monoxide was obtained with a band at 2140 cm^{-1}. Interdiffusion occurred after warming to 44 K, and strong bands appeared at 2070 and 1890 cm^{-1} due to chemisorbed species. This method gives relatively intense spectra and can be developed for operation under very clean ultra-high-vacuum conditions, but is inevitably restricted to use at very low temperatures.

An alternative way of reducing the support problem in transmission spectroscopy is to use very thin evaporated metal films deposited directly on to a transparent substrate. Ultra-high-vacuum conditions are essential and there are difficulties with bakable window seals. An ingenious method of avoiding a sealing medium has been devised[28] in which the ultra-high-vacuum cell is contained within a high-vacuum jacket. Adequate sealing of the ultra-high-vacuum windows is provided by pressing the windows onto polished steel rims on the cell. With co-evaporated sodium chloride and platinum films, a carbonyl band was obtained at 2053 cm^{-1}, in good agreement with previous work on platinum films.

The ultra-high-vacuum evaporated film technique previously used to study carbon monoxide adsorption on nickel[29] has been applied to a comparative study of adsorption at low temperatures on iron, cobalt, nickel, copper, silver, and gold films.[30] The best spectra were obtained from very thin films transmitting about 70% of the incident radiation, and the intensity of the resulting bands was necessarily low. The spectra on the transition metals resembled those given by silica-supported metals, but seemed to appear only after much of the carbon monoxide had been adsorbed. This result suggested that the strongly bound molecules may be in states which are i.r.-inactive, but Bradshaw and Vierle[31] subsequently reported that the electrical resistance and i.r. spectra of thin nickel films changed together as carbon monoxide was adsorbed. As it is well established that the resistance begins to change as soon as chemisorption occurs, the delay in the appearance of the spectra in the first experiments may have been an experimental artefact such as preferential adsorption by the film deposited on the cell walls. There was no delay with copper or gold films where the room-temperature film on the walls would not have adsorbed significantly. A single band was observed with films of each of these metals and with silver. Anomalous transmission effects[29] were very marked, causing the carbonyl absorption peak to resemble a derivative spectrum, presumably because of the sharpness and intensity of the bands. Copper at 113 K gave a band at 2105 cm^{-1} which increased

[27] G. Blyholder, M. Tanaka, and J. D. Richardson, *Chem. Comm.*, 1971, 499.
[28] A. Zecchina, C. Versino, S. Coluccia, and E. Borello, *J. Chim. phys.*, 1970, **67**, 1237.
[29] A. M. Bradshaw and J. Pritchard, *Surface Sci.*, 1969, **17**, 372.
[30] A. M. Bradshaw and J. Pritchard, *Proc. Roy. Soc.*, 1970, **A316**, 169.
[31] A. M. Bradshaw and O. Vierle, *Ber. Bunsengesellschaft Phys. Chem.*, 1970, **74**, 630.

linearly in intensity with coverage up to about half the maximum. At higher coverages the spectrum appeared unchanged, suggesting that the subsequent adsorption was i.r.-inactive. These two stages of adsorption correlated well with earlier surface potential measurements on polycrystalline copper films. On gold, which gave a band at 2120 cm^{-1}, there was no evidence for a second stage of adsorption. Very high peak absorption coefficients of 1.3×10^{-17} and 2.2×10^{-17} molecule^{-1} cm^2 were deduced for carbon monoxide on copper and gold films respectively. These apparent values were probably exaggerated by enhanced reflection losses. The band on silver at 2160 cm^{-1} was much less intense but because of the weakness of the adsorption the coverage was probably low even at 113 K. Thus Keulks and Ravi,[32] working with silica and alumina-supported silver at room temperature, could not detect a band due to carbon monoxide on fully reduced surfaces. After pretreatment with oxygen, a weak reversible adsorption was found which was attributed to chemisorption on surface oxygen ions. It gave a band at 2180 cm^{-1} in agreement with earlier work. Carbon monoxide adsorption on clean silver is extremely weak. Kavtaradze and Sokolova[33] have discussed the stability of surface carbonyls in relation to the catalytic activity of platinum metals. They report that the thermal stability increases in the sequence Ag, Au, Cu, Co, Pd, Rh, Pt, Ir. Bradshaw and Pritchard[30] have related the stretching frequencies, heats of adsorption, and surface potentials on copper, silver, and gold through a model of linearly adsorbed carbon monoxide in which the varying contributions of π- and σ-bonding are connected with the energy of the metal d-bands. Shopov, Andreev, and Palazov[34] have discussed the orbital energy requirements for effective π-bonding in molecular carbonyls, and, carrying over these ideas to chemisorption onto metal surfaces, have successfully correlated the force constants of the C—O stretching vibration with the third ionization potentials of the metal atoms.

2 Reflection Spectroscopy of Chemisorbed Carbon Monoxide

In 1964, Blyholder[14] rationalized the diverse results then available for the spectra of carbon monoxide on metals by using a simple model of linear carbonyl groups with degrees of π-bonding which depend on the co-ordination of surface metal atoms to other metal atoms and on the influence of other adsorbates on the availability of d-electrons for π-bonding. These ideas have provided a framework for rationalizing many of the spectra reported since. Transmission spectroscopy of metal samples requires that the metal be highly dispersed, as in supported catalyst or in very thin evaporated films, and therefore to possess a heterogeneous and usually ill-defined surface structure. Consequently, it has not yet been possible to directly relate

[32] G. W. Keulks and A. Ravi, *J. Phys. Chem.*, 1970, **74**, 783.
[33] N. N. Kavtaradze and N. P. Sokolova, *Russ. J. Phys. Chem.*, 1970, **44**, 93.
[34] D. Shopov, A. Andreev, and A. Palazov, *Izvest. Otdel. Khim. Nauk, Bulg. Akad. Nauk*, 1970, **3**, 335.

Infrared Spectra of Adsorbed Species on Metals

these bonding hypotheses to the structural information gained with techniques such as LEED from studies on single crystal surfaces. Reflection methods can, in principle, provide the necessary spectroscopic information and a number of exploratory studies of carbon monoxide adsorption have been reported recently.[35-38] Following earlier work,[39] an ultra-high-vacuum light pipe system has been used, giving 16 successive reflections from evaporated films of nickel and rhodium at 70° angle of incidence.[35] It is claimed that (110) oriented surfaces were produced by depositing the films in argon, but this was not shown experimentally. The nickel films slowly gave rise to a strong narrow band at 2056 cm^{-1} at 10 Torr pressure of carbon monoxide, while rhodium quickly gave two weaker bands at 2078 and 2060 cm^{-1}. The behaviour with nickel suggests that the spectrum may well be due to the slow formation of nickel tetracarbonyl but, as mentioned previously, there is evidence for a weakly adsorbed surface species on nickel absorbing at the same frequency. An additional band was observed at 191 cm^{-1} which could be the metal–carbon frequency for a weakly adsorbed state. The bands on rhodium remained after evacuation. Rhodium also gave bands at 278 and 253 cm^{-1} after exposure to oxygen which are ascribed to oxygen–rhodium stretching vibrations.

Pritchard and Sims[36] also used ultra-high-vacuum conditions to study carbon monoxide on evaporated copper films at room temperature by multiple reflections between parallel mirrors at angles of incidence greater than 80°. According to theoretical predictions[40] the optimum absorption per reflection should occur at 88° on a metal with the optical constants of copper. This prediction was verified experimentally, by measuring the absorption per reflection at ten angles between 82° and 90°, the peak absorption being 3.6% at 88°. Only the radiation polarized parallel to the plane of incidence was absorbed. The spectrum consisted of a sharp band at 2105 cm^{-1}, in agreement with transmission spectra of copper films and supported copper, together with a shoulder at about 2090 cm^{-1}. The shoulder band was not noticed in a previous measurement by the same technique. Drmaj and Hayes[37] used copper surfaces for reflection studies involving 37 reflections at 70°, but instead of copper films deposited under ultra-high-vacuum they employed electropolished copper plates which were reduced in flowing hydrogen at 473 K. These surfaces are unlikely to have been clean, but in one series of experiments a weak band at 2110 cm^{-1} was observed. Another band at 2180 cm^{-1} was thought to be due to adsorption or reaction on an oxide film.

In view of the relatively strong absorption by carbon monoxide in a single

---

[35] H. C. Eckstrom, G. G. Possley, S. E. Hannum, and W. H. Smith, *J. Chem. Phys.*, 1970, **52**, 5435.
[36] J. Pritchard and M. L. Sims, *Trans. Faraday Soc.*, 1970, **66**, 427.
[37] D. T. Drmaj and K. E. Hayes, *J. Catalysis*, 1970, **19**, 154.
[38] M. A. Chesters, J. Pritchard, and M. L. Sims, *Chem. Comm.*, 1970, 1454.
[39] H. C. Eckstrom and W. H. Smith, *J. Opt. Soc. Amer.*, 1967, **57**, 1132.
[40] R. G. Greenler, *J. Chem. Phys.*, 1966, **44**, 310.

reflection at 88° on copper films, a single reflection approach has been explored with a copper (100) single crystal surface,[38] on which simultaneous surface potential measurements could be made. After cleaning by ion-bombardment the crystal adsorbed carbon monoxide at 77 K. With increasing exposure the surface potential increased to a maximum positive value of 255 mV and then decreased to about 120 mV at saturation. The two stages of adsorption were reflected in the i.r. spectra; during the first stage the single sharp band at 2085 cm^{-1} grew in intensity, reaching its maximum intensity at the maximum surface potential; in the second stage the band remained at the same intensity but the peak moved to 2994 cm^{-1} as the surface potential fell. It is suggested that the second stage may involve strongly physically adsorbed molecules which do not absorb sufficiently strongly to be detected optically but which interact with the primary chemisorbed molecules. It is surprising that the band differs significantly in frequency from the sharp band observed on polycrystalline copper films, in which (100) surfaces may be expected to be present. The frequency is close to that in carbonyl complexes of Cu^I.[41]

3 Other Reflection Spectra

Poling[42] has reviewed applications of i.r. reflection spectroscopy to metal surfaces up to 1969. In most work the angle of incidence has been near 70°, for which the number of reflections for optimum sensitivity has been calculated[43] to be about 25, although for single reflections higher angles are more favourable.[40, 44, 45] According to Suëtaka[45] the optimum angle of incidence varies considerably, with high values preferred for strongly absorbing adsorbates. As already mentioned, experiment shows that 88° is the best angle for a single reflection with carbon monoxide on copper.

Applications of reflection spectroscopy to the adsorption of corrosion inhibitors on metals have been surveyed by Poling[42] and by Little.[46] Recently, the iron corrosion inhibitors diamyl sulphoxide and dihexyl sulphoxide have been studied.[47] After exposure to low concentrations of the sulphoxides the reflection spectra were identical with transmission spectra of the corresponding sulphides. At higher concentrations the band at 1020 cm^{-1} due to the sulphoxide group also appeared. It is concluded that the adsorbed sulphide confers the resistance to corrosion.

Poling[42, 48] has shown that copper oxalate films are produced on copper surfaces as an oxidation product of hydrocarbon oil. The spectrum of the

[41] G. Rucci, C. Zanzoterra, M. P. Lachi, and M. Camia, *Chem. Comm.*, 1971, 652.
[42] G. W. Poling, *J. Colloid Interface Sci.*, 1970, **34**, 365.
[43] R. G. Greenler, *J. Chem. Phys.*, 1969, **50**, 1963.
[44] A. A. Babushkin and L. A. Mukhitdinov, *Sbornik Trudov, Mosk. Inzh-Stroit. Inst.*, 1968, No. 58, 12.
[45] W. Suëtaka, *Sci. Reports, Tohoku Univ.*, Ser. A, 1966, **18** Suppl., 129.
[46] L. H. Little, *Australasian Corrosion Eng.*, 1970, **14**, 17.
[47] S. Thibault and J. Talbot, *Compt. rend.*, 1971, **272**, C, 805.
[48] G. W. Poling, *J. Electrochem. Soc.*, 1970, **117**, 521.

copper oxalate film, with seven reflections at 73°, disappeared when perpendicularly polarized radiation was used, confirming the general view that the intensity of the electric field of the radiation is nil at the metal surface and contrary to the theoretical conclusion of Eckstrom and Smith.[39]

The relationship between film thickness and absorption intensity is important if the spectrum is to be used for monitoring film growth. Poling [42, 49] has found that the oxide film growing on copper exhibits a band at 640 cm^{-1}, close to that of bulk cuprous oxide, the intensity of which is a linear function of film thickness. Rahn,[50] however, found a similar band which passed through a maximum intensity when the film thickness was about 200 nm. This result appears to be a consequence of the much higher angle of incidence, 85—90°, used in Rahn's work compared with the 73° used by Poling.

An important recent development which promises to solve many of the problems associated with reflection spectroscopy is the use of ellipsometric measurements in the i.r.,[51] which enable both the refractive index and absorption index of the surface film to be determined. Furthermore, the effects of gas-phase absorption are much less important. Preliminary results for the adsorption of methanol onto evaporated silver mirrors show well-resolved features in the C—H stretching region, and clearly demonstrate the potential power of this approach.

4 Hydrogen, Nitrogen, and Nitric Oxide

These diatomic adsorbates have received relatively little attention recently. Kavtaradze and Sokolova [52] investigated the spectra of hydrogen adsorbed onto alumina-supported iron, cobalt, nickel, rhodium, palladium, and iridium at 173, 213, and 303 K. Weak absorption bands in the region 1850—1940 cm^{-1} were found on all the metals, and it is concluded that M—H bonds are formed on all the metals. However, the band at 1940 cm^{-1} on iridium is much lower in frequency than those reported before [53] at 2050 and 2120 cm^{-1} and the assignments were not confirmed by isotopic shift measurements. In view of the further doubts as to whether the Pt—H band is observable on completely reduced platinum the interpretation of these bands requires caution.

Nitrogen adsorption onto thin nickel films deposited in the gas on cooled substrate at 115 K has been reported [54] to give a band at 2200 cm^{-1} similar to that on silica-supported nickel containing particles less than 7 nm in size. Vacuum-deposited films gave less intense and more complex spectra with additional bands at about 2255 and 2230 cm^{-1}. Evacuation removed the

---

[49] G. W. Poling, *J. Electrochem. Soc.*, 1969, **116**, 958.
[50] R. W. Rahn, *J. Electrochem. Soc.*, 1970, **117**, 1586.
[51] M. J. Dignam, B. Rao, M. Moskovits, and R. W. Stobie, *Canad. J. Chem.*, 1971, **49**, 1115.
[52] N. N. Kavtaradze and N. P. Sokolova, *Russ. J. Phys. Chem.*, 1970, **44**, 1485.
[53] F. Bozon-Verduraz, J. P. Contour, and G. Pannetier, *Compt. rend.*, 1969, **269**, C, 1436.
[54] A. M. Bradshaw and J. Pritchard, *Surface Sci.*, 1970, **19**, 198.

2200 cm^{-1} band. Comparison with the spectra of carbon monoxide suggested that the i.r.-active nitrogen is chemisorbed onto those sites which provide the best π-bonding for carbon monoxide and a similar chemisorption bond is proposed. The higher frequency bands were also obtained in later work [31] together with a band at 2180 cm^{-1}. Simultaneous electrical measurements showed a resistance increase smaller by an order of magnitude than that accompanying adsorption of carbon monoxide. A discussion of the relative roles of σ- and π-bonding in carbonyls and molecular nitrogen complexes has been presented.[55] Molecular orbital calculations [56] of the relative binding energy of molecular nitrogen in several alternative configurations on an iron surface lead to the conclusion that the strongest interaction will be with sites providing four adjacent iron atoms.

Batychko, Rusov, and Roev [57] have investigated the adsorption of nitric oxide on silica-supported nickel. They found bands at 1860 and 2190 cm^{-1} in approximate agreement with earlier work [58] but certain features have led them to reverse the association of these bands with specific kinds of site. Thus the 1860 cm^{-1} band was relatively insensitive to sintering compared with the removal of the 2190 cm^{-1} band, and it is attributed to adsorption onto close-packed planes. Because of the influence of sintering the 2190 cm^{-1} band is assigned to molecules adsorbed onto edge sites, *etc.*

The reaction between nitrogen and hydrogen on iron catalysts is the basis of ammonia synthesis and it has often been assumed that dissociative adsorption of nitrogen must be a first elementary reaction step. Evidence from several experimental approaches for non-dissociative adsorption leading to N_2H_x species has been discussed recently.[21, 59] Using magnesium oxide as a support, i.r. spectra could be recorded [60] over a wide range extending down to 600 cm^{-1}. The spectrum was much more complex than that of ammonia, but correlated well with that of hydrazine and its adsorption products. It is suggested that adsorbed species such as Fe...$\overset{+}{N}H_2$—NH_2 may be important precursors to ammonia formation. In previous work with silica-supported iron, only bands in the NH_2 stretching region (3200—3380 cm^{-1}) and bending region (1610 cm^{-1}) had been observed. With magnesium oxide, bands at 1525, 1225, 1140, 975, 815, 780, and 750 cm^{-1} were found as well. The inference that N_2H_x species are formed is well supported by field ionization mass spectrometry of ammonia on an iron tip.[21]

When hydrazine is adsorbed on to alumina-supported iridium, however, it has been concluded [61] that the N—N bond is readily broken, since the

---

[55] K. G. Caulton, R. L. DeKock, and R. F. Fenske, *J. Amer. Chem. Soc.*, 1970, **92**, 515.
[56] L. M. Roev, L. M. Dudkina, and M. T. Rusov, *Russ. J. Phys. Chem.*, 1971, **45**, 235.
[57] S. V. Batychko, M. T. Rusov and L. M. Roev, *Doklady Phys. Chem.*, 1970, **191**, 328.
[58] G. Blyholder and M. C. Allen, *J. Phys. Chem.*, 1965, **69**, 3998.
[59] R. Brill, *Ber. Bunsengesellschaft Phys. Chem.*, 1971, **75**, 455.
[60] R. Brill, P. Jiru, and G. Schulz, *Z. phys. Chem. (Frankfurt)*, 1969, **64**, 215.
[61] J. P. Contour and G. Pannetier, *Bull. Soc. chim. France*, 1970, **12**, 4260.

characteristic bands of ammonia adsorbed onto alumina appeared together with a band at 1220 cm^{-1} assigned to ammonia adsorbed on iridium.

5 Hydrocarbons and other Compounds

Acetylene adsorption on alumina-supported palladium led [62] to spectra very similar to earlier results with the silica-supported metal, with bands at 2960 and 3050 cm^{-1} assigned to C—H stretching vibrations in CH_3 and =CH groups. Similar evidence for self-hydrogenation occurs with γ-alumina-supported platinum [63] where an additional sharp band is reported at 1690 cm^{-1} corresponding to a C=C vibration. Hydrogen sulphide and carbonyl sulphide greatly reduced acetylene adsorption onto both the platinum and the support.

Much of the interpretation of the spectra of gases adsorbed onto metals has relied on comparison with the spectra of model compounds. With hydrocarbons a problem has existed because of the lack of model alkyl complexes. Morrow [64] has synthesised a series of triphenylphosphine platinum dialkyl compounds containing methyl, ethyl, [$^2H_3$]ethyl, n-propyl, n-butyl, and n-hexyl. The spectra of these compounds in the C—H stretching region generally confirms the assignments for chemisorbed species in previous work.[65]

Avery [5] has obtained spectra of several olefins adsorbed on silica-supported palladium. The branched chain mono-olefins, isobutene and 4,4-dimethylpent-l-ene, gave intense bands in the ν(C—H) region of saturated hydrocarbons. Hydrogenation caused some intensification and a slight shift to lower frequencies, and the hydrogenated species could be removed by evacuation. With linear mono-olefins, e.g. pent-l-ene, initial adsorption gave rather weak bands in the same region, but hydrogenation caused a major intensification and a considerable frequency shift. Following hydrogenation, the species responsible for the observed bands could be removed by pumping, but much of the original carbon skeleton remained, as shown by the re-appearance of a strong spectrum on rehydrogenation. This difference was emphasised with the di-olefin, buta-1,3-diene, which gave no spectrum at all when first adsorbed. Hydrogenation produced an intense spectrum which slowly diminished during evacuation but was completely restored by re-hydrogenation. In interpreting the spectra, it is postulated that C—H bands associated with carbon atoms directly attached to the metal (adsorbed carbon atoms) are too weak to be observed, and that all the recorded spectra arise from other C—H groups which may exist partly as terminal or branched alkyl groups and partly as sections of the carbon skeleton flanked by adsorbed

[62] N. P. Sokolova, L. A. Kazakova, and L. A. Borisenko, Russ. J. Phys. Chem., 1970, **44**, 1515.
[63] S. S. Randhava and A. Rehmat, Trans. Faraday Soc., 1970, **66**, 235.
[64] B. A. Morrow, Canad. J. Chem., 1970, **48**, 2192.
[65] N. Sheppard and J. W. Ward, J. Catalysis, 1969, **15**, 50; B. A. Morrow and N. Sheppard, Proc. Roy. Soc., 1969, **A311**, 391, 415.

carbon atoms. It is concluded that whereas the species produced by hydrogenation of branched chain mono-olefins are probably alkyl groups, the same interpretation is unsatisfactory for the much more strongly adsorbed linear olefins.

Alumina-supported platinum is used as a catalyst for the conversion of cyclohexane into benzene, but no i.r. bands characteristic of benzene were observed when cyclohexane was adsorbed [66] at 298 K. At higher temperatures bands due to adsorbed cyclohexene, methyl cyclopentane, methyl cyclopentene, and methylidene cyclopentene appeared. No band due to the aromatic C—H vibrations was observed even when benzene was adsorbed, but the same cyclopentane and cyclopentene derivatives appeared as in the case of cyclohexane adsorption. It is concluded that the aromatic character of the benzene ring cannot be maintained in the chemisorbed state. However, other workers [67] have deduced that benzene is adsorbed as a π-complex on nickel and platinum. The same authors [67] have also discussed the adsorption of olefins in terms of σ- and π-bonding with emergent d-orbitals of surface metal atoms.

Analogies with carbon monoxide appeared when butyro- and caprylonitrile were adsorbed on to silica-supported platinum and palladium.[68] Bands near 2200 cm^{-1} are assigned to nitriles in which the nitrogen atom is bound to the metal atom by a combination of σ- and π-bonding, the π-bonding component causing the C—N stretching frequency to fall below that in the free nitrile. On metal oxides the bonding is essentially σ in character with a frequency increase. In the presence of hydrogen, caprylonitrile on platinum is reduced to octylamine; i.r. spectra showed the same sequence of intermediate compounds, with —NH, =NH, and C=N groups, when octylamine was adsorbed on platinum [69] as when caprylonitrile was hydrogenated. A reversible reaction mechanism is proposed which involves species such as R—$\overset{*}{\text{C}}$H—$\overset{*}{\text{N}}$H where * denotes a bond to a platinum atom.

Several authors have investigated the interaction of organic oxygen compounds with metals. Silver did not adsorb dimethyl ether but it reacted with formaldehyde to give polyoxymethylene chains.[70] The strong interaction of formaldehyde with the silica-gel support complicated the spectra. However, bands at 2970 and 2900 cm^{-1} were taken as evidence for interaction between the silver and the oxygen atoms of polyoxymethylene.

Blyholder and Allen [71] used the oil film support method to study the interaction of evaporated manganese with a wide range of oxygen compounds. Again, ethers were not adsorbed. Alcohols were strongly adsorbed, the ex-

---

[66] P. Ratnasamy, *Chem. Age India*, 1970, **21**, 889.
[67] D. Shopov, A. Andreev, and A. Palazov, *Izvest. Otdel. Khim, Nauk. Bulg. Akad. Nauk*, 1969, **2**, 321.
[68] O. M. Oranskaya and V. N. Filimonov, *Doklady Phys. Chem.*, 1970, **194**, 675.
[69] O. M. Oranskaya, V. N. Filimonov, and Ya. E. Shmulyakovskii, *Kinetics and Catalysis*, 1970, **11**, 593.
[70] L. Kubelkova, P. Jiru, and P. Schürer, *Coll. Czech. Chem. Comm.*, 1969, **34**, 3842.
[71] G. Blyholder and M. C. Allen, *J. Catalysis*, 1970, **16**, 189.

tent of adsorption being greater for the primary alcohols methanol to butanol than the secondary and tertiary propanol and butanols. In all cases, alkoxides were formed with spectra very similar to those of the parent alcohols. Substitution of the hydroxy hydrogen by metal has little effect on the vibrational frequencies of the rest of the molecule. With methanol a band at 2810 cm^{-1} supported the alkoxide structure, and in the other cases a band at 540—550 cm^{-1} always appeared and was tentatively assigned to the oxygen –metal bond. The alkoxide structure also resulted from acetaldehyde and acetone adsorption, and even allyl alcohol gave an alkoxide rather than reacting at the unsaturated carbon centres. The same pattern of results was also found with evaporated chromium.[72] Evidently the surface alkoxides are very stable. Other potential modes of adsorption of oxygen compounds, such as co-ordination of the oxygen to the surface, C—H bond fission or addition of C=C to the surface, are all much less important.

Finally, Blyholder and Cagle [73] have investigated the adsorption of a number of sulphur compounds onto iron and nickel deposited in oil films. Hydrogen sulphide and carbon disulphide produced no observable i.r. bands but were sufficiently strongly adsorbed, presumably dissociatively, to prevent subsequent carbon monoxide adsorption. Ethyl and methyl mercaptans both gave spectra similar to the parent molecules and the adsorptions appear to be analogous with the alkoxide formation with alcohols. Thus adsorbed ethyl mercaptan gave bands at 1240, 1045, and 695 cm^{-1} compared with 1270, 1090, and 970 cm^{-1} for the gaseous molecule. Sulphur dioxide gave bands at 1055, 955, 865, and 630 cm^{-1} on iron and at 1060, 960, 880, and 640 cm^{-1} on nickel. These are well removed from the bands of gaseous sulphur dioxide and sulphur trioxide but correlate reasonably with bands for complexed sulphate groups. The authors make the interesting suggestion that the catalytic oxidation of sulphur dioxide to the trioxide may well involve surface sulphate groups as intermediates.

[72] G. Blyholder and M. C. Allen, *J. Colloid Interface Sci.*, 1970, **33,** 603.
[73] G. Blyholder and G. W. Cagle, *Environ. Sci. Technol.*, 1971, 5, 158.

9
Some Aspects of the Selective Action of Metal Catalysts

BY P. B. WELLS

1 Introduction

A catalyst is selective if it promotes the formation of a single product under conditions where the formation of additional products is thermodynamically feasible. This definition has the virtue of rigour but is too restricting for common usage. Its application would compel us to describe a catalyst for propylene hydrogenation ($\Delta G°_{298} = -20.6$ kcal mol^{-1}) as selective merely because it was inactive for hydrogenolysis of propylene to ethylene and methane ($\Delta G°_{298} = -10.8$ kcal mol^{-1}). Consequently, selective catalysis is usually defined more loosely and subjectively as a process whereby a given reaction provides high yields of desirable products and minimal yields of undesirable products. The selectivity of a reaction may then be expressed numerically as the yield of desired product(s) divided by the total product yield and may take values between zero and unity.

This definition applies, of course, to all catalysed reactions, regardless of whether the catalyst is a metal, an oxide, a charge-transfer complex, or an enzyme. In the natural world, catalysis by enzymes may be highly selective, not only in the sense of the above (rigorous) definition, but also in another sense, namely that these catalysts are substrate-selective. That is, they are selective as to reagents of a given type with which they will successfully interact. For example, it is well known that succinic acid dehydrogenase will convert succinate into fumarate, whereas the analogous conversion of malonate, which contains one less methylene group in the hydrocarbon chain, is not observed. Metal catalysts are seldom selective in this sense. If a given metal is known to catalyse, say, the hydrogenation of simple olefins, then it is likely that it will also catalyse the saturation of oleic acid to give stearic acid. For this reason, metals have sometimes been called 'class catalysts'.

This Report is concerned with the factors that govern the selective action of metal catalysts. The subject has been widely studied in recent years. It was explicitly a topic for discussion at the Second International Congress on Catalysis and was discussed under other headings at the Third and Fourth Congresses. In addition, there is an extensive literature consisting in great measure of descriptions of selective catalysts but containing little interpretation of their selective action. It is not the author's intention to provide a complete survey of this literature. We shall not be concerned with descriptive material that may be found in standard texts, nor with the effect of diffusion upon selectivity which is adequately dealt with elsewhere.[1] Rather, the intention is to discuss, by the consideration of selected examples, those structural and mechanistic features of the catalytic situation which enable the highly versatile metal surface to catalyse reaction selectively.

2 Selectivity and Catalyst Structure

Reaction occurs at a metal surface following chemical interaction of an adsorbing molecule with a *site*. The necessary characteristics of such a site have been the subject of much debate. Taylor, in his well-known paper of 1925 concerning active centres,[2] envisaged that the amount of surface which is catalytically active will be determined by the reaction that is catalysed. Thus, the site must befit the reaction. At one extreme a single metal atom in a close-packed two-dimensional array may constitute a site. Alternatively, precise groupings of two, three, or more surface atoms may be required, as in the five-neighbour $B5$ site considered by van Hardeveld and van Montfoort.[3] Or at the other extreme, the requirement might be for surface atoms situated at certain edges, or steps, or at particular crystal defects.[4]

Most frequently such types of site are considered from the standpoint of the specific activity that they impart to the catalyst, but there are important consequences also for selectivity. Suppose that molecules of a substance RAB have two functional groups A and B, the first of which may be converted into X by process 1 and the second into Y by process 2. If processes 1 and 2 occur at sites of the same type, then the selectivity of the reaction of RAB will be independent of the detailed topography of the surface, and hence upon the mode of catalyst preparation. However, if a typical site for process 1 is a specific cluster of surface atoms, whereas virtually any atom in the surface may constitute a site for process 2, then the selectivity of the reaction will be very sensitive to the structure of the surface and to variables such as the mode of catalyst preparation. Thus, in this section, we shall consider instances in which selectivity has been investigated as a function of the nature of the sites available at the catalyst surface.

Normally, catalytically active metal is polycrystalline and possesses a particle size distribution. Direct experimental determination of the distribution can often be achieved by electron microscopy [5] using the direct transmission technique, provided either that a suitably thin support has been chosen (*e.g.* 'Aerosil' or 'Cabosil' silica) or that embedding of the sample in a resin and subsequent sectioning is feasible. The limit of resolution of this

[1] A. Wheeler, *Adv. Catalysis*, 1951, **3**, 250.
[2] H. S. Taylor, *Proc. Roy. Soc.*, 1925, **A108**, 105.
[3] R. van Hardeveld and A. van Montfoort, *Surface Sci.*, 1966, **4**, 396.
[4] J. M. Thomas, *Adv. Catalysis*, 1969, **19**, 293.
[5] T. A. Dorling and R. L. Moss, *J. Catalysis*, 1967, **7**, 378.

technique is in the region of 15 Å. (Normally, the results obtained by electron microscopy are supported by X-ray diffraction measurements and by determinations of metal surface area by the selective chemisorption of, say, hydrogen or oxygen.) The ability to produce such small assemblies of metal atoms prompts the thought that it may be possible to prepare supported catalysts in which each cluster of metal atoms present is simply that which constitutes a site. This is the concept that prompted Kobosev to formulate his theory of active ensembles long before these techniques became available;[6] indeed, efforts should now be made to test some of his predictions.

The geometric consequences of the aggregation of metal atoms to give microcrystals has been discussed by Poltorak, Boronin, and Mitrofanova.[7-9] Following these workers, we take as a model a face-centred cubic metal which forms octahedral crystallites with regular faces. The surface atoms will have a co-ordination number of 9 unless they occur at edges (co-ordination number = 7) or at corners (co-ordination number = 4). Thus, a microcrystal containing, say, twelve atoms, and possessing a maximum number of metal-metal bonds, has an average co-ordination number less than 7. Such particles are expected to exhibit activity typical of the structural defects in macrocrystals (*e.g.* edges of incomplete planes). Particles containing about fifty atoms have an average co-ordination number of about 7, and such particles are expected to portray activity typical of atoms situated at edges (on the octahedral model). Finally, crystallites having an edge-length as large as 40 Å (containing about 2200 atoms), and which have been annealed at temperatures sufficiently high to reduce the concentration of defects to the equilibrium value at that temperature (*e.g.* for Pt, 753 K) possess a large majority of surface atoms in an environment which is indistinguishable from the ideal face of an infinitely large crystal. Thus Poltorak concluded that, provided metal atoms form close-packed crystallites—and not, for example, rafts—an examination of catalytic properties of microcrystals in the range 10—20 Å would provide information on the catalytic activity of atoms of abnormally low co-ordination number. Furthermore, experiments with crystallites in the so-called mitoedrical region [7] of 20—50 Å would make it possible to distinguish the catalytic properties of atoms located at crystal edges and faces.

These important conclusions are not greatly dependent upon the model chosen. van Hardeveld *et al.*[3,10] and Bond [11] have discussed the nature of sites present at the surface of cubo-octahedral microcrystals, especially in

---

[6] N. I. Kobosev, *Zhur. fiz. Khim.*, 1939, **13**, 1; *Russ. J. Phys. Chem.*, 1945, **19**, 48, 142; 1947, **21**, 1413; *Uspekhi Khim.*, 1956, **25**, 545.
[7] O. M. Poltorak and V. S. Boronin, *Russ. J. Phys. Chem.*, 1965, **39**, 1329.
[8] O. M. Poltorak and V. S. Boronin, *Russ. J. Phys. Chem.*, 1966, **40**, 1436.
[9] O. M. Poltorak, V. S. Boronin, and A. N. Mitrofanova, Paper No. 68, Fourth International Congress on Catalysis, Moscow, 1968.
[10] R. van Hardeveld and F. Hartog, Paper No. 70, Fourth International Congress on Catalysis, Moscow, 1968.
[11] G. C. Bond, Paper No. 67, Fourth International Congress on Catalysis, Moscow, 1968.

relation to the concentration of $B5$ sites, and have concluded that these sites will be abundant only for crystallites of less than 70 Å diameter.

Thus, to examine the dependence of activity and selectivity upon the average co-ordination number of metal atoms which may function as surface sites, well-characterized catalysts must be prepared containing crystallites in a narrow size range and of the order of 10—50 Å in diameter.

Accordingly, Poltorak prepared and characterized (albeit, without electron microscopy) samples of platinum supported on silica which corresponded to the three types of co-ordination described above.[12] In accordance with expectation, virtually all of the platinum atoms present in the catalysts containing metal particles < 10 Å in edge-lengths (average co-ordination number = 6) were available for the chemisorption of hydrogen; 80% were so available when the edge-length was in the region of 10 Å (average co-ordination number = 7), and 30% when the edge-length was about 40 Å (average co-ordination number = 8.6). Table 1 shows typical values for the specific

Table 1 *Specific activities for hydrogen–deuterium exchange and for hex-1-ene hydrogenation over 5.5% platinum–silica at 298 K*

Catalyst	Pt atoms exposed at surface (%)	Specific activities [μmol reactant min^{-1} (Pt atom in crystal)$^{-1}$]	
		H_2–D_2 exchange	C_6H_{12} hydrogenation
1	100	0.11	0.33[a]
2	80	0.09	0.29
3	30	0.05	0.13

[a] The reader of the English translations is warned that the relevant graph is numbered wrongly in reference 7, but correctly in reference 8.

activities of the hydrogen–deuterium exchange reaction, and of hex-1-ene hydrogenation catalysed by these three samples of platinum-silica.[7] Within experimental error, each reaction occurs at the same rate at sites of each type irrespective of the average co-ordination number of the surface metal atoms. Such reactions have been described as *facile*, or *structure-insensitive* by Boudart, who has emphasised the importance of these and other analogous results to our understanding of the kinetics of surface reactions.[13] However, structure insensitivity of single-step reactions is not the concern of this Report, and we merely note that the reactions listed in Table 2 are of this type, and

Table 2 *Examples of structure-insensitive reactions*

Reaction	Reference
Hydrogenation of cyclohexene and of allyl alcohol	8
Hydrogenation of benzene	14, 15
Dehydrogenation of cyclohexane and of isopropanol	8
Hydrogenolysis of cyclopropane	16
Hydrogenolysis of cyclopentane	8

[12] O. M. Poltorak and V. S. Boronin, *Russ. J. Phys. Chem.*, 1965, **39**, 781.
[13] M. Boudart, *Adv. Catalysis*, 1969, **20**, 153.

that the concept of the structure-insensitive reaction is in the process of being established.

This concept may now be applied to reactions in which a number of products are formed simultaneously, and in which the dependence of selectivity upon crystallite size (in the critical range) can be determined. Wells and Oliver [17] have observed that the initial product of buta-1,3-diene hydrogenation over nickel-silica (*e.g.* at 373 K, 57% but-1-ene, 28% *trans*-but-2-ene, 14% *cis*-but-2-ene, 1% n-butane) was independent of the particle size distribution over the range < 20 Å to 100 Å. Since the butenes are formed as primary products by parallel routes,[18] this reaction appears to be structure-insensitive for the formation of each product. This provides a more basic understanding of the fact that the product composition of alkadiene or alkyne hydrogenation over a Group VIII metal is virtually independent of the form of the catalyst.[19] For example, buta-1,3-diene hydrogenation over palladium at room temperature affords a product composition of about 60% but-1-ene, 35% *trans*-but-2-ene, and 5% *cis*-but-2-ene irrespective of whether the metal is supported on silica, alumina, or carbon, or unsupported, or in wire form. It is now reasonable to suppose that the reaction may be structure-insensitive, and hence the precise physical nature of the metal—although important in terms of the efficiency of metal usage—is unimportant in determining the product composition obtained.

The converse of a facile or structure-insensitive reaction is a *demanding* or *structure-sensitive* reaction.[13] No examples of single-step structure-sensitive reactions have yet been established unequivocally. The report by Sinfelt and Yates,[20] that the relative specific activities of the noble Group VIII metals for the hydrogenolysis of ethane at 478 K increase linearly with the percentage *d*-character of the metal for catalysts of particle size ranging from 14 Å for iridium to 106 Å for palladium, has been interpreted by Boudart [13] as a demonstration of structure insensitivity. However, the variation of acti-

Table 3 *The dependence on crystallite size of the specific activity of rhodium for ethane hydrogenolysis at* 526 K

% Rh on silica	0.1, 0.3	1	5	10	5	5	(wire)
Mean crystallite size (Å)	11	12	20	23	41	127	2,560
Specific activity[a]	4.3	16.0	12.2	13.5	8.1	0.41	0.79

[a] units = mmol C_2H_6 converted $h^{-1}(m^2Rh)^{-1}$ at certain specific reactant pressures.

[14] P. C. Aben, J. C. Platteuw, and B. Stouthamer, Paper No. 31, Fourth International Congress on Catalysis, Moscow, 1968.
[15] T. A. Dorling and R. L. Moss, *J. Catalysis*, 1966, **5**, 111.
[16] M. Boudart, A. Aldag, J. E. Benson, N. A. Dougharty, and C. G. Horkins, *J. Catalysis*, 1966, **6**, 92.
[17] R. G. Oliver, Ph.D. thesis, University of Hull, 1971; R. G. Oliver and P. B. Wells, unpublished results.
[18] J. J. Phillipson, P. B. Wells, and G. R. Wilson, *J. Chem. Soc.* (*A*), 1969, 1351.
[19] P. B. Wells, *Platinum Metals Rev.*, 1963, **7**, 18; see also ref. 34(*b*).
[20] J. H. Sinfelt and D. J. C. Yates, *J. Catalysis*, 1968, **10**, 362.

vity from the least active metals (Pt, Pd) to the most active (Os) spans seven orders of magnitude, and the scatter in the results is such that a structure sensitivity effect involving one or two orders of magnitude in the specific rate would not seriously upset the correlation.[20] Indeed, the results shown in Table 3 indicate that the reaction at the rhodium surface is structure sensitive,[21] there being a marked decrease in specific rate when the crystallite size exceeds 41 Å. Boudart attributes [13] this effect to a perturbation of the metallic structure of small rhodium particles by adherent carbonaceous residues which appear to participate in the mechanism of hydrogenolysis. However, it is difficult to envisage how such residues can *enhance* the activity of small metal crystallites. Such a perturbation is more easily understood for the case of hydrogen peroxide decomposition catalysed by platinum–silica [7] (Table 4) where *diminished* activity of catalysts containing very small crystallites is attributed to a perturbation of the metallic character of the small platinum crystallites by the extensive surface coverage of oxygen atoms.

Table 4 *Specific activities for hydrogen peroxide decomposition over 5.5% platinum–silica at 298 K*

Catalyst (see Table 1)	1	2	3
Specific activities [min^{-1} (Pt atom in crystal)$^{-1}$]	0.1	0.1(5)	0.7

Clearly, structure sensitivity will be most clearly revealed if a system is chosen in which a reaction which is structure-sensitive proceeds in competition with one that is structure-insensitive. In this case, marked changes in selectivity will be expected to occur as particle size is varied within the critical range 10—50 Å. A well-authenticated example is the report by van Hardeveld and Hartog [10] of the reaction of benzene with deuterium catalysed by nickel–silica. The two competing reactions in this case are the hydrogenation of benzene to cyclohexane (specific rate = k_H) and the exchange of protium atoms in benzene for deuterium (specific rate = k_E). Table 5 shows values for k_H and k_E calculated from data in reference 10; (the original designations of the catalysts have been retained). It is seen that hydrogenation is structure-insensitive, the activity per unit surface area of nickel being approximately the same for all of the catalysts, whereas isotope exchange was structure-sensitive and was mostly confined to catalysts containing the larger crystallites.

In 1966, Anderson and Avery reported [22] that the reaction of neopentane with hydrogen over platinum at elevated temperatures results in skeletal isomerization to isopentane and hydrogenolysis to isobutane and methane. Dissociative chemisorption of the neopentane occurred giving both di- and tri-adsorbed intermediates; the former was crucial to the process of hydro-

---
[21] D. J. C. Yates and J. H. Sinfelt, *J. Catalysis*, 1967, **8**, 348.
[22] J. R. Anderson and N. R. Avery, *J. Catalysis*, 1966, **5**, 446.

Table 5 Specific activities of nickel–silica for benzene hydrogenation (k_H) and exchange (k_E) at 298 K

Catalyst	B	A_f	C	D
Metal content (wt % Ni)	6.8	7.4	23.3	25.3
Range of crystallite size (Å)	~200	Mostly <70	All <50	All <50
$10^5 k_H$ (mol h^{-1} m^{-2})	4.7—5.3	11.0	9.0—12.5	9.5
$10^5 k_E$ (mol h^{-1} m^{-2})	77—90	20	0.7—3.8	0.28

genolysis. Boudart et al. have more recently obtained the dependencies of the specific activities of these two processes upon the dispersion of the metal;[23] (Table 6, columns 1—6). Isomerization appears to be more structure-sensitive than hydrogenation, but the interpretation of the results is somewhat confused by the wide variety of physical forms of catalyst used, and an uncertainty as to whether the nature of the support influences the activity of the metal. If we equate dispersion as defined by Boudart with the percentage of atoms present in the octahedral crystallites of the Poltorak model, then the likely mean particle sizes for the seven catalysts of Table 6 are respectively (reading from left to right): very large; 200Å ; 90 Å; 40 Å; 20 Å; 15 Å; 40 Å. (Such an equation assumes a narrow band of crystallite size, and such may not have been the case.) It then appears that marked changes in specific activity are occurring in this system when crystallite size exceeds 100 Å, and are not restricted to the mitoedrical region.

Table 6 Specific activities of various platinum catalysts for neopentane isomerization (N_I) and hydrogenolysis (N_H) at 573 K

Catalyst support	none (powder)	η-Al$_2$O$_3$	silica	carbon (spheron)	η-Al$_2$O$_3$	γ-Al$_2$O$_3$	carbon (spheron)
Metal content (wt % Pt)	100	1.96	4.30	1.00	1.96	0.60	1.00
Dispersion (%)	0.028	7.6	17	35	64	73	35
$N_I{}^a$	0.73	11	3.8	0.86	3.4	2.8	0.26
$N_H{}^a$	0.08	4.5	13	0.34	2.3	1.9	0.02
Selectivity = N_I/N_H	9.1	2.4	0.3	2.5	1.5	1.5	13.0

^{a} units = (10^3 molecules converted) (surface Pt atom)$^{-1}$ s^{-1}.

Morikawa et al. have reported [24] that nickel present as large crystallites in a nickel–alumina preparation is active both for the hydrogenation of toluene to methylcyclohexane and for the hydrogenolysis of toluene to benzene and methane. However, if these large crystallites are removed by dissolution in acid, and if the nickel aluminate present at the former nickel–alumina boundary is reduced to give an amorphous nickel, then the newly-formed catalyst is completely selective for hydrogenation, i.e. no hydrogenolysis occurs. Although no detailed information on crystallite size is given, there is *prima facie* evidence that this hydrogenolysis is structure-sensitive.

[23] M. Boudart, A. W. Aldag, L. D. Ptak, and J. E. Benson, *J. Catalysis*, 1968, **11**, 35.
[24] K. Morikawa, T. Shirasaki, and M. Okada, *Adv. Catalysis*, 1969, **20**, 98.

Thus far, emphasis has been laid on crystal size. However, this emphasis must not obscure the fact that the calculations of Poltorak [8] and of van Hardeveld [3, 10] are based on models that possess a greater degree of geometrical perfection than the real crystallites present in supported catalysts. Crystallites of 40 or 50 Å in edge-lengths may, under certain circumstances, take up special configurations in preference to geometrically symmetrical shapes. And the act of chemisorption of reactant may bring about surface reconstruction. Each of these factors may effect the selectivity of a reaction containing a structure-sensitive step. An example of this aspect of selectivity is contained in Table 6. The selectivity of carbon-supported platinum for the isomerization and hydrogenolysis of neopentane was 2.5 when the catalyst was prepared by reduction at 773 K (column 4). However, a second catalyst, reduced similarly and subsequently fired at 1173 K *in vacuo* before use, exhibited a selectivity of 13 (column 7), the gain resulting mainly from a loss of hydrogenolysis activity. It is particularly notable that these two catalysts have the same degree of dispersion. The authors present evidence [23] that the firing at 1173 K caused crystallites to develop (111) faces, and these faces contain the trios of atoms that constitute the site for the triadsorption of neopentane and its subsequent rearrangement to isopentane. Thus, selectivity depends upon the nature of the exposed planes when one of the catalysed reactions is structure-sensitive.

Wells and co-workers have reported [18, 25] that cobalt and nickel powders formed by reduction of oxides below 623 K catalyse preferential 1,2-addition of hydrogen to buta-1,3-diene at 373 K, whereas reduction above 623 K gives catalysts that are highly selective for 1,4-addition. Subsequent work [17] has shown that this is not a particle size effect, as was first thought, and it is possible that crystal planes particularly suited to the production of but-2-ene by 1,4-addition are formed at the higher temperatures. For reasons unknown, the presence of Ni^{3+} ions in the oxide before reduction is necessary in order to obtain catalysts which are selective for 1,4-addition.

The message is clear. Where two or more primary reactions occur in parallel, the dependence of selectivity upon metal crystallite size and mode of preparation will be governed by the structure sensitivity or otherwise of the various independent processes. If all processes are facile, then selectivity will be independent of these parameters. If just one of the processes is structure-sensitive, or if all are demanding but require differently constituted sites, then selectivity will vary with crystallite size and hence with those variables that modify crystallite size. There is a great need for further studies, and it is too early to make generalizations with confidence. However, evidence so far suggests that hydrogenations and dehydrogenations may normally be structure-insensitive, and that for the hydrogenolysis of carbon–carbon bonds the situation will vary from case to case, some being structure-sensitive and others not.

[25] B. J. Joice, J. J. Rooney, P. B. Wells, and G. R. Wilson, *Discuss. Faraday Soc.*, 1966, No. 41, 223.

3 Selectivity and Reaction Mechanism

Not only may the geometry of a catalyst surface influence selectivity, as shown above, but so may the geometrical and energetic aspects of reaction mechanism. Molecules of reactant in the free state have certain geometrical characteristics and their geometry will normally be modified upon chemisorption. The products that are observed are those which have followed an energetically favourable route, for which the geometries of the chemisorbed states of reactants and intermediates have been allowed. Examples of selective processes abound, and are as diverse as hydrogenation and oxidation. In this section, the factors influencing selective hydrogenation of di-unsaturated hydrocarbons will be discussed in detail, but the general principles are of wider relevance and it is hoped that this presentation will enable them to be applied *mutatis mutandis* to other systems.

$$A \xrightarrow{+X} B \xrightarrow{+X} C$$

Scheme 1a

$$A \xrightarrow{+X} B \xrightarrow{+A+X} C \xrightarrow{+A+X} \ldots$$

Scheme 1b

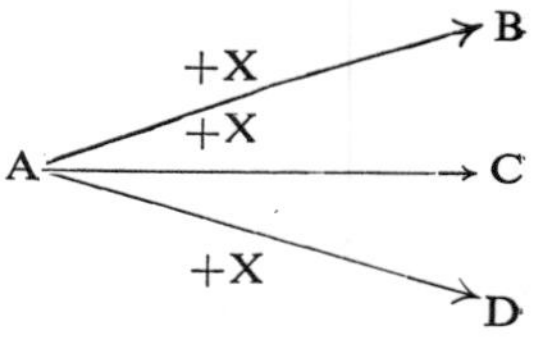

Scheme 2

For present purposes, potentially selective reactions may be divided into the two general types shown in Schemes 1 and 2. Scheme 1a represents a reaction consisting of two consecutive steps, for example the hydrogenation of an alkadiene to give alkene and alkane. The system becomes of interest from the standpoint of selectivity if the intermediate B is the desired product, as would normally be the case in the example quoted. The copper-catalysed hydrogenation of nitrobenzene to aniline is an example of a commercial process of this type.[26] Furthermore, the necessity to remove small quantities of acetylene from olefin streams in processes subsequent to the steam-cracking of naphtha is of continuing importance. It will become even more important as the production of olefins from crude oil is increased, as the latter process gives higher yields of acetylenes.[27] Scheme 1b represents a long-chain consecutive process such as the Fischer–Tropsch reaction for the production of high molecular weight alcohols and acids from hydrogen and carbon monoxide. Although this particular process is of little commercial importance at the present time, the reactions involved are still receiving considerable attention,[28]

---

[26] W. L. Faith, D. B. Keyes, and R. L. Clark, 'Industrial Chemicals', Wiley, New York, 1957, p. 118.
[27] R. K. Goldstein and A. L. Waddams, 'The Petroleum Chemicals Industry', 3rd edition, Spon, London, 1967, p. 107.
[28] H. Kölbel, G. Patzschke, and H. Hammer, *Brennstoff-Chem.*, 1966, **47**, 4; T. Kunugi, T. Sakai, H. Ose, and Y. Hamada, *J. Chem. Soc. Japan, Ind. Chem. Sect.*, 1966, **69**, 2244; T. Kunugi, T. Sakai, H. Ose, and S. Ho, *ibid.*, 1967, **70**, 810; P. N. Mukherjee,
(continued)

apparently because the selectivities that can be obtained are not well understood. Scheme 2 represents any situation in which substance A reacts with X to give isomeric products. An important example is the hydrogenation of natural oils to give edible fats,[29] where A might be linoleic acid and B and C isomeric linoleic acids, or alternatively, A might be a linoleic acid and B and C isomeric oleic acids.

The key to the achievement of high mechanistic selectivity is simple: the system must contain a constraint which greatly reduces the rate of all but one of the many possible reaction paths. The various subsections below provide examples of these constraints.

Consecutive Reactions.—*The Thermodynamic Factor.* All of the Group VIII metals catalyse the hydrogenation of acetylene and, above about 348 K, ethylene is invariably the major (Ru, Rh, Os, Ir, Pt) and sometimes the only (Fe, Co, Ni, Pd) initial product.[30] Even when a stoicheiometric excess of hydrogen is employed, the selectivity for ethylene formation is often retained until the near-removal of acetylene. This situation is achieved because one reactant (acetylene) is more strongly chemisorbed than (*i*) the other reactant (hydrogen) and (*ii*) the products (ethylene and ethane). In general, we may say that if substances A and B have pressures in the gas phase of P_A and P_B, and if equilibrium is achieved between the gaseous and adsorbed phases such that the surface coverages θ_A and θ_B are described by the appropriate Langmuir equations for competitive associative adsorption, then

$$\theta_A/\theta_B = (P_A/P_B) \exp(-\delta\Delta G/RT) \qquad (1)$$

where $\delta\Delta G$ is the difference between the Gibbs free energy changes upon chemisorption of A and B. A value of $\delta\Delta G$ of only a few kilocalories per mole is sufficient to ensure that reactant A attains a very high surface coverage. Thus, in the example above, once ethylene is formed as a gaseous product, its re-entry to the chemisorbed layer is greatly inhibited.

The relative surface coverages of acetylene and hydrogen may be determined from an equation analogous to (1) which allows for dissociative chemisorption of hydrogen. Again, for chemisorption at Group VIII metal surfaces, $\delta\Delta G$ takes values such that the surface coverage of hydrogen atoms attainable by competitive chemisorption is very low. Thus, selective hydrogenation

---

[28] *(continued)*
N. C. Ganguli, and A. Lahiri, *Indian J. Technol.*, 1965, **3**, 15; N. C. Ganguli, *ibid.*, 1966, **4**, 326; B. K. Nefedov and Ya. T. Eidus, *Uspekhi Khim.*, 1965, **34**, 630; V. M. Vlasenko, G. E. Yuzefovich, and M. T. Rusov, *Kinetika i Kataliz*, 1965, **6**, 688; Yu. B. Kagan, A. N. Bashkirov, L. A. Morozov, and N. A. Orlova, *Neftekhimiya*, 1965, **5**, 536; R. B. Anderson, F. S. Karin, and J. F. Schultz, *J. Catalysis*, 1965, **4**, 56.
[29] C. R. Scholfield, R. O. Butterfield, V. L. Davison, and E. P. Jones, *J. Amer. Oil Chemists' Soc.*, 1964, **41**, 615; A. E. Johnston, C. A. Glass, and H. J. Dutton, *ibid.*, 1964, **41**, 788; L. F. Albright, *ibid.*, 1965, **42**, 250; S. Koritala and H. J. Dutton, *ibid.*, 1966, **43**, 86; R. R. Allen, *ibid.*, 1967, **44**, 466; T. Hashimoto and H. Shina, *Reports Govt. Chem. Ind. Res. Inst., Tokyo*, 1964, **59**, 295.
[30] P. B. Wells, *Chem. and Ind.*, 1964, 1742.

always proceeds under conditions of hydrogen starvation. The importance of hydrogen starvation during selective hydrogenation of fats and oils has been stressed by Bailey.[31]

Similarly, the principle embodied in equation (1) and steric considerations together determine which of a member of similar substances (*e.g.* olefins) will undergo reaction when a mixture is admitted to a catalyst.[32]

It may be noted, in passing, that the poor selectivity of Group VIII metal catalysts for the oxidation of alkenes to aldehydes, ketones, and acids (or alternatively, the high selectivity for carbon dioxide formation) is attributable to the high surface coverages of oxygen that are invariably achieved. Because conditions of oxygen starvation are not obtained, intermediate products cannot undergo desorption before complete oxidation occurs.

The Kinetic Factor. The relative yields of alkene and alkane are determined in Scheme 3 at the point where the routes to products become independent. Thus, the catalyst is very selective for the formation of alkene if $k_2 \gg k_3 f(\theta)_H$.

$$C_nH_{2n-2}(ads) \xrightarrow[k_1]{+2H(ads)} C_nH_{2n}(ads) \xrightarrow[k_3]{+2H(ads)} C_nH_{2n+2}(g)$$

$$C_nH_{2n}(ads) \xrightarrow{k_2} C_nH_{2n}(g)$$

Scheme 3

The yield of alkane will clearly depend in part upon the value of k_3, *i.e.* upon the inherent activity of the catalyst for the hydrogenation of the alkene in question. The situation with respect to the value of k_2 is more complex. It is likely that alkyne or alkadiene may chemisorb by causing the displacement of alkene from a surface site. Thus k_2 may not be a simple desorption rate constant, but instead its value may be determined by the magnitude of $\delta \Delta G$ [equation (1)]. Consequently, the thermodynamic and kinetic factors which determine selectivity are interrelated. Nevertheless, the alkane yield will clearly depend upon the value of k_3 and upon the surface coverage of chemisorbed hydrogen.

A Selectivity Pattern and its Interpretation. The theoretical framework described above will now be applied to interpret the selectivity pattern observed in the hydrogenation of di-unsaturated hydrocarbons catalysed by the Group VIII metals. The variation of selectivity (defined as $S_M = P_{C_nH_{2n}}/(P_{C_nH_{2n}} + P_{C_nH_{2n+2}})$ with hydrogen pressure and with temperature is shown in Figures 1(a) and (b). Although these Figures refer to the hydrogenation of acetylene [33]

---

[31] A. E. Bailey, *J. Amer. Oil Chemists' Soc.*, 1945, **26**, 644.
[32] V. N. Rozhkova, and A. L. Markman, *Uzbek. khim Zhur.*, 1964, **2**, 38; R. Mauriel and J. Tellier, *Bull. Soc. chim. France*, 1968, 4191; I. Kh. Freidlin, E. F. Litvin, and S. K. Tilyaev, *Neftekhimiya*, 1968, **8**, 155; A. S. Hussey, R. H. Baker, and G. W. Keulks, *J. Catalysis*, 1968, **10**, 258.
[33] G. C. Bond and P. B. Wells, *J. Catalysis*, 1965, **4**, 211; 1966, **5**, 65, 419; 1966, **6**, 397; G. C. Bond, G. Webb, and P. B. Wells, *ibid.*, 1968, **12**, 157.

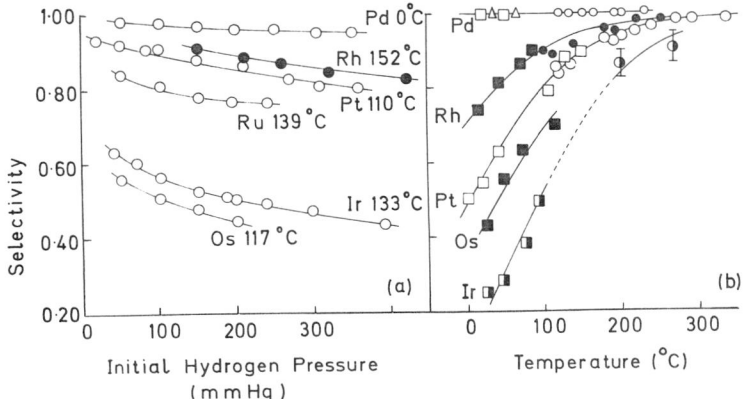

Figure 1 (a) *Acetylene hydrogenation. The dependence of selectivity upon initial hydrogen pressure observed using alumina-supported metal catalysts. Static system: Initial acetylene pressure = 50 mmHg, conversion = 10%.* (b) *Buta-1,3-diene hydrogenation. The dependence of selectivity upon temperature. Values for ruthenium–alumina resemble closely those for rhodium–alumina, and are not shown. (Squares: alumina-supported metal; circles: metal wire; triangles: metal film). For detailed experimental conditions see reference 34. For each reaction, the selectivities afforded by iron, cobalt, and nickel catalysts resemble that of palladium.*

and of buta-1,3-diene,[18, 34] the form of the graphs is normally found to be independent of the di-unsaturated hydrocarbon undergoing reduction. The information is summarized as the *selectivity pattern* shown in Scheme 4.

$$S_{Fe} \approx S_{Co} \approx S_{Ni} \approx S_{Cu} \approx 1.00$$
$$\vee \quad \vee \quad \text{?}$$
$$S_{Ru} \leqslant S_{Rh} < S_{Pd}$$
$$\vee \quad \vee \quad \vee$$
$$S_{Os} \approx S_{Ir} < S_{Pt}$$

Scheme 4

Under comparable experimental conditions selectivity increases on ascending each vertical triad of Group VIII and there is a tendency for it to increase on passing from left to right. Thus, the least selective metals are osmium and iridium, whereas nickel and palladium are among the most selective. An interpretation of this selectivity pattern for the case of acetylene hydrogenations is as follows. The relative specific activities for ethylene hydrogenation over a number of Group VIII metals have been reported by Beeck[35] and by Schuit and van Reijen.[36] The values of $\log_{10} k_{C_2H_4}$, quoted relative to

---

[34] (a) G. C. Bond, G. Webb, P. B. Wells, and J. M. Winterbottom, *J. Chem. Soc.*, 1965 3218; (b) P. B. Wells and A. J. Bates, *J. Chem. Soc. (A)*, 1968, 3064; (c) A. J. Bates Z. K. Leszczynski, J. J. Phillipson, P. B. Wells, and G. R. Wilson, *ibid.*, 1970, 2435

[35] O. Beeck, *Discuss. Faraday Soc.*, 1950, No. 8, 118.

[36] G. C. A. Schuit and L. L. van Reijen, *Adv. Catalysis*, 1958, **10**, 242.

Table 7 *The logarithm of the relative values of the velocity constant for ethylene hydrogenation ($k_{Rh} = 1$)*

	Fe	Ni	Ru	Rh	Pd	Ir	Pt
Metal films[35]	−3.0	−2.8	—	0.0	−0.8	—	−1.6
Supported metal[36]	—	—	−0.3	0.0	−0.9	−2.0	−1.5

rhodium as the standard, are given in Table 7. Thus, the high selectivity of the first row elements of Group VIII is related to their low activities for ethylene hydrogenation (*i.e.* k_3 in Scheme 3 has a relatively low value). The lower selectivities of ruthenium and rhodium are then attributed to the higher activities of these metals for ethylene hydrogenation. Thus, for these catalysts, the effect of the kinetic factor is predominant. The question then arises as to why the third row elements are the least selective, since the specific activities for ethylene hydrogenation over iridium and platinum are reported to be low. Investigations of selectivity as a function of conversion show that the thermodynamic factor operates with least severity in reactions of this type catalysed by osmium, iridium, and platinum.[33] For example, only over these metals does selectivity diminish appreciably with increasing conversion. If this occurs because ethylene is mostly strongly chemisorbed on these metals (which would be consistent with trends in the stabilities of ethylene complexes of these metals) then the rate constant for ethylene desorption (k_2 in Scheme 3) would be low. For this situation, a low value of k_3 does not preclude k_2 being comparable in magnitude to $k_3\theta_H$, and a low selectivity results.

The capacity of palladium, unique among the Group VIII metals, to occlude hydrogen exothermically may further enchance its ability to catalyse these hydrogenations selectively. There is evidence from studies of ethylene hydrogenation over palladium [37] that the surface coverage of hydrogen atoms during hydrogenation is unusually low by comparison with the situation at the surfaces of the other noble metals under comparable conditions. Thus, high selectivity ($k_2 \gg k_3\theta_H$) in alkadiene and alkyne hydrogenation may be attributed not only to an intermediate value of k_3 (kinetic factor) and a high value of k_2 (thermodynamic factor) but in addition to an unusually low value of θ_H.

Attention was first drawn to this selectivity pattern in 1963, although at that time the available data were rather restricted.[19, 38] The pattern has been amply confirmed by evidence that has accumulated since that date, although (*i*) iron, ruthenium, and osminum catalysts have been little studied and (*ii*) some hydrogenations over nickel have exhibited lower selectivities than indicated in Scheme 4. This further evidence concerns the hydrogenation

---

[37] G. C. Bond, J. J. Phillipson, P. B. Wells, and J. M. Winterbottom, *Trans. Faraday Soc.*, 1966, **62**, 433.

[38] G. C. Bond, P. B. Wells, and J. M. Winterbottom, *J. Catalysis*, 1962, **1**, 74; G. C. Bond and P. B. Wells, *Adv. Catalysis*, 1964, **15**, 171.

[39] Y. Kabe and I. Yasumori, *J. Chem. Soc. Japan*, 1964, **85**, 410; 1965, **86**, 39, 385.

Table 8 *Extent of hydrogen occlusion in metal powders formed by reduction of chlorides in hydrogen at one atmosphere pressure at* 473 K

		Ru	Rh	Os	Ir	Pt
x in MH_x	by exchange with D at 293 K	—	0.03	0.11	0.10	0.005
	by titration with butene at 373 K	0.04	0.02	0.07	0.13	0.007

of acetylene,[33, 39, 57] propadiene [40] and substituted 1,2-dienes,[41] propyne,[42] but-1-yne,[43] but-2-yne,[44] buta-1,3-diene [18, 34] and other conjugated alkadienes,[45] non-conjugated alkadienes,[46] and cyclic alkadienes.[47] Hence, it would appear that the factors which have been discussed above with particular reference to acetylene hydrogenation are similarly relevant to the hydrogenation of higher di-unsaturated hydrocarbons.

The Role of Occluded Hydrogen. Recently, it has become apparent that the above interpretation is incomplete, and that it is necessary to take a further factor into account. Selectivity appears to be related to the extent of hydrogen occlusion in the Group VIII metals. Catalytically active metal is normally prepared by a reductive procedure and in consequence the metal contains occluded hydrogen. Mellor, Smith, and Wells have shown, by hydrogen-deuterium exchange and by titration of occluded hydrogen with butene, that the capacity of the noble metals to occlude hydrogen is as shown in Table 8.[48] Occlusion in iron, cobalt, and nickel under these circumstances is minimal; palladium is not considered in this context because, as mentioned above, it occludes hydrogen exothermically.

It is striking that the order of occlusion, $Ir \geqslant Os > Ru > Rh > Pt$ is the

[40] R. S. Mann and D. E. To, *Canad. J. Chem.*, 1968, **46**, 161; R. S. Mann and D. E. Tiu, *ibid.*, 1968, **46**, 3249.
[41] L. Crombie, P. A. Jenkins, D. A. Mitchard, and J. C. Williams, *Tetrahedron Letters*, 1967, 4297.
[42] R. S. Mann and S. C. Naik, *Canad. J. Chem.*, 1967, **45**, 1023; R. S. Mann and K. C. Khulbe, *ibid.*, 1967, **45**, 2755; 1968, **46**, 623; 1969, **47**, 215; *J. Phys. Chem.*, 1969, **73**, 2104; *Indian J. Technol.*, 1967, **5**, 65.
[43] E. F. Meyer and R. L. Burwell, *J. Amer. Chem. Soc.*, 1963, **85**, 2881; R. S. Mann and K. C. Khulbe, *J. Catalysis*, 1968, **10**, 401; 1969, **13**, 25.
[44] G. Webb and P. B. Wells, *Trans. Faraday Soc.*, 1965, **61**, 1232; J. Phillipson, P. B. Wells and D. W. Gray, *Proc. 3rd Internat. Congr. Catalysis*, 1964, **2**, 1250.
[45] L. Kh. Freidlin and E. F. Litvin, *Neftekhimiya*, 1964, **4**, 374; L. Kh. Freidlin, E. F. Litvin, and L. M. Krylova, *ibid.*, 1964, **4**, 185; L. Kh. Freidlin, E. F. Litvin and R. N. Shafran, *ibid.*, 1964, **4**, 552, 669; L. Kh. Freidlin, N. V. Borunova, and L. I. Gvinter, *Doklady Akad. Nauk S.S.S.R.*, 1965, **163**, 1173; L. Kh. Freidlin, E. F. Litvin, and L. Guan-Khun, *Izvest. Akad. Nauk S.S.S.R., Ser. khim.*, 1965, 134; N. L. Sanina, P. S. Kogan, and S. N. Kazarnovskii, *Trudy Khim. i khim. Tekhnol*, 1966, **2**, 326.
[46] L. Horner and I. Grohmann, *Annalen*, 1964, **670**, 1; L. Kh. Freidlin, E. F. Litvin, and L. M. Krylova, *Neftekhimiya*, 1965, **5**, 468.
[47] L. Kh. Freidlin and I. L. Popova, *Izvest. Akad. Nauk S.S.S.R., Ser. khim.*, 1968, 1514; A. A. Balandin, A. I. Kukina, C. Hou-Sheng, and I. Ya. Kosinskaya, *Zhur. fiz. Khim.*, 1963, **37**, 2504; T. J. Katz and N. Acton, *Tetrahedron Letters*, 1967, 2601.
[48] S. D. Mellor, Ph.D. Thesis, University of Hull, 1968; S. D. Mellor, N. C. Smith, and P. B. Wells, unpublished results.

Table 9 *Dependence of selectivity in buta-1,3-diene hydrogenation at* 373 K *upon the extent of hydrogen occlusion in iridium powder*

x in MH_x before experiment	0.19	0.17	0.05	0.01
Selectivity $[C_4H_8/(C_4H_8+C_4H_{10})]$	0.45	0.55	0.78	0.95

reverse of the order of selectivity. Iridium, which reversibly occludes the most hydrogen, catalyses the most extensive formation of alkane during alkadiene hydrogenation, and the reverse is the case for iron, cobalt, and nickel. This correlation suggests that at least some of the alkane may be formed selectively at regions of the surface that are particularly rich in hydrogen. To test this, the selectivity for buta-1,3-diene hydrogenation was measured as a function of the extent of hydrogen occlusion in iridium, using the occluded material itself as the sole source of hydrogen. It was found that selectivity was clearly related to the extent of hydrogen occlusion at the commencement of the experiment (Table 9). The relative importance of hydrogen occlusion and the other factors mentioned above in the determination of the general pattern of selectivity has yet to be established.

This correlation of alkane yield with extent of hydrogen occlusion prompts three comments. First, the reproducibility of the selectivity pattern as the alkyne or alkadiene is varied is more readily understood if the pattern is partly determined by a constitutional property of the catalyst. Secondly, deuterium tracer experiments have shown that the butene and butane formed during buta-1,3-diene hydrogenation over platinum and rhodium were not wholly formed at common sites.[34c] The surface of a metal–hydrogen phase may now be regarded, tentatively, as a region where butane may be selectively formed. Thirdly, the dependence of selectivity upon temperature [Figure 1(b)] is too complex for analysis in terms of the thermodynamic and kinetic factors mentioned above. However, if it is shown in the future that selectivity is governed predominantly by the extent of hydrogen occlusion in the catalytically active metal, then an increase in selectivity with increasing temperature will be seen as consistent with the endothermic nature of hydrogen occlusion in these metals.

Unresolved Problems. It is sometimes observed with palladium catalysts that selectivity diminishes abruptly as hydrogen pressure is increased beyond a critical point;[33] the mechanism of this selectivity breakdown is not understood.

It is to be expected that, as the chain length of the alkyne or alkadiene is increased, so random chemisorption will provide increasingly inefficient packing of reactant molecules and a point should be reached where gaseous hydrogen is able to chemisorb without competition on sites which are not available for further hydrocarbon adsorption. Such uncompetitive adsorption of hydrogen would give rise to rapid and probably non-selective hydrogenation. Models of monolayers of alkynes chemisorbed at idealized surfaces suggest that this uncompetitive adsorption of hydrogen should be important

Some Aspects of the Selective Action of Metal Catalysts 251

in the hydrogenation of C_5 and higher alkynes and alkadienes. The available data to test this prediction are sparse and contradictory. The hydrogenation of gaseous penta-1,3-diene is analogous to that of buta-1,3-diene as regards selectivity,[49] whereas the hydrogenation of penta-1,3-diyne over palladium at 293—353 K was non-selective, and afforded products of all stages of hydrogenation simultaneously.[50] This prediction assumes immobile chemisorption of reactant which may be incorrect, especially at higher temperatures.

Parallel Reactions.—As indicated in Scheme 2, A may be so constituted that its reaction with X leads to isomeric products B, C, and D. In this situation it would be justifiable to refer to the *stereoselectivity* of the reaction but here the term 'selectivity' will be retained and used to embrace this concept. The hydrogenations of alkynes and alkadienes will again be used as illustrative examples.

Once more, the simplicity of the mechanism is crucial to the attainment of selective reaction. Examples of simple and of complex mechanisms are afforded by but-2-yne hydrogenation and buta-1,3-diene hydrogenation respectively; the former is always highly selective for the formulation of *cis*-but-2-ene, whereas the latter always provides a mixture of all three n-butenes.

First, consider the processes that occur in the metal-catalysed hydrogenation of but-2-yne.[43, 44] Gaseous molecules of the reactant are symmetrical and, on associative chemisorption, they may take up only one conformation which is shown as species (I) of Scheme 5. Addition of one hydrogen atom gives the half-hydrogenated state, species (II). Hydrogen atom loss from (II) usually regenerates (I) but may alternatively form chemisorbed buta-1,2-diene (III), whereas further hydrogen-atom addition to (II) gives the product *cis*-but-2-ene. Thus completely selective reaction is only jeopardized by the

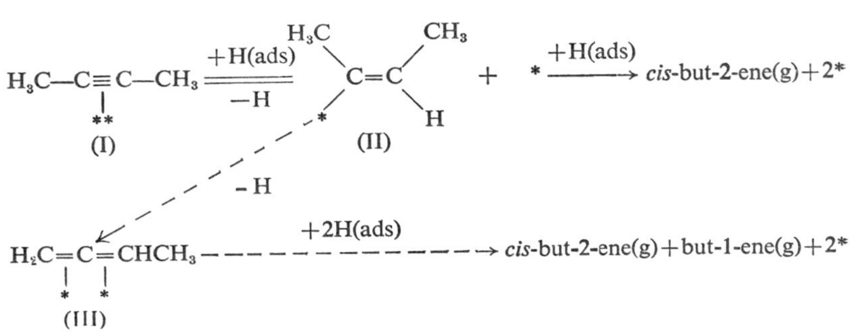

Scheme 5†

---

[49] P. B. Wells and G. R. Wilson, *J. Chem. Soc.* (A), 1970, 2442.
[50] P. B. Wells and G. R. Wilson, *Discuss. Faraday Soc.*, 1966, No. 41, 237.

† Asterisks symbolize catalyst sites.

possible isomerization of the reactant (and, coincidentally, the eventual hydrogenation of the buta-1,2-diene is likely to provide equimolar proportions of cis-but-2-ene and but-1-ene). Thus reaction is always found to be highly selective for cis-but-2-ene, the yields in reactions catalysed by alumina-supported metals being: Fe(473 K) 76%; Ru(361 K) 80%; Os(393 K) 74%; Co(413 K) 88%; Rh(427 K) 85%; Ir(433 K) 88%; Ni(425 K) 95%; Pd(287 K) 100%; Pt(431 K) 86%; Cu(397 K) 100%. The importance of reactant isomerization decreases on moving from left to right across Group VIII, and selectivity increases accordingly. [Small yields (4—7%) of trans-but-2-ene are also formed over some metals by trans-addition of hydrogen to the triple bond via a free-radical form of the intermediate (II)].

Analogously, the hydrogenation of α-acetylenes is highly selective for the formation of α-olefins.[43]

In contrast, the hydrogenation of buta-1,3-diene is a complex process,[18, 34c] as shown in Scheme 6. The simpler mechanism B contributes to reaction at the surfaces of palladium, and the so-called Type B cobalt and nickel, but, at the surfaces of the other elements of Group VIII, over Type A cobalt and nickel, and over copper, mechanism A prevails. Molecules of the reactant in the gas phase are not symmetrical. On chemisorption, interaction of one double bond with the surface gives species (II) whereas if interaction of both double bonds occurs a given molecule must adopt one of two conformations, (I) or (III). The relative abundance of each conformation is important, since 1,4-addition of hydrogen gives distinguishable products, trans-but-2-ene being formed from (I) and the cis-isomer from (III). There are no less than six half-hydrogenated states derivable from the species (I), (II), and (III), and they may interconvert. It is therefore not surprising that buta-1,3-diene hydrogenation is not selective for the formation of any one butene when mechanism A occurs. Butene compositions in the region: but-1-ene 50%, trans-but-2-ene 25%, cis-but-2-ene 25% are commonly observed. The reactions show a preference for but-1-ene formation because the routes to this product are more numerous than those to cis- and to trans-but-2-ene.

When some special constraint operates, the number of important product-forming steps may be diminished, and the reaction may become more selective. The copper-catalysed hydrogenation of buta-1,3-diene is highly selective for the formation of but-1-ene [18] (yields of but-1-ene range from 85 to 95% depending upon experimental conditions) because chemisorption of the reactant is so weak in this system that, in the primary chemisorption step, species (II) is formed almost exclusively. The majority of reaction then follows steps 4 and 10, and the high selectivity for but-1-ene formation is thereby achieved. Another example is buta-1,3-diene hydrogenation catalysed by Type B cobalt, in which up to 80% of the product may be trans-but-2-ene.[18, 25] At this surface, species appear able to attain a π-allylic mode of chemisorption (possibly due to the particular crystal faces exposed), and hence the reactant adsorbs as species (IA) and (IIIA) of mechanism B. The characteristics of rotation about the single bond for buta-1,3-diene molecules

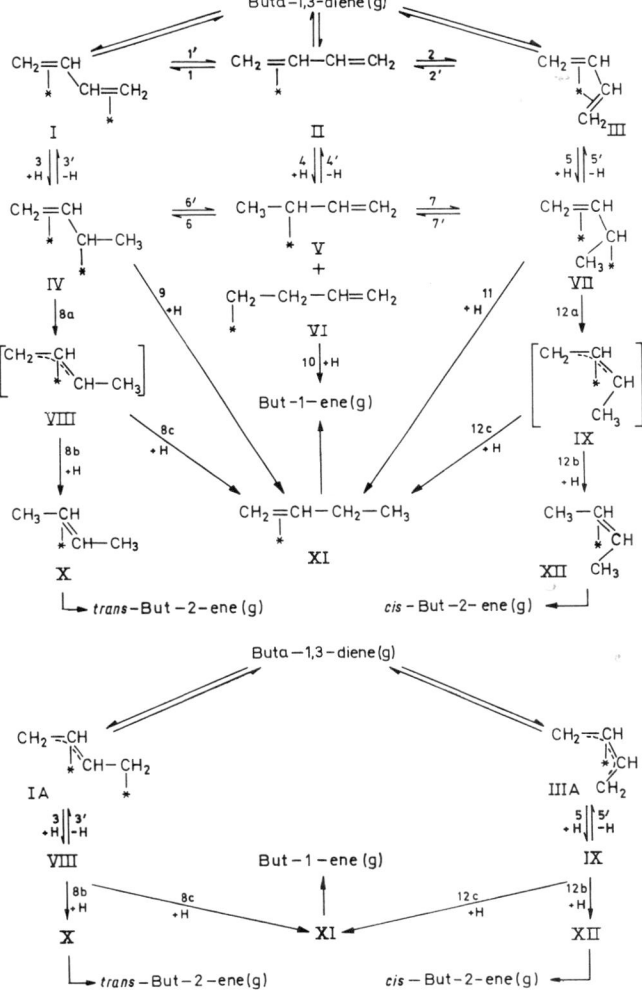

Scheme 6 Mechanisms of buta-1,3-diene hydrogenation

in the gas phase are such that chemisorption as conformation (IA) is favoured in comparison with (IIIA). Thus, after 1,4-addition of hydrogen, the *trans*: *cis* ratio in the product is high. Furthermore, but-1-ene formation by hydrogen atom addition to (VIII) and (IX) is sterically less favourable than but-2-ene formation, and hence the yield of the former is also low. The net result is a substantial selectivity for *trans*-but-2-ene formation. Thus, chemical reactivity of the surface may play a critical role in determining the complexity or otherwise of the mechanism, and thereby the selectivity.

There have been no reports, to date, of the selective formation of cis-but-2-ene by the heterogeneously catalysed hydrogenation of buta-1,3-diene. Such selectivity has been observed for reactions catalysed by pentacyanocobaltate(II) complexes in homogeneous solution.[51] Here the catalytically active entities are mononuclear complexes, and the only co-ordinated states of the buta-1,3-diene that can be formed are analogous to (II) and (III) or (IIIA). Where the latter occurs, 1,4-addition provides cis-but-2-ene very selectively, the cis : trans ratio being as high as 7 in methanol solution and 11 when ethylene glycol is used as solvent. The achievement of this situation in a heterogenously catalysed reaction would require the preparation of catalysts containing isolated Group VIII metal atoms. The possibility of suitably alloying say, cobalt, with a metal of Groups IB or IIB, to achieve such a catalyst, has not been investigated.

Substituent Effects in Selectivity.—Frequently, a substance to be hydrogenated may contain two functional groups of the same nominal type, which are dissimilarly located within the molecule. For example, penta-1,3-diene contains two double bonds, one terminal and the other internal. In the hydrogenation of this hydrocarbon catalysed by cobalt, nickel, copper, palladium, and platinum the internal double bond (i.e. the one more sterically hindered) is reduced more rapidly than the terminal double bond.[49] Over copper, where the effect is most marked, 3,4-addition accounts for about 70% of reaction, 1,2-addition for 13%, and 1,4-addition for 17%. Clearly the electron-releasing effect of the methyl group activates the internal double bond and renders its reduction more rapid than that of the terminal double bond.

Isomerization of Product before Desorption.—It was noted above that selectivity may be diminished by the isomerization of the reactant before hydrogenation. Equally, the isomerization of a chemisorbed product before its desorption may have a similar effect. It used commonly to be held that buta-1,3-diene hydrogenation occurred by 1,2-addition, yielding but-1-ene as a primary product, and that this was followed by isomerization of a proportion of the but-1-ene to cis- and trans-but-2-ene.[52] This is equivalent to classifying the reaction under Scheme 1(a) rather than Scheme 2. However, deuterium tracer studies have shown that, at the surfaces of Co, Ni, Cu, Rh, Pd, and Pt, the three butenes are formed with virtually identical deuterium distributions, and hence they are formed by parallel routes.[18, 34c] Indeed, there is no well-documented example in the literature of alkadiene or alkyne hydrogenation in which isomerization of primary products occurs before desorption, provided that the reactant has not been seriously depleted.

[51] T. Sasaki, S. Eguchi, and T. Toru, *Tetrahedron Letters*, 1971, 1109.
[52] R. L. Augustine, 'Catalytic Hydrogenation', Arnold, London, 1965, p. 60.

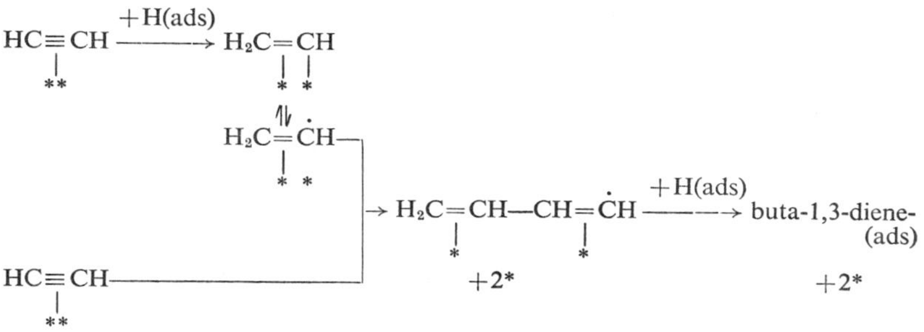

Scheme 7

$$HC\equiv CH \xrightarrow{+H(ads)} H_2C=CH$$

$$H_2C=\overset{\cdot}{C}H$$

$$\rightarrow H_2C=CH-CH=\overset{\cdot}{C}H \xrightarrow{+H(ads)} \text{buta-1,3-diene-(ads)}$$

$$HC\equiv CH \qquad\qquad +2* \qquad\qquad +2*$$

Reduction of Selectivity by Polymerization.—The chemisorption of unsaturated hydrocarbons at metal surfaces may result in dimerization, trimerization, or polymerization of hydrocarbon units, thus diminishing the selectivity of the reaction for the primarily required products. For example, terminal alkynes and allene undergo polymerization during hydrogenation.[33, 40, 53] Sheridan proposed that this occurs as a result of the formation of half-hydrogenated states having free-radical character.[53] Thus buta-1,3-diene may be formed from acetylene by the process shown in Scheme 7. The alkadiene so formed then undergoes hydrogenation. In the reaction of acetylene with hydrogen over palladium–alumina at 273 K, almost the whole of the C_2-product is ethylene (and from this standpoint the reaction is highly selective) but 38% of the acetylene is converted into butenes. The fraction of acetylene appearing as C_4-products during acetylene hydrogenation over the other platinum metals was found to be:[33] Pt–Al_2O_3(409 K) 28%; Os–Al_2O_3(396 K) 16%; Ir–Al_2O_3(409) 15%; Rh–Al_2O_3(405 K) 12%; Ru–Al_2O_3 (439 K) 8%.

4 Selective Poisoning and Shape-selective Catalysts

The preceding sections have demonstrated how the characteristics of the surface and of the reacting molecules each contribute to the determination of the selectivity of a reaction. The results that have been quoted are for reactions in which the catalyst surfaces have not been unduly contaminated by substances other than the reactants or species derived from them. A subject now ripe for further study is the poorly understood phenomenon of the effect of selective poisons upon catalyst activity; that is, the effect upon selectivity of intentionally introducing foreign substances on to metal surfaces. Present knowledge on this subject will not be rehearsed here, but the implications of the principles discussed in Section 2 will be pointed out.

Selective poisoning of catalysts is an old art, the practical objective being to suppress the activity of the catalyst for the formation of unwanted products. The application of a so-called selective poison frequently deactivates a cata-

[53] J. Sheridan, *J. Chem. Soc.*, 1945, 305.

lyst in an absolute sense, the relative deactivation of the unwanted products being greatest. An understanding of such deactivation would be advantageous. At the same time, the search for the ideal should not be overlooked, namely the ability so to modify a surface that activity for the formation of unwanted products may be re-channelled into activity for formation of the premier product.

It is well known that selective poisoning may be achieved by contaminating the surface with an electron-donating substance.[54a] Such poisons include compounds having lone pairs of electrons *e.g.* thiophen and alkyl sulphides, pyridine, ammonia, quinoline, and metals such as mercury and lead. In the well-known Lindlar catalyst [54b, 55] the poisons lead and quinoline deprive palladium of its activity for alkene hydrogenation, so that alkynes and alkadienes may be selectively reduced to alkenes. Similar results can be obtained using mercury as a poison.[56] The extent to which such poisons act simply by steric blocking of selected sites, and the extent to which their effect is electronic, by virtue of the donation of electrons into the d-band of the metal, is uncertain. It has long been known that a poison may reduce activity by a larger factor than the fraction of the surface that it occupies, but such results can frequently be interpreted either using the concept of active centres or by invoking an electronic effect. Two new modes of attack on this problem are now available.

First, future investigators should determine whether or not the reactions to be modified by a poison are structure-sensitive. If the (unpoisoned) reaction consists of two structure-insensitive parallel steps it is unlikely that any poison, however it operates, will selectively deactivate the catalyst for one of the steps. However, if, say, the conversion of A into B is structure-sensitive, whereas the conversion of A into C is not, then selective deactivation of the process giving B should be achievable if a poison can be found which is selectively chemisorbed at the special sites at which B is formed. This poisoning would be achieved by the simple steric blocking of these particular sites. By similar reasoning, for the system of consecutive reactions shown in Scheme 1(a), a poison which simply blocks sites would selectively deactivate the second stage if the conversion of B into C was the only structure-sensitive step. If both processes are structure-insensitive, as may well prove to be the case for alkadiene and alkyne hydrogenation, then selective poisoning will only be achievable by the modification of the electronic nature of the metal so as to weaken drastically the chemisorption of B. Thus, a knowledge of the structure-sensitivity of the unpoisoned reaction should allow a correct choice to be made of the type of poisoning mechanism to be attempted, and a new framework for thoughtful experimentation is hereby provided.

Secondly, more information is required about the *reactivity* of poisons.

---

[54] G. C. Bond, 'Catalysis by Metals', Academic Press, London and New York, 1962 (*a*) p. 99; (*b*) p. 297.
[55] H. Lindlar, *Helv. Chim. Acta*, 1952, **35**, 446.
[56] G. C. Bond and P. B. Wells, *Proc. 2nd Internat. Congr. Catalysis*, 1962, **1**, 1159.

Some Aspects of the Selective Action of Metal Calatysts

Not infrequently it is assumed that thiophen or pyridine acting as a poison is simply an unreactive adsorbate. This is unlikely to be true. Studies of the reactivities of heterocyclic compounds in the chemisorbed state, which are now in their infancy,[57] may assist in providing an understanding of their mode of operation as poisons. Is it possible that poisons themselves will be divisible into structure-sensitive and structure-insensitive types?

In the introduction it was noted that metals are generally class catalysts, whereas enzymes are not infrequently substrate-selective. An important recent development has been the preparation of reactant-selective metal catalysts. This has been achieved by causing the reaction to occur within the intracrystalline space of a molecular sieve. The molecular sieve may be a zeolite or a carbon. Straight-chain organic compounds can penetrate the channels of certain sieves, whereas the branched-chain isomers cannot.[58] Consequently, if catalytically active metal is present in the interior cavities of the structure, then metal-catalysed reaction of selected compounds can be achieved. The really novel aspect of this work has been the manner in which catalytically active metal has been located in the interior of the catalyst particles. Chen and Weisz[59] enlarged the channels of sodium mordenite by base exchange with hydrochloric acid; the pores where then large enough to allow passage of tetrammineplatinum(II) ions into the internal cavities of the zeolite. After incorporation of 0.2% platinum, the pores were narrowed, that is, restored to their original size, by back-exchange of hydrogen for sodium ions. The resulting catalyst selectively hydrogenated ethylene present in ethylene–propylene mixtures. Such shape-selective catalysts have already found commercial application[60, 61] for the selective elimination of normal hydrocarbons from gasolines, thus improving the anti-knock quality of the product.

Catalysts containing platinum dispersed in amorphous carbons which exhibit molecular sieve properties have been described by Cooper.[62] One method of preparation involves the polymerization of an alcoholic medium containing a colloidal suspension of platinum, and the resin so obtained is then carbonized. Alternatively, conventional charcoal-supported metal catalysts may be coated with polymer, and carbonized, so that a skin of the carbon molecular sieve is formed over the original material. Such catalysts hydrogenate propylene in preference to but-1-ene, isobutene, or substituted butenes.

A further valuable feature of shape-selective catalysts is the immunity of the

---

[57] G. E. Calf, J. L. Garnett, and V. A. Pickles, *Austral. J. Chem.*, 1968, **21**, 961; C. G. Macdonald and J. S. Shannon, *Tetrahedron Letters*, 1964, 3351; K. Kishi and S. Ikeda, *J. Phys. Chem.*, 1969, **73**, 2559; R. B. Moyes and P. B. Wells, *J. Catalysis*, 1971, **21**, 86.
[58] P. B. Weisz and V. J. Frilette, *J. Phys. Chem.*, 1960, **64**, 382; P. B. Weisz, V. J. Frilette, R. W. Maatman, and E. B. Mower, *J. Catalysis*, 1962, **1**, 307.
[59] N. Y. Chen and P. B. Weisz, *Chem. Eng. Progr., Symp. Ser. 73*, 1967, **63**, 86.
[60] N. Y. Chen, J. Mazuik, A. B. Schwartz, and P. B. Weisz, *Oil Gas J.*, 1968, Nov. 18 issue, 154.
[61] P. B. Weisz, *Ann. Rev. Phys. Chem.*, 1970, **21**, 193.
[62] B. J. Cooper, *Platinum Metals Rev.*, 1970, **14**, 133.

metal to poisoning by substances such as thiophen which cannot penetrate the channels, and so cannot chemisorb at the catalytically active centres.

Procedures of this sort, which involve the utilization of intracrystalline space as a reaction zone, and its modification as a means of obtaining selective reaction, have been described as 'molecular engineering'.[61]

5 Summary

In Section 2 of this Report, the effect of metal crystallite size upon selectivity was discussed.

Section 3 demonstrates that the selectivity of a reaction may be influenced:

(*i*) by the conformational simplicity of complexity of a reactant before and after chemisorption;

(*ii*) by electronic effects of substituent groups in the reactant;

(*iii*) by the mode of bonding of chemisorbed species;

(*iv*) by the number and/or reactivity of intermediates, particularly half-hydrogenated states;

(*v*) by the propensity of the reactant to isomerize before reaction;

(*vi*) by the propensity of a product to isomerize before desorption;

(*vii*) by the relative surface coverages of reactants, and in particular the necessity to achieve low surface coverage of hydrogen during hydrogenation;

(*viii*) by the interplay of thermodynamic and kinetic factors where consecutive reactions occur;

(*ix*) by the formation of more than one catalyst phase, as occurs when hydrogen is occluded in certain metals.

A more basic understanding of the selective poisoning of metal catalysts, and the development of substrate-selective catalysts are mentioned in Section 4 as topics which, along with further careful work on the effects of crystallite size, merit careful study in the immediate future.

Author Index

Abbatista, F., 39
Aben, P. C., 240
Abkin, A. D., 113
Abrahams, S. C., 140
Ackermann, R. J., 14
Acton, N., 249
Adams, I., 134
Adler, G., 97, 99, 100, 106, 118, 124
Aika, K., 212
Albon, N., 135
Albright, L. F., 245
Alcock, C. B., 21
Aldag, A. W., 240, 242
Allen, F. G., 145, 163
Allen, M. C., 232, 234, 235
Allen, R. R., 245
Allpress, J. G., 13, 14, 16, 46, 47, 48
Altham, J. A., 220
Amelinckx, J., 16
Amelinckx, S., 25, 79
Amelio, G. F., 186
Ander, P., 95
Anderson, J., 214
Anderson, J. S., 2, 11, 24, 27, 33, 36, 48, 241
Anderson, R. B., 245
Andersson, G., 13, 26
Andersson, S., 10, 22, 23, 33, 41, 47
Andreev, A., 228, 234
Andrew, E. R., 115
Apker, L., 145, 148
Arakawa, E. T., 192
Ariya, S. M., 4, 27
Arnat, R. A., 132
Arnold, B., 103, 107
Ashmeed, D. R., 211
Augustine, R. L., 254
Avery, N. R., 222, 241
Azaki, A., 211

Babushkin, A. A., 230
Baddour, R. F., 222
Bagchi, S. N., 68
Bailey, A. E., 246
Baker, A. D., 172, 174
Baker, B. G., 157
Baker, C., 79, 172, 174
Baker, R. H., 246
Balandin, A. A., 249
Ballantine, D., 99, 100
Ballantine, D. S., 95, 119
Balwit, J. S., 121
Bamford, C. H., 96, 99, 102, 107, 109, 111, 129, 133

Ban, L. L., 56, 59, 68, 71, 82, 91, 133
Bando, Y., 30
Barbanel, V. I., 23
Bardeen, J., 145
Barkalov, I. M., 113, 114
Barker, W. W., 15
Bartholomew, R. F., 30
Bashkirov, A. N., 245
Bassett, D. C., 125, 135
Bates, A. J., 247
Batt, R. J., 168
Batychko, S. V., 232
Baukus, J., 31
Baysal, B., 99, 100
Becker, G. E., 180
Beeck, O., 247
Beer, M., 94
Bensasson, R., 100, 112
Benson, J. E., 240, 242
Berak, J., 16
Berglund, C. N., 156, 169, 190
Bergner, D., 39
Bergsnov-Hansen, B, 208
Berkes, J. S., 30
Berlin, A. A., 114
Bevan, D. J. M., 2, 4, 15
Bibby, A., 102, 109, 111
Biscoe, J., 66
Bishop, H. E., 183
Blomberg, B., 15
Bloomer, R. N., 164
Blum, H., 132
Blumenthal, R. N., 31, 50
Blyholder, G., 224, 226, 227, 232, 234, 235
Bogdanova, N. I., 27
Bogomolov, V. N., 23, 30
Bohn, G. K., 184
Bond, G. C., 206, 220, 238, 246, 247, 248, 256
Boon, C., 215
Bordass, W. T, 192
Borello, E, 227
Borelly, R., 20
Borisenko, L. A., 233
Born, M., 55
Boronin, V. S., 238, 239
Borunova, N. V., 249
Bouas-Laurent, J., 134
Boudart, M., 211, 239, 240, 242
Bouwman, R., 158
Bowden, M. J., 104, 106
Bozon-Verduraz, F., 223, 231
Bradley, D. E., 80
Bradshaw, A. M., 227, 231

Brattain, W. H., 145
Brauer, G., 21
Brerger, A. Kh., 117
Brill, R., 232
Brodskii, I. A., 226
Brooks, J. D., 68
Broudy, R. M., 163, 167
Brown, I. F., 117
Bruche, E., 54
Bruk, M. A., 113
Brundle, C. R., 172, 174, 201
Brunie, S., 2
Buckland, A., 222
Bühl, H., 222
Bühl, A., 84
Bunn, G. W., 89
Burdese, A., 39
Burgers, J. M., 133
Burlant, W., 121
Bursiu, L. A., 7, 18, 20, 23, 24, 25
Burton, J. A., 147
Burwell, R. L., 249
Buseck, P., 36
Butterfield, R. O., 245

Cagle, G. W., 235
Calas, R., 138
Calf, G. E., 257
Camia, M., 230
Cannavo, C., 125
Cant, N. W., 226
Carazollo, G., 122, 123
Carlson, T. A., 179, 198
Cashion, J. K., 190
Castles, J. R., 2
Caulton, K. G., 232
Cha, D. Y., 217
Chachaty, C., 107
Chang, C. C., 173, 186
Chapiro, A., 99, 121
Charlesby, A., 96, 119, 121
Chase, J. D., 32
Chatani, Y., 97, 117, 122
Chen, C. S. H., 99
Chen, N. Y., 257
Chen, W. K., 50, 51
Chernyak, I. V., 113
Cherov, A. A., 134
Cheselke, F. J., 209
Chesters, M. A., 229
Chevreton, M., 2
Chirico, V., 68
Chu, C. W., 31
Chukin, G. D., 222
Chukov, V. I., 31
Chung, M. F., 183

Church, E. L., 183
Clapp, P. C., 25
Clark, A. J., 211
Clark, R. L., 244
Clarke, J. K. A., 226
Clement, G., 222
Coad, J. P., 186
Coburn, J., 31
Cockayne, D. J. H., 62
Cohen, J. B., 2
Cohen, M. D., 116, 133, 134
Cole, T., 107
Collén, B., 22
Colombo, P., 99
Coluccia, S., 227
Connick, W., 134
Contour, J. P., 231, 232
Cooper, B. J., 257
Corke, N. T., 132, 134
Costachuk, F. M., 105
Cottrell, A. H., 101, 133
Coutures, J. P., 21
Cowley, J. M., 25, 43
Cox, B. M., 164
Craig, D. P., 138
Cranstoun, G. K. L., 221
Crewe, A. V., 57
Crick, F. H. C., 92
Crombie, L., 249
Crowell, A. D., 217
Cullis, C. F., 68
Curry, C., 161
Curti, R., 141
Cuthill, J. R., 168

Dalrymple, M., 157
Damask, A. C., 132
Danley, W. J., 31
Danon, J., 113
David, C., 121
Davison, V. L., 245
Dawson, I. M., 132
DeKock, R. L., 232
Delavignette, P., 16
Delchar, T. A., 164, 168
Delgass, W. N., 196
Deschamps, J., 125
Desvergne, J. P., 134
Dey, B. N., 134
Dickey, J., 145
Dienes, G. J., 95
Dignam, M. J., 231
Di Stefano, T. H., 157
Dooley, G. J., 183
Dorgelo, G. J. H., 158
Dorling, T. A., 237, 240
Doughatty, N. A., 240
Dove, D. B., 80
Dowden, D. A., 145
Dowell, W. C. T., 56
Drmaj, D. T., 229
Dudkina, L. M., 232
Dümbgen, G., 23
Dukova, E. D., 134
Dunning, W. J., 135
Dupouy, G., 55
Durigon, D. D., 218
Dutton, H. J., 245

Dworkin, A., 112, 113

Earnshaw, J. W., 215
Eastman, D., 161, 174, 190, 191, 192
Eastmond, G. C., 96, 99, 102, 103, 104, 107, 109, 111, 125, 129, 133
Eckert, F. J., 68
Eckstrom, H. C., 229
Eden, R. C., 156
Eguchi, S., 254
Ehrlich, G., 213
Ehrlich, P., 21
Eidus, Ya. T., 245
Eikum, A., 23
Eischens, R. P., 203
Eisenhandler, C. B., 57
Ejima, T., 22
Eley, D. D., 205, 211
Elo, R., 51
Emerson, D., 139
Enikolopyan, N. S., 113
Ergun, S., 67
Ertl, G., 224
Estrup, P. J., 214
Evans, J. R. N., 134
Evans, S., 174
Evans, W. C., 134
Everhart, T. E., 61
Eyring, L., 2, 4, 5, 27

Fadley, C. S., 168, 195, 196, 199
Fadner, T. A., 99
Faith, W. L., 244
Fan, H. Y., 148
Farnsworth, H. E., 207, 209
Farren, G. M., 226
Faucitano, A., 118
Fender, B. E. F., 2, 5
Fenske, R. F., 232
Fernandez-Moran, H., 56
Ferreira, L. C., 224
Filimonov, V. N., 234
Fischer, T. E., 163, 190
Fisher, G. B., 191
Fisher, R. M., 55
Flack, H. D., 129
Flörke, O. W., 33
Foëx, M., 21
Førland, K. S., 22
Ford, R. R., 223
Forstmann, F., 192
Fourdeaux, A., 86
Fowler, R. H., 148
Frank, F. C., 132
Frankl, D. R., 30
Franklin, R. E., 67, 92
Freidlin, I. Kh., 246, 249
Frennet, A., 220
Frilette, V. J., 257
Fujimoto, M., 106
Fukui, F., 122
Fukui, K., 122
Fydelor, P. J., 119

Gado, P., 10, 14

Gallon, T. E., 184, 200
Galy, J., 10
Ganguli, N. C., 245
Gareyeva, D. A., 114
Garnett, J. L., 257
Gasser, R. P. H., 207, 211, 212, 214, 215, 216, 218
Gaylord, N. G., 117
Gebert, E., 14
Geil, P. N., 89, 135
Gerlach, R. L., 201
Gesell, T. F., 192
Geutsch, H., 210
Gevers, R., 133
Gibb, R. M., 33
Gibson, D. T., 140
Gift, J., 161
Gillbro, T., 106
Gillin, L. M., 79
Gilson, D. F. R., 105
Glass, C. A., 245
Glassler, F. P., 114
Glassler, L. S. D., 114
Glazer, A. M., 129
Glemser, O., 13
Gobeli, G. W., 145, 163
Gol'danskii, V. I., 113, 114
Gol'din, V. A., 117
Goldsmith, R. L., 222
Goldstein, R. K., 244
Goodsel, A. J., 226
Gordon, G., 23
Gordon, R. B., 132
Goshing, R. G., 92
Goto, T., 30
Grabar, D. G., 99
Graeme-Barber, A., 141
Grant, J. T., 183, 185, 186
Gray, D. W., 249
Green, J. C., 174
Green, M. L. H., 174
Greenler, R. G., 229, 230
Griffiths, E., 134
Grohmann, I., 249
Gromov, V. F., 113
Grubb, W. T., 121
Gruehn, R., 36, 39, 40
Gruenwald, T. B., 23
Grunberg, L., 168
Guan-Khun, L., 249
Guarini, G., 134
Gvinter, L. I., 249

Haas, T. W., 183, 185, 186
Hädicke, E., 117
Hagstrum, H. D., 174, 180, 192
Haigh, E., 104
Hale, A., 214
Hall, C. E., 67, 91
Hall, W. K., 209, 226
Hamada, Y., 244
Hammer, H., 244
Hamming, R. W., 57
Hanawa, T., 80
Hannum, S. E., 228
Hansen, R. S., 213, 214
Hardy, G., 99, 127
Harrick, N. J., 203
Harris, L. A., 185

Author Index

Hartley, B. M., 203
Hartog, F., 238
Hasegawa, M., 115, 116
Hashimoto, H., 62
Hashimoto, T., 245
Hasiguti, R. R., 31
Haul, R., 23
Hawthorne, H. M., 132, 135
Hayashi, K., 95, 120, 121, 122, 124, 125, 127
Hayes, K. E., 229
Heckman, F. A., 68
Heidenreich, R. D., 55, 56, 57, 59, 61, 133
Heine, V., 192
Heller, H. C., 107
Hendricks, S. B., 140
Henfrey, A. W., 2
Herickx, C., 86
Herz, J. E., 122, 127
Hess, W. M., 48, 56, 59, 71, 82
Hess, W. W., 133
Heyne, Hj., 216
Hibi, T., 56
Hill, G. J., 32
Hill, R. D., 183
Hirsch, P. B., 55, 133
Hirthe, W. M., 31
Ho, S., 244
Hobert, H., 225
Hobson, J. P., 215
Hockey, J. A., 223
Holland, V. F., 134
Holmberg, B., 14
Horkins, C. G., 240
Horner, L., 249
Hornyak, J., 119
Hoseman, R., 68
Hou-Sheng, C., 249
Houston, J. E., 201
Howe, A. T., 51
Howie, A., 55, 62, 133
Howling, D. H., 157
Huber, E. E., 168
Hughes, O. H., 157
Hughes, T. R., 196
Hugo, J. A., 86
Hull, D., 133
Humphreys, C. J., 55
Huntingdon, H. B., 22, 148
Hurlen, T., 22
Hussey, A. S., 246
Hutchinson, J. L.,
Hyde, B. G., 2, 4, 7, 11, 18, 23, 24, 25, 27

Ignat'eva, L. A., 222
Iguchi, M., 115
Ikeda, S., 257
Imre, L., 14
Ino, T., 80
Isaacson, M., 57
Ishigaki, I., 121
Israelsson, M., 39
Ito, A., 121, 122
Ito, Y., 99, 106, 119
Ivanov, V. S., 117
Iwai, T., 121

Iwasaki, M., 103, 106
Izu, M., 122

Jackson, D. A., 51
Jackson, P. J., 134
Jakabhazy, S. Z., 115
James, W. J., 22
Janak, J. F., 190
Janninck, R. F., 50
Jeannin, Y. P., 2
Jenkins, G. M., 84
Jenkins, L. H., 183
Jenkins, M. S., 5
Jenkins, P. A., 249
Jennings, T. J., 167
Jiru, P., 225, 232, 234
Johannson, H., 54
Johnson, B. B., 157
Johnson, C. D., 57
Johnson, D., 57
Johnson, D. J., 86
Johnson, O. W., 22
Johnston, A. E., 245
Joice, B. J., 243
Jones, E. P., 245
Jones, G. R., 174
Jongepier, R., 158
Joris, G. G., 211
Jostons, A., 2
Joyner, R. W., 185

Kabe, Y., 248
Kagan, Yu. B., 245
Kagiya, T., 122
Kaiser, J., 117
Kakinoki, J., 80
Kane, E. O. 145
Kanter, H., 182
Karin, F. S., 245
Katada, K., 80
Kato, Y., 30
Katsura, S., 106, 119
Katz, T. J., 249
Kautaradze, N. N., 223, 228, 231
Kawada, A. A., 134
Kawamiya, N., 31
Kawamura, K., 84
Kay, D. H., 91
Kazakova, L. A., 233
Kazarnovskii, S. N., 249
Keith, H. D., 134
Keller, A., 89, 134
Kelly, A., 79
Kemball, C., 219
Keulks, G. W., 228, 246
Kevorkian, H. K., 129
Keyes, D. B., 244
Keys, L. K., 31
Khan, A. S., 27
Khomikovskii, P. M., 113
Khulbe, K. C., 249
Kihlborg, L., 15, 18, 20, 39
Kinell, P. O., 106
King, D. A., 218
King, M. F., 214
Kingsbury, P. I., 22
Kirk, C. T., 168
Kirk, J. C., 31
Kishi, K., 257

Kiss, L., 117
Kitahama, K., 122
Kitaigorodskii, A. I., 142
Kitanishi, Y., 95, 120, 121
Kleber, W., 31
Klemperer, O., 178
Klier, K., 217
Knözinger, H., 222
Knoll, M., 54
Kobosev, N. I., 238
Koch, F., 2
Koch, J., 224
Kochendorfer, A., 132
Kölbel, H., 225, 244
König, H., 20
Kössler, I., 117
Kofstad, P., 22, 50, 51
Kogan, P. S., 249
Kohlschütter, H. W., 120
Kolnais, J., 32
Komaki, A., 124
Komoda, T., 56
Kondo, M., 121
Konstantinova, V. P., 135
Konvalinka, J. A., 205
Koritala, S., 245
Kosinskaya, I. Ya., 249
Koslovskii, M. I., 134
Kosuge, K., 32
Krakowski, K. A., 210
Krauch, C. H., 117
Krause, M. O., 179
Krieger, K. A., 221
Krishnan, K. S., 142
Kröger, F. A., 2
Krolikowski, W. F., 169, 191
Kronick, P. L., 129, 132
Krylova, L. M., 249
Kubelkova, L., 222, 234
Kukina, A. I., 249
Kunakura, M., 121
Kunugi, T., 244
Kupperman, A., 173
Kurita, Y., 106
Kuylenstierna, U., 22

Labes, M. M., 129, 132, 134
Lachi, M. P., 230
Lahiri, A., 245
Lalande, R., 138
Lally, J. S., 55
Lambert, R., 222
Lander, J. J., 202
Lando, J. B., 104, 122
Lang, A. R., 133, 135
Lawrence, C. P., 212
Lawton, E. J., 121
Lea, C., 156
Le Brusq, H., 50
Lee, C. W., 33
Leghissa, S., 123
Leidheiser, H., jun., 217
Leisegang, E. C., 224
Lemmlein, G. G., 134
Leszczynski, Z. K., 247
Lienard, G., 220
Ligotti, A., 121
Lindenmeyer, P. H., 134

Lindlar, H., 256
Linnett, J. W., 192
Lipscomb, T., 99
Lipson, H., 71
Littke, W., 21
Little, L. H., 230
Litvin, E. F., 246, 249
Locchi, S., 141
Logan, S. R., 211
Lonsdale, K., 129, 135, 142
Ludmer, Z., 133, 134
Lund, A., 106
Lutz, D. A., 115

Maatman, R. W., 257
McAllister, A. J., 168
McColm, I. J., 51
Macdonald, C. G., 257
McGarvey, B., 106
McGhie, A. R., 129, 134
McGregor, D. R., 140
McKee, C. S., 185
McKee, D. W., 219, 220
MacRae, A. K., 202
Madey, T. E., 212, 214, 215, 216, 218
Magneli, A., 14, 15, 22, 23, 33
Maire, G. L. C., 157
Malin, A. S., 2
Mammi, M., 122, 123
Mann, R. S., 249
Mannami, M., 62
Marans, N. S., 121
Marion, F., 50
Markinson, R. E. B., 148
Markman, A. L., 246
Marsay, C. J., 216
Marsh, H., 133
Marsh, P. A., 79
Martin, A. E., 129
Martin, R. L., 15
Marx, R., 112
Massey, C. J., 214
Mathew, J. A. D., 187
Mathews, L. D., 157
Matsumoto, T., 124
Matsushita, K., 213, 214
Mattern, P. L., 132
Matthews, L. D., 217
Mauriel, R., 246
May, F. G. J., 134
May, J. W., 171
Mazuik, J., 257
Medana, R., 209
Medvedev, Yu. V., 117
Mee, C. H. B., 156, 168
Meesen, A., 145
Meleshina, V. A., 135
Mellor, S. D., 249
Mel'nik, P. V., 198
Menter, J. W., 55, 132
Merritt, R. R., 27
Mertin, W., 36
Mesrobian, R. B., 95
Meye, W., 222
Meyer, E. F., 249
Meyer, H., 145
Mez, E. C., 117
Michell, D., 134

Milledge, H. J., 139, 141
Miller, K., 202
Minelich, J. W., 183
Minkoff, I., 134
Mitchard, D. A., 249
Mitrofanova, A. N., 238
Miyairi, T., 106, 119
Miyake, Y., 122
Moddeman, W. E., 179
Modell, M., 222
Moll, J. L., 156
Moodie, A. F., 43
Moore, G. E., 208, 209, 214
Morawetz, M., 96, 99, 104, 106, 115
Morikawa, K., 242
Morikawa, Y., 211
Morosoff, N., 115
Morozov, L. A., 245
Morozova, M. P., 4
Morris, G. C., 129
Morrison, J., 202
Morrison, R. A., 221
Morrow, B. A., 233
Moser, J. B., 50
Moskovits, M., 231
Moss, R. L., 237, 240
Moudrianakis, E. N., 94
Mower, E. B., 257
Moyes, R. B., 257
Mukherjee, P. N., 244
Mukhitdinov, L. A., 230
Mulay, L. N., 31
Mullens, J. J., 79

Nagasawa, K., 30
Nagy, L., 127
Naik, S. C., 249
Naiki, T., 62
Nakamura, Y., 95
Nakanishi, H., 115, 116
Nakatani, S., 117
Nakodkin, N. G., 198
Nashakio, S., 121
Nefedov, B. K., 245
Neumann, D., 174
Newman, D. G., 212
Nicholson, R. B., 55
Niehrs, H., 62
Niizeki, N., 36
Nimmo, K. M., 48
Nishii, M., 95, 120, 121, 122, 127
Nitta, I., 97, 104
Nordling, C., 185
Norin, R., 40
Norton, F. J., 219, 220
Norton, P. R., 205
Nyburg, S. C., 89

Ochi, H., 121
Odajima, A., 103
O'Donnell, J. H., 104, 106
Oehlig, J. J., 50
Ogilvie, G. J., 129
Ogilvie, J. L., 196
Ohlsen, W. D., 22
Ohnishi, S. I., 104

Okada, M., 242
Okada, T., 30
Okamura, S., 95, 120, 121, 122, 124, 127
O'Keefe, D. R., 210
Olander, D. R., 210
Oliver, R. G., 240
Olsen, H., 59
Oranskaya, O. M., 234
Orchard, A. F., 174
Orlova, N. A., 245
Ose, H., 244
Oshima, K., 99, 106, 113, 119
Osipov, V. B., 117
Ozaki, A., 211, 212

Pabst, H., 31
Palazov, A., 228, 234
Palmberg, P. W., 183, 184, 186, 189, 200
Palmer, H. B., 68
Palmer, R. L., 210
Pannetier, G., 231, 232
Park, R. L., 201
Parker, H. S., 39
Parks, T. C., 5
Parravano, G., 217, 220
Pashley, D. W., 55
Pasternak, R. A., 208
Patzschke, G., 244
Pearce, D. R., 211
Penczek, S., 121
Pernoux, E., 20
Perret, R., 86
Peterson, J. M., 135
Petropoulos, J. H., 106
Phillips, V. A., 86
Phillipson, J. J., 240, 247, 248, 249
Philp, J., 211
Pickles, V. A., 257
Pierce, D. T., 157
Pines, D., 57, 182
Pipenbring, F. J., 152
Piper, T. C., 203
Pirogovskaya, G. P., 27
Platteuw, J. C., 240
Pliskin, W. A., 203
Poling, G. W., 230, 231
Pollard, J. H., 185
Poltorak, O. M., 238, 239
Pope, M., 129
Popov, Y. G., 4
Popova, I. L., 249
Possley, G. G., 229
Post, B., 115
Powers, H. E. C., 135
Predecki, P., 134, 135
Presland, A. E. B., 83
Price, D., 79
Price, W. C., 174
Prince, M., 119
Pritchard, J., 222, 227, 229, 231
Propst, F. M., 203
Provcost, F., 121
Ptak, L. D., 242
Pullen, B. P., 179

Author Index

Quinn, C. M., 157, 164
Quinn, J. J., 154, 180

Raccah, P., 5
Rahn, R. W., 231
Ralek, M., 225
Ramsbotham, J., 222
Ramsey, J. A., 168
Rand, B., 133
Randhava, S. S., 233
Rao, B., 231
Rao, H., 119
Rapaport, V. G., 113
Ratnasamy, P., 234
Ravi, A., 228
Ray, I. L. F., 62
Reams, W., 97, 100
Redfield, D., 151
Redhead, P. A., 216, 218
Rehmat, A., 233
Rest, A. J., 224
Reucroft, P. J., 129, 134
Rhodin, T. N., 183, 200
Ricca, F., 209
Rice, J. K., 173
Rice, S. A., 129
Richardson, J. D., 227
Richardson, P. C., 211
Riganti, R., 141
Riley, F. D., 5
Riviere, J. C., 183, 186
Roberts, B. W., 86
Roberts, J. K., 132
Roberts, K., 207
Roberts, M. W., 157, 164, 168, 185, 217
Robertson, J. M., 135, 141
Robin, M. B., 172, 174
Robinson, P. M., 129, 134
Rochester, C. H., 222
Roev, L. M., 232
Ron, I., 133
Rood, A. P., 139
Rooney, J. J., 243
Roscoe, C., 71
Roth, J. F., 196
Roth, R. S., 15, 39, 40, 47
Roussel, M., 113
Roy, R., 30
Rozhkova, V. N., 246
Rubalcava, H. E., 226
Rubin, I., 99
Rucci, G., 230
Rudham, R., 211
Ruland, W. O., 67, 86
Ruska, E., 54
Rusov, M. T., 232, 245
Rymer, T. B., 57

Sabine, T. M., 134
Sachtler, W. M. H., 158, 220
Sahl, K., 34
Saini, G., 209
St. Pierre, L. E., 105
Sakai, T., 244
Sakai, Y., 103, 104, 106, 109
Sakamoto, M., 121

Sakata, Y., 97
Saltsburg, H., 210
Sancier, K. M., 210
Sander, Y. L., 218
Sanders, J. V., 46
Sanina, N. L., 249
Sarti-Fantoni, P., 129, 138
Sasada, Y., 116
Sasaki, T., 254
Sauer, H., 13
Sauer, J. A., 103
Schäfer, H., 36, 39, 40, 47
Scheer, J. J., 145, 162
Scheibner, E. J., 186
Scherzer, O., 57
Schlick, S., 107
Schmidt, G. M. J., 116, 133
Schmidt, L. D., 209
Schmidt, M. C., 107
Scholfield, C. R., 245
Scholten, J. J. F., 205
Schroeder, W., 31
Schürer, P., 222, 234
Schuit, G. C. A., 247
Schultz, J. F., 245
Schulz, G., 232
Schwartz, A. B., 257
Scott, H. G., 134
Scurrell, M. S., 222
Sears, G. W., 134
Seib, D. H., 161
Sella, C., 100, 125
Septier, A., 54
Shafran, R. N., 249
Shannon, J. S., 257
Shavlova, E. Y., 27
Shearer, H. M., 141
Sheets, R., 226
Shepherd, J. P. G., 178
Sheppard, N., 233
Sheridan, J., 255
Sherwood, J. N., 129, 132, 134
Shimanouchi, H., 116
Shimizu, A., 121, 127
Shina, H., 245
Shinagawa, H., 56
Shioji, Y., 104
Shirasaki, T., 242
Shirley, D. A., 168, 195, 199
Shirota, K., 56
Shmulyakovskii, Ya. E. 234
Shockley, W., 145
Shooter, D., 207, 209
Shopov, D., 228, 234
Shu, S., 113
Siegbahn, K., 172, 185
Siegel, B. M., 57
Sienko, M. J., 16
Silverton, J. V., 140
Sim, G. A., 141
Simmons, G. W., 203
Sims, M. L., 229
Sinfelt, J. H., 240, 241
Singer, N., 222
Smallman, R. E., 23
Smith, A. P., 134

Smith, H. A., 209
Smith, J. N., jun., 210
Smith, N. C., 249
Smith, N. V., 169, 180, 191, 202
Smith, W. H., 229
Smrkovska, M., 222
Snyder, L. C., 174
Sokolova, N. P., 223, 228, 231, 233
Sokolowski, E., 185
Speakman, J. C., 140
Spicer, W. E., 145, 147, 156, 161, 168, 169, 174, 180, 190, 202
Sprengel, L., 120
Spyridelis, J., 16
Stanevich, A. E., 226
Stannett, V., 122, 127
Statton, W. O., 134, 135
Steadman, R., 51
Steele, B. C. H., 21
Stephan, V., 117
Stephenson, N. C., 39
Stern, R. C., 215
Stevens, A. J., 207
Stobie, R. W., 231
Stokes, A. R., 92
Stolz, H., 222
Stone, F. S., 167
Stouthamer, B., 240
Straumanis, M. E., 22
Strivastava, S. W., 141
Sturkey, L., 61, 62
Suétaka, W., 230
Sugano, T., 156
Suhrman, R., 216
Sukhikh, T. A., 117
Sullivan, G. A., 22
Sundholm, A., 33
Suzuki, F., 115, 116
Suzuki, T., 119
Suzuki, Y., 115, 116
Swalin, R. A., 50, 51
Swan, J. B., 203
Sworakowski, J., 129
Sykes, K. W., 217
Symons, M. C. R., 109
Sze, S. M., 156

Tabata, Y., 99, 106, 113, 119
Tadokoro, H., 117, 122
Taft, E., 145
Takada, T., 30
Talbot, J., 230
Tamaki, T., 116
Tamm, P. W., 209
Tanaka, M., 227
Tannhauser, D. S., 22
Taylor, B., 104
Taylor, C., 121
Taylor, C. A., 71
Taylor, E. H., 209
Taylor, G. H., 68
Taylor, H. F. W., 114
Taylor, H. S., 211, 237
Taylor, N. J., 173
Tellier, J., 246
Terasaki, O., 2

Terasaki, S., 23
Terekhova, S. F., 113, 114
Teroni, R., 129
Thibault, S., 230
Thieme, J., 225
Thomas, D., 107
Thomas, H., 145, 155, 182
Thomas, J. M., 71, 104, 129, 132, 133, 134, 138, 237
Thomas, L. E., 55
Thomas, T. D., 178
Thomson, B. J., 161
Thomson, S. J., 129, 221
Thon, F., 64
Thornber, M. R., 2
Tilley, R. J. D., 16, 24, 51
Tilyaev, S. K., 246
Tiu, D. E., 249
Tkachenko, E. V., 39
To, D. E., 249
Tompkins, F. C., 164
Toru, T., 254
Tovey, H., 215
Tracey, J. C., 184, 189
Trajmar, S., 173
Traum, M. M., 169
Trillat, J. J., 100
Trofimova, G. M., 113, 114
Tsuchita, H., 121
Tsukamoto, H., 122
Turcotte, R. P., 5
Turner, D. W., 172, 174
Turner, J. J., 224
Turton, L. M., 129

Uchida, T., 122
Ueda, H., 106
Ueno, K., 122
Uhlaner, C. J., 218
Unterwald, F. C., 208, 214
Uyeda, R., 55

Vaight, P. R., 214, 216
Vainshtein, E. E., 31
Valle, G., 122
Vallet, P., 5

Vand, V., 132
Van der Paaren, J., 121
Van der Plank, P., 220
Van Hardeveld, R., 237, 238
Van Laar, J., 145, 162
Van Landuyt, J., 25
Van Montfoort, A., 237
van Reijen, L. L., 247
Van Ruyven, L. J., 32
Vasil'eva, I. A., 27
Versino, C., 227
Vierle, O., 227
Vlasenko, V. M., 245
Vodhenal, J., 117
Von Hippel, A., 32
Voshcenkov, A. M., 134

Waddams, A. L., 244
Wadsley, A. D., 4, 11, 41, 46
Wallace, W. E., 209
Ward, J. C., 99
Ward, J. W., 233
Waring, J. L., 15, 39, 40
Warren, B. E., 66
Watanabe, D., 2, 23
Watanabe, H., 120
Watanabe, M., 56
Watson, D. G., 141
Watson, J. D., 92
Webb, G., 220, 246, 247, 249
Weber, E. C., 99
Wedler, G., 216
Weertman, J., 133
Weertman, J. R., 133
Wegner, G., 116, 117
Wei, P. S. P., 200, 203
Weisz, P. B., 257
Wells, B. R., 168
Wells, P. B., 220, 240, 243, 245, 246, 247, 248, 249, 251, 256, 257
Wessells, F. A., 121
Westphal, W. B., 32
Wheeler, A., 237
Whelan, M. J., 55, 62, 133

White, D. M., 117
White, J. R., 83
White, W. B., 30
Whitehead, P., 211
Whitmore, D. H., 50
Wilkins, M. H. F., 92, 93
Williams, A. R., 190
Williams, J. C., 249
Williams, J. O., 104, 129, 133, 134, 135, 138
Williams, M. L., 168
Willis, B. T. M., 2
Wilson, G. R., 240, 243, 247, 251
Wilson, H. R., 92
Winterbottom, J. M., 247, 248
Wise, H., 210
Wishlade, J. L., 221
Witnamer, L. P., 115
Wittke, J. P., 22
Wolberg, A., 196
Wolf, E., 55
Woodward, A. E., 103
Wright, K. H. R., 168

Yada, K., 56
Yagi, E., 31
Yamada, M., 121
Yamashino, H., 121
Yamanchi, T., 122
Yao, H. C., 107
Yaroslavskii, N. G., 226
Yasumori, I., 248
Yates, D. J. C., 240, 241
Yates, J. T. jun., 212, 214, 215, 216, 218
Yenikolopyan, N. S., 114
Yuzefovich, G. E., 245

Zador, S., 21, 27
Zanzoterra, C., 230
Zaukelies, D. A., 134
Zecchina, A., 227
Zeitler, E., 57, 59
Zettlemoyer, A. C., 217
Zhuza, V. P., 30
Zurakowska-Orszagh, J., 99

QC
176.8
E4
S87
v.1

AUG 11 1977